D1535146

Public Infrastructure Asset Management

About the Authors

Waheed Uddin is Professor of Civil Engineering and Director of the Center for Advanced Infrastructure Technology (CAIT) at the University of Mississippi. Previously a pavement expert for the United Nations, he has contributed to highway and airport infrastructure management projects in the United States and many countries abroad. Dr. Uddin is a visionary in airborne laser mapping and satellite imagery applications in transportation and disaster impact assessment.

W. Ronald Hudson is the Dewitt C. Greer Centennial Professor Emeritus in Transportation Engineering at the University of Texas at Austin and Senior Consultant for AgileAssets, Inc. Dr. Hudson is a pioneer in developing and applying management systems to pavements, bridges, and other civil infrastructure assets.

Ralph Haas is the Norman W. McLeod Engineering Professor and Distinguished Professor Emeritus in the Department of Civil Engineering at the University of Waterloo in Ontario, Canada. He holds the country's highest civilian honor, the Order of Canada, and the highest academic honor, Fellow of the Royal Society of Canada. Dr. Haas is an innovator in pavement design and civil infrastructure management systems.

Public Infrastructure Asset Management

Waheed Uddın

W. Ronald Hudson

Ralph Haas

Second Edition

New York Chicago San Francisco
Lisbon London Madrid Mexico City
Milan New Delhi San Juan
Seoul Singapore Sydney Toronto

Library of Congress Cataloging-in-Publication Data

Hudson, W. Ronald.
  [Infrastructure management]
  Public infrastructure asset management / Waheed Uddin, W. Ronald Hudson,
  Ralph Haas.—2nd ed.
      p. cm.
  Edition of: Infrastructure management / W. Ronald Hudson, Ralph Haas,
  Waheed Uddin. c1997.
  ISBN 978-0-07-182011-0 (hardback)
  1. Engineering—Management. 2. Infrastructure (Economics)—Management.
  3. Life cycle costing. 4. Engineering economy. I. Uddin, Waheed. II. Haas, R. C. G.
  (Ralph C. G.) III. Title.
  TA190.H82   2013
  363.6068—dc23   2013018493

McGraw-Hill Education books are available at special quantity discounts to use as premiums and sales promotions or for use in corporate training programs. To contact a representative, please visit the Contact Us page at www.mhprofessional.com.

Public Infrastructure Asset Management, Second Edition

1 2 3 4 5 6 7 8 9 0   DOC/DOC   1 9 8 7 6 5 4 3

ISBN 978-0-07-182011-0
MHID 0-07-182011-6

This book is printed on acid-free paper.

The first edition of this book was published as *Infrastructure Management: Integrating Design, Construction, Maintenance, Rehabilitation, and Renovation*, copyright © 1997.

Sponsoring Editor	**Project Manager**	**Indexer**
Michael McCabe	Charu Khanna, MPS Limited	Robert Swanson
Editing Supervisor		**Art Director, Cover**
Stephen M. Smith	**Copy Editor**	Jeff Weeks
	Erica Orloff	
Production Supervisor		**Composition**
Pamela A. Pelton	**Proofreader**	MPS Limited
	Surendra N. Shivam,	
Acquisitions Coordinator	MPS Limited	
Bridget L. Thoreson		

This book is dedicated to Rose Haas, wonderful wife, mother, and friend, who has inspired us all in life and in our pursuit of improved infrastructure asset management over these 40-plus years.

Contents

Foreword

Until recently *infrastructure* was an esoteric word rarely heard or written in public discourse. That has dramatically changed as awareness of our "deteriorating infrastructure" has increased through political conversations involving candidates for public office and commentary by members of our media.

This second edition of *Public Infrastructure Asset Management* effectively addresses the role and importance of civil infrastructure across the spectrum of quality-of-life concerns including a clean and adequate water supply, energy needs and consumption, and multiple means of transportation. Advocating public attention to this critical subject is increasing in importance globally as our nation addresses significant allocation-of-resources challenges. Infrastructure improvement cannot be delayed if we are to continue as a vital nation.

Professors Waheed Uddin, Ronald Hudson, and Ralph Haas have rendered great service to professional engineers and to our society by producing a well-organized, thoughtful guide for moving forward as we address urgent needs that affect all of us.

Robert C. Khayat
Chancellor Emeritus
University of Mississippi

Dr. Robert C. Khayat served as the 15th Chancellor of the University of Mississippi (Ole Miss) from 1995 until he retired in 2009. Educated at Yale and Ole Miss where he was an Academic All-American football and baseball player, Dr. Khayat received the NFL Lifetime Achievement Award and the National Football Foundation Distinguished American Award. Under his outstanding leadership, the university experienced amazing change, increased enrollment, renovation of campus infrastructure, and addition of new facilities. Ole Miss hosted the first presidential debate between Barack Obama and John McCain in 2008.

Preface to the Second Edition

We preserved the preface to the first edition of this book for your information because we feel it correctly captures the technical situation at that time (1997). However, subsequent rapid changes indicate the need for a second edition preface. General usage has adopted the term "asset management" instead of "infrastructure management," which we originally proposed for the systems and components described. We now yield to this common usage and the term "asset management" is widely used in the second edition. For clarity we have broadened the title of the book to *Public Infrastructure Asset Management* because the concepts are most widely used in publicly owned facilities. These concepts are equally applicable to privately owned assets such as railroads, bus fleets, shopping centers, hospitals, etc. You will see in the book the terms "infrastructure management," "asset management," and "infrastructure asset management" used almost interchangeably, because that is a fact of life in the practical world today.

Since the first edition was published in 1997, computing and data storage technology have seen quantum advancements, enhancing the breadth and depth to which users can do asset management. But the foundation and the principles presented in this book remain the same as those in the first edition.

Since many current asset management systems are proprietary, we have left in place Chapter 17, which still illustrates how the principles of asset management are implemented. However, we have added Chapter 18, which provides details of proprietary systems available in 2013 with guidance to where the reader can get more complete information and demonstrations on any specific system from the vendor. We also provide a suggested procedure for evaluating and obtaining the best possible system for your needs. History makes it clear that trying to develop your own system in-house is as expensive and time-consuming as trying to build your own computer. The future of asset management is bounded only by your imagination. In 1997 when the first edition was published, the primary asset management systems were pavements and bridges. Since that time, the concept has grown immensely, including pavements, bridges, safety, fleet, facilities, congestion, and other areas. A number of public agencies have combined their resources with two or more of these systems into a true asset management system using data from the bottom-up to create effective top-down financial decisions.

Asset management is now also being used by many smaller public agencies, cities, and countries for their facilities. We have included a section in Chapter 16 showing the implementation of asset management systems by agencies such as cities, counties, metropolitan transportation authorities, and others. We are indebted to Alan Kercher, Kercher Engineering, Civil/Municipal Engineers and Surveyors, for providing detailed information on the ways in which he has implemented asset management in such smaller agencies as indicated in the references.

We may sound like we are preaching in this book, but our intent is to urge you into action on asset management and to guide you toward knowing the many basic principles necessary. We may have missed the mark on some factors, so searching elsewhere is always useful. But since many have encouraged us to update the first edition and there seems to be no other comprehensive text, we offer this second edition. As with any book some flaws and some errors may well exist, for which we take full responsibility.

There are a number of national publications on asset management, including a recent one from the U.S. DOT Federal Highway Administration that outlines the "concepts" of asset management and considers the "big picture" [FHWA IF-10-009]. This is important but only a starting point. To accomplish what is needed requires detailed planning, nuts and bolts technologies, and models, such as mathematical techniques, optimization methods, etc. Just as President John F. Kennedy voiced the goal of putting a man on the moon in a speech, it took money, design of rockets, spacesuits, moon rovers, calculations of earth orbits, and much coordination to do this through systems engineering. In the same way, this book describes the details (the *system*) required to flesh out the skeleton of asset management concepts. While we applaud the progress being made in sustainability and environmental stewardship there is still much that is in the concept stage, and it may well take a decade to fully expand asset management to adequately incorporate these important factors.

Special thanks are due to a number of people who helped in the preparation, proofreading, and editing of this book, starting with Dr. Rukhsana "Juhee" Uddin and Mr. Asad Uddin whose help, understanding, and encouragement to the lead author were invaluable. Also, Ms. Jan Zeybel of AgileAssets and Ms. Shelley Bacik of the University of Waterloo deserve special thanks for their hard work throughout. We are also indebted to Dr. Roger Smith of Texas A&M

University and his colleagues at MTC for providing details of StreetSaver® in Chapter 16 as a long-running, successful IAMS implementation.

Finally, we are grateful to the many graduate students who have contributed so much to our work over the years, and we have tried to acknowledge them as best we can in the references.

Waheed Uddin
W. Ronald Hudson
Ralph Haas

Preface to the First Edition

A great deal of thought went into titling this book. We considered *Infrastructure Management*, *Asset Management*, and *Facilities Management* because all of these terms are occasionally used to describe the kinds of activities that are covered in this book. It should be noted that these terms are similar and are often used interchangeably to describe the process of life-cycle cost integration. That is, the process of integrating design, construction, maintenance, rehabilitation, and renovation to maximize benefits to the user and minimize total cost to the owners and users.

There are, however, some subtle differences among the terms which led to the final choice for this book. The term "asset management" has arisen primarily from use in the private sector. Private corporations in the profit-making business try to manage their "assets," which include physical infrastructure as well as capital, personnel, technology, and technology ideas.

"Facility management" is most often used with a physical entity broadened to include operations and furnishings. In fact, the term facility management is generally meant to describe operation and use of a facility, such as a hotel, including both the infrastructure or building itself as well as furniture, ballrooms, banquet facilities, dishes, linen, and so forth. The activity therefore often includes scheduling, setting up furniture in a ballroom for a banquet, and similar operations.

The term "infrastructure management" has been coined to generalize the concepts of pavement management, bridge management, and building management and has most often been applied to public civil engineering infrastructure such as water, wastewater, bridges, airports, parks, pavements, and the like.

Much of the technology used in the three areas described is the same, that is, good data inputs, economic models and analysis, benefit cost studies, good maintenance, and rehabilitation are all needed to provide adequate protection of the investment. Given the similarities, but keeping in mind the subtle differences of these three terminologies, the term Infrastructure Management Systems (IMS) was selected for use. We felt that IMS was more descriptive of the process that covers public infrastructure assets and facilities than the other two terms, and we will use this term throughout this book, *Infrastructure Management*. The reader should not be confused by others who may choose to use the word *assets* or *facilities*, they are not "new and different." The concepts presented herein are applicable no matter which choice of wording you select.

This book is intended to provide an overview for those interested in all aspects of civil engineering infrastructure, especially those interested in learning how to procure and preserve it more efficiently and economically. The book can be used by students or practicing engineers to enlighten themselves about the broader aspects of IMS. Each section of the book stands alone, and the book can be read or studied in parts, by those active in one part of the process (i.e., design) who want to learn more about another part of the process, such as maintenance and rehabilitation. Someone in a peripheral area of infrastructure, such as a small town or an airport authority not now using infrastructure management in their agency, can use the book to excite their interest in new applications and developments in their workplace.

Perhaps most importantly, the authors hope that this book will provide a compilation of course materials and an impetus to the development of courses and a change in university curricula to specifically teach engineering students the real need to integrate maintenance and rehabilitation into their future activities. Current curricula too often stress primarily design and construction when, in fact, most engineering work in the next two decades will involve a large component of preservation, maintenance, and rehabilitation, as well as improved use of infrastructure facilities.

This book is dedicated to our respective wives, Martha Hudson, Rose Haas, and Rukhsana (Juhee) Uddin, and to our many colleagues known and unknown who have for the past 30 years helped us to advance the technology and use of better management concepts in all phases of civil engineering infrastructure. Thanks are due to Dr. Zhanmin Zhang, Omar Uddin, and Usman Uddin for their contributions in the preparation of the manuscript. The manuscript was carefully typed by Julie Wickham, Jan Zeybel, and Loretta McFadden, and reviewed by Betty Brown, who are due special thanks for their perseverance.

We are specially indebted to Dr. Zhanmin Zhang, who prepared Chapter 17. He has done pioneering work in developing working IMS computer programs. Dr. Zhang is currently Associate Professor at the University of Texas in Austin.

The roots of the IMS process lie in pavement management, for which we make no apology because none is necessary. There is proven application of the concepts already in the areas of highways, bridges, water, and wastewater facilities. Only the imagination and the ability of you, the reader, limit the future use of this book and the concepts presented.

W. Ronald Hudson
Ralph Haas
Waheed Uddin

CHAPTER 1
The Big Picture

1.1 Infrastructure and Society

The success and progress of human society depends on physical infrastructure for distributing resources and essential services to the public. The quality and efficiency of this infrastructure affects the quality of life, the health of the social system, and the continuity of economic and business activity. A nation's economic strength is reflected in its infrastructure assets. Many examples can be cited from history. The Romans built a strong empire by constructing all-weather roads and viaducts throughout Europe, North Africa, and the Middle East to move people, goods, and water. In the colonial era of the 16th to 19th centuries, European nations emerged as strong shipbuilders and explorers. This was followed by the products of the Industrial Revolution, particularly in the use of steam engines for ships and railroad transport. Since then migration of rural-area population to cities and urban areas has been continuing because of higher-education institutions, diverse job opportunities, proximity to built infrastructure, and better quality of life [Uddin 12a]. Currently about 80 percent of the population in industrialized countries lives in cities and metropolitan areas. Urban population is also vulnerable to natural disasters, especially in coastal areas where about 60 percent of the U.S. population lives [Uddin 12b].

In the United States and other regions of the world, historical development of the economic and social systems closely parallels phases of infrastructure development and urban growth. Demands on infrastructure and related services increase as people expect a higher standard of life and public services. But, more importantly, good infrastructure facilitates a higher quality of life.

1.1.1 Energy Infrastructure and Economic Development

Energy demands in cities and urban areas are large and higher for industrialized countries compared to developing countries. A country's gross domestic product (GDP) highly correlates with its energy consumption. Figure 1-1, prepared using the U.S. Energy Information Administration (EIA) data [EIA 11], compares GDP per capita (pc)

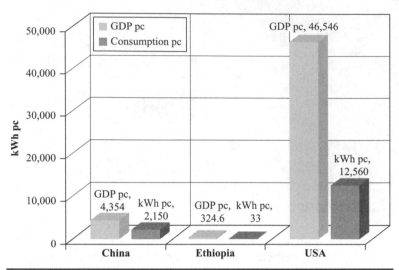

Figure 1-1   GDP versus energy consumption per capita comparison of high-, medium-, and low-GDP countries.

with kilowatt hour (kWh) pc for a high-GDP country (United States), a medium-GDP country (China), and a low-GDP country (Ethiopia).

1.1.2   Transportation Infrastructure and Economic Development

1.1.2.1   Road Infrastructure and Economic Indicators

Clarification of the relationship between infrastructure and economic development is provided by Queiroz in his World Bank study [Queiroz 92]. It shows a very strong association between economic development, in terms of per capita gross national product (GNP), and road infrastructure. GNP is the measure of a nation's total market value of the goods and services that are produced annually. GNP per capita is a country's gross national product divided by its population. Road infrastructure can be characterized by spatial density, which is a country's road length per land area, and road density, which is the length of the road network per capita. Road transport is important to economic activity, especially in developing countries, where it plays an essential role in marketing agricultural products and providing access to health, education, and other services. A good road system gives a country a competitive edge in moving goods efficiently and economically.

Considering the United States as illustrative of a highly developed country, a vast amount of historical data is available on the road network and the economy [FHWA 91, Abstract 91]. A time-series analysis of this data from 1950 to 1988 shows a significant positive relationship between per capita GNP (*PGNP,* in U.S. $1,000 per inhabitant, using 1982 constant dollars) and density of paved roads (length of paved roads, *LPR*, in km per 1,000 inhabitants):

$$PGNP = -3.39 + 1.24 \ (LPR) \tag{1.1}$$

A similar type of analysis for 98 developing countries was carried out by Queiroz and Gautam, with the resulting equation:

$$PGNP = 1.39 \ (LPR) \tag{1.2}$$

Figure 1-2 shows this relationship and the associated statistics. To make a 1988 constant dollar comparison with the developing countries, the time-series equations from the U.S. data were converted using the GNP implicit price deflator between 1982 and 1988 [Queiroz 92]:

$$PGNP_{88} = -4.1 + 1.50 \ (LPR) \tag{1.3}$$

A similar analysis of 1950 to 1988 Canadian data [Queiroz 94] yielded the following equation:

$$PGNP_{88} = 0.86 + 1.33 \ (LPR) \tag{1.4}$$

This equation is also plotted on Figure 1-2 (i.e., a comparison is provided of Eqs. (1.2), (1.3), and (1.4), according to their inference space). There is relatively good consistency between the two equations [Eqs. (1.2) and (1.4)] for 98 countries and Canada. The United States has about 13 percent greater road density for any given *PGNP* value.

An interesting time-lag analysis shows that the highest correlation exists for *PGNP* of a given year compared with *LPR* four years earlier. It suggests that a paved-road investment today will result in an

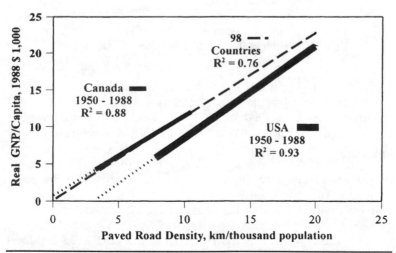

FIGURE 1-2   Comparison of data from the United States, Canada, and 98 developing countries [after Queiroz 94].

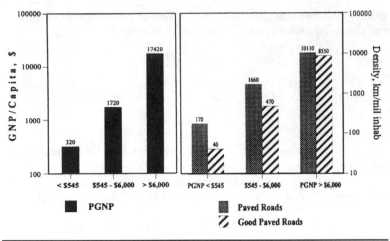

Figure 1-3 Average road density in low-, middle-, and high-income economies [after Queiroz 92].

increased GNP, about four years later. This four-year time lag is in broad agreement with the "half a decade" lag period observed by Aschauer [Aschauer 89]. A second comparison analysis between the supply and condition of paved-road networks in 98 developing and developed countries is shown in Figure 1-3 [Queiroz 92]. Figure 1-3 shows that the supply of good paved-road infrastructure in high-income economies is dramatically higher than that in middle- and low-income economies. The countries are grouped as defined by the World Bank [World 90]:

1. Low-income economies are those with a *PGNP* of $545 or less in 1988 (42 countries).

2. Middle-income economies are those with a *PGNP* of more than $545 but less than $6,000 in 1988 (43 countries).

3. High-income economies are those with a *PGNP* of $6,000 or more in 1988 (13 countries).

A recent analysis of paved-road density per capita for 173 countries worldwide based on 2006 data [CIA 09, World 09] showed a significantly high correlation of 0.76 between *PGNP* and paved-road density per capita.

A decline in maintenance and capital improvement investments may significantly affect the condition and adequacy of the infrastructure. This has been examined in detail for 23 states by the University of Colorado at Boulder in *Hard Choices: A Report on the Increasing Gap between America's Infrastructure Needs and Our Ability to Pay for Them* [Colorado 84]. The report shows that state and local infrastructure outlays have declined from 2.2 percent of GNP in 1961 to 1.9 percent in 1982 and have reduced across all regions of the United States. For the United States as a whole, U.S. $450 billion (in 1982 dollars) was the estimated future infrastructure funding gap between available and needed outlays. The most dominant need was identified for highway and bridge infrastructure. The 1998 to 2004 United States Transportation Equity Act for the 21st Century (TEA-21) was enacted on June 9, 1998; and the 2005 to 2009 Safe, Accountable, Flexible, Efficient Transportation Equity Act: A Legacy for Users (SAFETEA-LU) was signed into law on August 10, 2005. These federal transportation bills of the U.S. Congress provided funds to preserve and improve the nation's roads, bridges, and transit facilities. The combined annual government investment of U.S. $147.5 billion in highway and bridge infrastructure assets increased by 44.7 percent between 1997 and 2004. The infrastructure rehabilitation expenditure was 51.8 percent of capital investments [DOT 07].

1.1.2.2 Air-Transport and Economic Indicators

The air-transport industry and airport infrastructure in the United States has evolved in parallel with the nation's economic, industrial, and social development [DOT 94]. The 1998 to 2002 National Plan of Integrated Airport Systems (NPIAS) listed 5,174 airports in the United States [FAA 08]. The U.S. NPIAS inventory includes more than 2,400 general-aviation airports. The world airport infrastructure and air-transportation system have also continually grown in recent years to meet the air-transport demands of passengers and freight. Between 1979 and 1988, world air traffic in terms of passenger kilometers grew annually at the rate of 5.5 percent and the number of freight kilometers grew at 7.5 percent annually, as compared to the 2.5 percent annual growth of GNP and 4 percent growth of international trade in the Organisation for Economic Cooperation and Development (OECD) countries [Veldhuis 92]. In 2006, 2.1 billion passengers departed on scheduled airline flights worldwide and 734.7 million passengers enplaned on U.S. airlines, which is expected to reach an annual 1 billion passengers in 2023 [Uddin 09]. According to the International Air Transport Association (IATA) the 2011 top-ranked freight airline was FedEx at 15,939 million ton-km annually followed in turn by UPS, Cathay Pacific, Korean Airlines, Emirates, and Lufthansa [IATA 12]. These indicate the trend of increased air transport in the global economy.

1.1.2.3 Ports Infrastructure, Global Supply Chain, and Economy

Marine ports infrastructure is an integral part of both imports and exports through container ships for bulk freight traffic between countries. Marine freight markets have been steadily growing in shipping volumes worldwide, although in 2008 to 2012 freight-shipping growth in Europe declined as a result of slow world trade. The Far East and Middle East have had the strongest growth in containerized sea transport, the Northern European market has contracted, and sea freight from the European Union (EU) to Asia continues to increase [IATA 12]. An unprecedented growth in container freight traffic is expected for many U.S. ports in the future as even more companies (such as Boeing, auto plants, and computer manufacturers) obtain equipment and parts from around the world.

The top-ranked 50 ports on the U.S. coasts handled a total of 3,240 million dead weight tons (DWT) in 2010 [DOT 12]. The following nine ports handled more than 100 million DWT each: Houston, TX; Los Angeles/Long Beach, CA; New Orleans, LA; New York/New Jersey, NY and NJ; San Francisco, CA; Virginia Ports, VA; Savannah, GA; Columbia River, OR; and Philadelphia, PA. Many of these U.S. ports are striving to deepen their navigation channels and port facilities to accommodate supersize container ships expected to pass through the expanded Panama Canal. This clearly increases the need for an infrastructure asset management system (IAMS).

1.1.3   Urban Infrastructure and Economic Growth

Business communities, commercial enterprises, and industries know the benefits of a well-maintained and well-served infrastructure system. According to *Fortune* magazine [Perry 89], in a 1988 Lou Harris survey of 250 companies conducted for Cushman & Wakefield, a real-estate firm, most respondents rated "easy access to domestic markets, customers, or clients" as the most important factor in choosing an office location. The survey ranked Atlanta, which boasts an impressive array of freeways, mass transit, and air-transportation facilities, as the number-one city in which to put a business. Says George Berry, Georgia Commissioner of Industry and Trade, "Transportation services are to Atlanta what gambling is to Las Vegas" [Perry 89]. Most urban areas suffer from congestion and related air-quality degradation. Operational traffic flow in U.S. urban areas deteriorated from 6.2 hours of traffic congestion each day in 1997 to 6.6 hours in 2004 [DOT 07]. These problems affect productivity, economic loss, emissions, and public health issues [Uddin 06].

A recent survey of 568 respondents from a broad range of businesses in the United Kingdom describes infrastructure impacts on business investment decisions [CBI 12]. Some highlights are: (1) Over 80 percent of companies see the quality and reliability of transport infrastructure as significant considerations in investment decisions; (2) over 90 percent of manufacturing companies rated energy costs as important; and (3) 65 percent of companies feel the condition of local road networks declining, and cited congestion and lack of investment as the main concerns.

1.2   Infrastructure Definition

Infrastructure has been defined in many ways; for example, "those physical facilities that are sometimes called public works" [Grigg 88]. Public works have been defined by the American Public Works Association (APWA) as follows [Stone 74]:

Public works are the physical structures and facilities that are developed or acquired by the public agencies to house governmental functions and provide water, power, waste disposal, transportation, and similar services to facilitate the achievement of common social and economic objectives.

The National Science Foundation report on civil infrastructure systems states [NSF 94]:

A civilization's rise and fall is linked to its ability to feed and shelter its people and defend itself. These capabilities depend on infrastructure—the underlying, often hidden foundation of a society's wealth and quality of life. A society that neglects its infrastructure loses the ability to transport people and food, provide clean air and water, control disease, and conduct commerce.

The following infrastructure definition is given by the Associated General Contractors of America (AGCA), for all long-lived assets owned by local, state, and federal governments and utilities owned by businesses [Kwiatkowski 86]:

The nation's infrastructure is its system of public facilities, both publicly and privately funded, which provide for the delivery of essential services and a sustained standard of living. This interdependent, yet self-contained, set of structures provides for mobility, shelter, services, and utilities. It is the nation's highways, bridges, railroads, and mass transit systems. It is our sewers, sewage, sewage treatment plants, water supply systems, and reservoirs. It is our dams, locks, waterways, and ports. It is our electric, gas, and power-producing plants. It is our court houses, jails, fire houses, police stations, schools, post offices, and government buildings. America's infrastructure is the base upon which society rests. Its condition affects our life styles and security and each is threatened by its unanswered decay. [AGCA 82]

During the 1980s the U.S. government accounting statements considered infrastructure to be the fixed assets of value to the government, which included roads, bridges, curbs and gutters, streets and sidewalks, drainage systems, lighting systems, and similar assets that are immovable and of value only to the governmental unit [Kwiatkowski 86]. Such a definition, while useful for accounting, ignores the value of these assets to the citizens.

The Government Accounting Standards Board (GASB) in the United States promulgated its GASB Statement 34 (GASB 34) on June 15, 1999, which was implemented by the U.S. Department of Transportation (DOT) for reporting of infrastructure asset valuation and condition beginning after June 15, 2001 [DOT 00]. GASB 34 established financial and condition reporting methods for capital assets that required government agencies to be transparent to bond-market analysts, legislators, citizens, the media, and others who may be interested in public finance and infrastructure performance. State and local governments accounted for $175 billion expenditure on highways and bridges in the United States. GASB 34 considers these expenditures on long-lived assets as investments and defines *infrastructure assets*, as follows [NCHRP 08]: "Infrastructure assets are long-lived capital assets that normally can be preserved for a significantly greater number of years than most capital assets and that normally are stationary in nature. Examples of infrastructure assets include roads, bridges, tunnels, drainage systems, water and sewer systems, dams, and lighting systems." In essence, the GASB 34 definition is limited to public infrastructure assets built and owned by government agencies. Other similar asset valuation standards include PSAB3150 in Canada and similar standards in Australia.

In this book, "public infrastructure" refers to all these combined facilities that provide essential public services of energy, transportation, roads, airports, water supply, solid waste disposal, parklands, sports and athletic fields, recreational facilities, and housing. Infrastructure also provides the physical systems used to provide other services to the public through economic and social actions. These infrastructure facilities and services are provided by both public agencies and private enterprises. Public utilities function as enterprises providing vital infrastructure assets for electric power and renewable energy, solid waste disposal, wastewater treatment, water supply, and wireless and telecommunications to serve communities and businesses.

1.3   Historical Overview of Infrastructure Development

The trend toward major infrastructure development in the industrial nations, especially in road and bridge construction, began in the 18th century. In the 19th century, major infrastructure development paralleled economic development. Street-construction technology was improved by using asphalt and concrete to facilitate all-weather automobile traffic. Most modern roadways have been constructed since 1900. Prior to that, railroads were the primary mode of long-distance surface transportation during the 19th century and early-20th century. Philadelphia was the first city to initiate the construction of a large-scale municipal water supply system in 1798, followed by New York and Boston by the mid-19th century [Grigg 88].

A chronological summary listing of infrastructure milestones in the past and present century follows, based on published literature [Hudson 97, Horonjeff 94] and Internet sources.

1900–1909.   Construction of a ship canal in Michigan; initiation of the Panama Canal project; construction of the 320-km-long (200-mi) Owens River aqueduct to bring water to Los Angeles; introduction of gasoline and diesel power to rollers and excavators; Wright Brothers' first airplane flight; opening of New York City's subway system; first concrete road paved in Wayne County, Michigan; 1906 earthquake that ravaged San Francisco.

1910–1919. The Roosevelt Dam was constructed near Phoenix, Arizona; construction of the first transcontinental highway (Lincoln Highway U.S. 30); introduction of the motor grader; start of World War I in 1914 and conclusion in 1918; first U.S. federal-aid highway program legislation in 1916, with 50 percent federal participation; gasoline tax to pay for roads; post-construction boom of roads, rural highway network grown to 4 million km (2.5 million mi) with 10 percent surfaced, including 16,800 km (10,500 mi) of asphalt, 3,680 km (2,300 mi) of concrete, and 2,560 km (1,600 mi) of brick roads; first air-mail service between Washington, DC, and New York in 1918; Panama Canal project completed in 1914.

1920–1929. Use of pneumatic rubber tires; 9 million autos and trucks damage the unbound roads; introduction of earth-scrapers, mobile cranes, and tractors; construction of the Sukkur Barrage started in 1923 over Indus River and opened in January 1932 as the Lloyd Barrage in British India (now Pakistan), forming the World's largest water canal network for agriculture; assignment of national road-numbering system (even numbers to east-west routes and odd numbers to north-south; main transcontinental routes numbered in multiples of 10); completion of the Holland Tunnel under the Hudson River between Manhattan and Jersey City; opening of intercontinental air travel; completion of Newark's international airport; 1928 Air Commerce Act; stock market crash in 1929 and start of the Great Depression.

1930–1939. President Roosevelt's National Recovery Act (NRA), which included a large program of public works construction and $400 million to the states for highway construction; completion of the Detroit-Windsor tunnel, New York City's George Washington Bridge, Hoover Dam, Bonneville Dam and hydroelectric power station in Oregon, San Francisco's Trans Bay Bridge; Golden Gate suspension bridge; the Blue Ridge Parkway; establishment of the Tennessee Valley Authority (TVA); 1938 Civil Aeronautics Act defining the federal government's role in aviation; Germany's invasion of Poland and start of World War II.

1940–1949. Creation of the Pentagon in Washington, DC; construction of military airfields; construction of the 2,270 km (1,420 mi) Alaska Highway, the Pennsylvania Turnpike, and the Grand Coulee Dam (concrete construction) and hydroelectric power station; nuclear research leading to the atomic bomb and later to nuclear-power generation; slip-form concrete paver introduced; Federal Aid Highway Act of 1944 authorizing more than $1.6 billion over three years and the designation of a National System of Interstate Highways; Federal Airport Act of 1946; end of World War II in 1945; postwar construction of the Maine

Turnpike and conversion of 500 military airports to civilian use in the United States.

1950–1959. Large-scale toll-road construction started with the 189-km (118-mi), $225 million New Jersey Turnpike; construction of the first prestressed concrete bridge (the Walnut Lane Bridge in Philadelphia) and the first segmented concrete bridge (the Griffith Road Bridge in New York State); start of hauling of truck trailers on rail flat-bed cars; construction of the nation's first commercial nuclear power plant in Pennsylvania; the first transcontinental telephone cable in operation; Federal Aid Highway Act of 1956 authorizing 65,600 km (41,000 mi) of the National System of Interstate and Defense Highways at an estimated $41 billion and establishment of the Highway Trust Fund; construction and testing of the $27 million AASHO Road Test to produce data for improved highway design; Federal Aviation Act of 1958 establishing the Federal Aviation Agency.

1960–1969. Space Age initiated with the first American in orbit in 1962, the first moonwalk in 1969, and construction of National Aeronautical Space Administration (NASA) facilities; construction of the 28-km-long (17.6-mi), $200 million Chesapeake Bay Bridge-Tunnel; construction of the Oroville Dam (earthen-fill dam) in Northern California; start of construction on 26 nuclear power plants; passage by Congress of the Water Quality Act in 1965, the National Environmental Protection Act in 1969, and creation of a cabinet-level U.S. Department of Transportation (DOT) in 1967, which included the Federal Highway Administration (FHA), the Urban Mass Transportation Administration, the Federal Aviation Administration (FAA), the Federal Railroad Administration, and the Maritime Administration; collapse of the 39-year-old Silver Bridge over the Ohio River and, as a result, development of national bridge-inspection standards and conception of a bridge-replacement program.

1970–1979. Authorization by Congress of $250 million to start the bridge-replacement program; establishment of the Occupational Safety and Health Administration (OSHA), U.S. Environmental Protection Agency (EPA); passage by Congress of the Clean Air Act; Federal Water Pollution Act authorizing $18 billion for construction of wastewater treatment plants; Airport and Airway Development Act of 1970, Airline Deregulation Act of 1978, and Aviation Safety and Noise Abatement Act of 1979; authorization by Congress of the 1,280-km-long (800-mi) Alaskan Oil pipeline and tanker terminal; construction of JFK Causeway in Corpus Christi, a cable-stayed bridge in Alaska, and the first cast-in-place segmented Pine Valley bridge near San Diego; the Eisenhower–Johnson Tunnel for four-lane Interstate 70 and approximately 80-km (50-mi) west of Denver, Colorado, was the world's highest elevation tunnel when opened initially in 1973 at maximum elevation of 11,158 ft (3,401 m) above sea level; opening of the Bay Area Rapid Transit (BART) System in

San Francisco; Federal Aid Highway Act permitting limited use of federal funds for urban mass-transit projects; first trial of asphalt recycling for road construction; first Earth Day in 1970.

1980–1989. Rebuilding highways without shutting down traffic; Congress expands the Interstate Resurfacing, Restoration, Rehabilitation, and Reconstruction (4R) program with the federal match of 90 percent; the Surface Transportation Assistance Act of 1982 raising the fuel tax on gasoline from 4 to 9 cents per gallon, with 1 cent earmarked for mass transit; 72,000-km-long (45,000-mi) interstate system near completion; Airport and Airway Improvement Act of 1982; construction of the first roller-compacted Willow Creek Dam in Oregon; earthquakes causing major damage in Mexico and California.

1990–1999. Updating of Clean Air Act in 1990 with strict regulation on reduced vehicle emissions; construction decline due to recession; Airway Safety and Capacity Expansion Act of 1990; the 1991 Intermodal Surface Transportation Efficiency Act (ISTEA) authorizing $155 billion over six years with emphasis on maintenance, service-life analysis, and management systems; 1992 Chicago flood and Hurricane Andrew in South Florida; rebuilding infrastructure in Kuwait after the 1991 Gulf War, attracting U.S. and other international contractors; other significant global changes, including the collapse of the Berlin Wall, opening of East European countries to the West, breakup of Soviet Union into independent republics and end of Cold War; post-1992 recession in the U.S. and world economies; continuation of highway rehabilitation projects and establishment of a high-quality National Highway System (NHS) consisting of nearly 256,000 km (160,000 mi) of roads in the United States; opening of the Denver International Airport; construction of Olympic facilities in Atlanta; the 1994 ice-storm damage to electric-power stations and water-supply systems in the Midsouth region; enormous infrastructure damages during the 1994 Los Angeles earthquake in the United States, and the 1995 Kobe earthquake in Japan; 1-m high-resolution satellite imagery through IKONOS satellite missions by a private company in the United States in September 1999, which marked the first significant commercial spaceborne imagery of the Earth.

2000–2009. High-resolution, 61-cm commercial satellite imagery through QuickBird2 satellite successfully launched in the United States on October 18, 2001; construction of facilities in Salt Lake City for the 2002 Winter Olympics at a public investment of $2.7 billion; the 2004 Summer Olympic–associated infrastructure construction in Athens, Greece, at an investment exceeding $10 billion; $40 billion public investment for the 2008 Beijing Summer Olympics. This decade saw construction of record-breaking tall buildings: Taiwan's Taipei 101 in Taipei set a record height of 509.2 m (1,671 ft) in 2004, superseding the 452-m-high (1,482-ft) Kuala Lumpur's Petronas Twin Towers of 1998 in Malaysia.

Post–2009. The world's tallest-built structure, the 829.8-m (2,722-ft) Burj Khalifa in Dubai, United Arab Emirates, opened on January 4, 2010, exceeding the 2007 Petronius offshore oil platform that stands 610 m (2,000 ft) off the sea floor in the Gulf of Mexico. Continued investment in transportation included the Mike O'Callaghan–Pat Tillman Memorial Bridge, the first concrete-and-steel-composite arch bridge and part of the $240 million Hoover Dam Bypass in the United States. Expansion of the Panama Canal is in progress and scheduled to be completed in 2014.

1.4   Infrastructure Assets

1.4.1   Categories of Infrastructure Facilities

Physical facilities and related services can be categorized in seven groups based on their primary functions and services, as shown in Table 1-1.

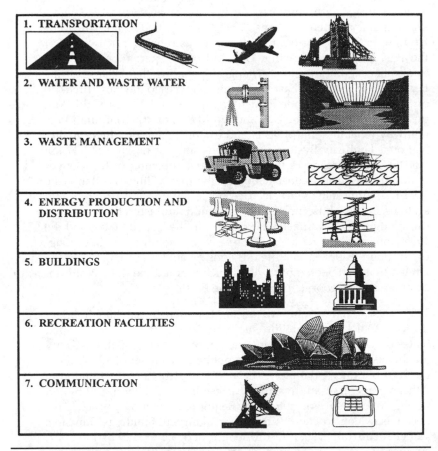

TABLE 1-1   Categories of Infrastructure Assets

In the United States, most of the ports, airports, roads, urban mass-transit infrastructure facilities, water and wastewater facilities, and waste-management facilities are built and maintained by public agencies. However, there is a growing trend, particularly internationally, to finance, build, maintain, and operate these facilities as public-private partnerships. Both public agencies and private enterprises are involved in energy production and distribution infrastructure, buildings, and recreational facilities. Railroad facilities, manufacturing and supply chain infrastructure, and communications/cyber-infrastructure assets are historically built and managed by the private sector in the United States. The following subgroups of infrastructure assets include most of the publicly and privately owned and managed facilities used to provide the essential services to support and sustain a nation's civilian life.

1.4.1.1 Transportation

- Ground transportation (roads, bridges, tunnels, railroads)
- Air transportation (airports, heliports, ground facilities, air-traffic control)
- Waterways and ports (inland waterways, shipping channels, terminals, dry docks, sea ports)
- Intermodal facilities (rail/airport terminals, truck/rail/port terminals)
- Mass transit (subways, bus transit, light rail, monorails, platforms/stations)
- Pipelines (natural gas, crude oil)

1.4.1.2 Water and Wastewater

- Water supply (pumping stations, treatment plants, main water lines, wells, mechanical/electric equipment)
- Structures (dams, diversion, levees, tunnels, aqueducts)
- Agricultural water distribution (canals, rivers, weir, gates, dikes)
- Sewers (main sewer lines, septic tanks, treatment plants, stormwater drains)
- Stormwater drainage (roadside gutters and ditches, streams, levees)

1.4.1.3 Waste Management

- Solid waste (transport, landfills, treatment plants, recycling facilities)
- Hazardous waste (transport, storage facilities, treatment plants, security)
- Nuclear waste (transport, storage facilities, security)

1.4.1.4   Energy Production and Distribution

- Fossil fuel-based electric power production (gas-, oil-, and coal-fueled power generation)
- Electric power distribution grid networks (high-voltage power-transmission lines, substations, distribution systems, energy-control centers, service and maintenance facilities)
- Gas pipelines (gas production, pipeline, computer stations and control centers, storage tanks, service and maintenance facilities)
- Petroleum/oil production (pumping stations, oil/gas separation plants, roads)
- Petroleum/oil distribution (marine and ground tanker terminals, pipelines, pumping stations, maintenance facilities, storage tanks)
- Nuclear power stations (nuclear reactors, power-generation stations, nuclear-waste disposal facilities, emergency equipment and facilities)
- Renewable energy and non-fossil fuels (infrastructure for solar power, wind power, hydro-electric power, biofuels)

1.4.1.5   Buildings

- Public buildings (schools, hospitals, government offices, police stations, fire stations, postal offices, prison systems, parking structures)
- Other buildings and structures—public/residential/commercial/offices (public housing, structures, utilities, swimming pools, security, ground access, parking) Multipurpose and sports complexes (coliseums, amphitheaters, convention centers)
- Housing facilities (public, private)
- Industrial, manufacturing/warehouse, and supply chain facilities (private)

1.4.1.6   Health Infrastructure Facilities

- Hospitals and public health facilities (public, private)
- Veterans hospitals (public)
- University teaching and research hospitals and medical centers (public, private)
- Nursing homes and assisted-living facilities (private)

1.4.1.7   Recreational Facilities

- Parks and playgrounds (roads, parking areas, recreational facilities, office buildings, restrooms, ornamental fountains, swimming pools, picnic areas)

- Lake and water sports (roads, parking areas, picnic areas, marinas)
- Theme parks/casinos (access roads, buildings, restaurants, security facilities, structures)

1.4.1.8   Communications and Information Technology Assets

- Telecommunications networks (land telephone/fiber-optic line networks, telephone-exchange stations, cable distribution, power supplies, switching and data-processing centers, buildings, transmission towers, and repeat stations)
- Television/cable networks (production stations, transmission facilities, cable distribution, power supplies, buildings)
- Wireless/satellite networks (satellites, ground-control centers, wireless cell towers, communication systems, receivers, buildings, services and maintenance facilities)
- Information technology (IT) infrastructure and information highway networks (computer networks, cable distribution, data-processing hardware/software, on-line and off-line cyber infrastructure including wireless and telecommunication networks, systems, information sources, buildings, backup and recording mediums, cloud computing infrastructure)

1.4.2   Examples of Infrastructure Assets

The road network in the United States exceeds 6 million km (3.9 million mi), as shown in Figure 1-4 by functional classification for 2006 [BTS 12]. Because the paved-road networks are already well developed compared to most other countries, there is only slight increase in road length in each class in 2006 compared to 1992 [Abstract 94], such as a 9.1 percent increase in arterial road length. While the interstate highway system is about 1 percent of the total road network, it carries over 20 percent of the heavy-truck traffic [Choate 81]. As a part of the ISTEA legislation, Congress designated about 256,000 km (160,000 mi) of road as the National

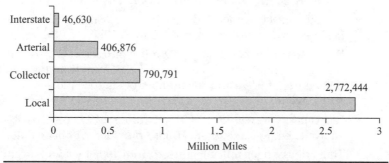

FIGURE 1-4   U.S. national road network by functional class, 2006 [BTS 12].

Highway System (NHS) in 1995. Although the NHS comprises only 4 percent of the nation's roads, it carries 40 percent of auto travel and 75 percent of truck traffic, and connects 95 percent of the businesses and 90 percent of the households in the United States [FHWA 96, DOT 07]. The national bridge inventory in 1990 contained 577,000 bridges of which 70 percent were constructed prior to 1935, and 40 percent were, by reasons of their condition or appraisal, eligible candidates for rehabilitation or replacement [Golabi 92]. As of September 11, 2007 the DOT reported 597,876 bridges in the national inventory, and 25 percent were structurally deficient and functionally obsolete [DOT 07]. In 2012 many of these bridges were more than 70 years old. Roads and highway networks represent the largest public infrastructure investment in the United States at a cost of more than U.S. $175 trillion.

A World Bank study [Faiz 87] provides an overview on the status of main roads in 85 developing countries that have received $9.4 billion in road-assistance programs from the World Bank from 1974 to 1985. The aggregate length of the main road systems in these countries is about 1.8 million km, of which slightly more than 1 million km are paved. The main roads carry about 70 to 80 percent of the interurban traffic.

Transit system coverage, capacity, and use in the United States have been steadily increasing since 1992 [Abstract 94]. The TEA-21 transportation bill provided record levels of federal investment with the highest level of U.S. $28.4 billion in 2002, a 62.6 percent increase from $17.5 billion in 1997 [DOT 07]. The milestones of the status of national transit ways by mode in 2004 follow:

- Out of 640 transit agencies serving urban areas, 600 were public agencies. There were 1,215 transit operators in rural areas.
- Transit agencies had 120,659 vehicles in their inventory (18 percent more than the 1997 level) of which 92,520 (76.7 percent) were in urban and metropolitan areas of more than 1 million people. A total of 19,185 transit vehicles were operated in 2000 in rural areas. The 2002 survey shows 37,720 special-service transit vehicles for older adults and persons with disabilities.
- Transit agencies operated 793 bus- and rail-maintenance facilities, an increase of 8.8 percent from 1997.
- Rail systems operated on 10,892 miles of track and 2,961 rail stations. The rail miles increased by 9.8 percent from the 1997 track miles of 9,992 and the number of stations increased by 10.4 percent.
- Rail capacity increased in 2004, providing 2.4 billion capacity-equivalent miles, and non-rail transit provided 2.1 billion miles.
- Transit passenger miles increased from 1997 by 15.8 percent to 46.5 billion (41 percent on motor bus, 21 percent on commuter rail, 3 percent on light rail, and 4 percent other).

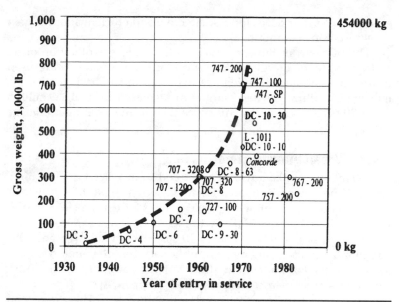

FIGURE 1-5   Trends in the gross weight of transport aircraft [after Horonjeff 94].

Historically airports have been reliable and stable and have kept pace with the technological developments in air-traffic-control systems and aircraft fleets. Figure 1-5 shows the advances in an aircraft fleet as a result of travel demand [Horonjeff 94]. Average passenger capacity per aircraft was 106 seats in 1969, which grew to 162 seats by 1978, and 181 seats by 1989. New wide-bodied aircraft have a capacity of 300 to 500 passengers. In 1978, about 250 million passengers in the United States traveled by air; by 1989 this figure had already reached 445 million [Hamiel 92]. The worldwide airline fleet in 1975 to 1978 was somewhere in the vicinity of 3,500 to 4,000 aircraft; by 1989 the fleet grew to 7,441 aircraft, as estimated by Boeing and Airmark. New generations of regional jets and new large aircraft (NLA) from Boeing (B-777) and Airbus (A-380) in the post-2000 era have improved fuel-efficient jet engines, enabling airlines to meet increased travel demands while reducing emissions.

The railroad network in the United States has been used largely for bulk freight over long distances since the 1960s. Light-rail trains have become effective in some large urban areas for mass public transit. In Europe, Japan, and other parts of the world, railroads are a popular mode of passenger travel. Major technological advances in speed and equipment have been made in Europe, Japan, and most recently in China by introducing high-speed magnetic levitation (maglev) trains. The 2008 to 2009 U.S. economic stimulus bill allocated billions of dollars to develop high-speed passenger-train corridors. In recent years intermodal railroad facilities have emerged in the United States, especially at major ports, for cost-effective and efficient supply-chain transportation operations.

The oil crisis of the early 1970s spurred new exploration of oil and gas all over the globe. Super oil-tanker fleets are being used to transport oil and liquefied gas across oceans. The U.S. pipeline mileage of natural gas and petroleum increased to total 1.7 million mi in 2009 [BTS 12] up by 310 percent from the 1992 total mileage [Abstract 94].

1.4.3   Urban Infrastructure

There continues to be a worldwide trend to increased urbanization. About 50 percent of the population in developing countries now lives in cities, and 80 percent of the population of industrialized countries lives in cities [Uddin 12a]. In the United States, approximately 90 percent of the population inhabits 10 percent of the land [Bragdon 95]. Urban infrastructure and related public services are therefore crucial to our society. For illustration, the following summary of New York City's infrastructure assets is presented [Wagner 84].

New York City's capital needs are enormous. The extent and variety of New York's infrastructure are extraordinary:

- Forty-seven waterway bridges and 2,057 highway bridges and elevated structures.

- A water-supply system that delivers 1.45 billion gallons of water a day from a reservoir system of 1,956 sq. mi. It delivers water through two tunnels (a third is under construction), 32 billion linear feet of trunk and distribution mains, and 20,000 trunk valves.

- 6,100 mi of sewers, 12 operating water-pollution-control plants, and 450 combined sewer-overflow regulators.

- 6,200 mi of paved streets that cover approximately 30 percent of the city's land.

- 6,700 subway cars that ride on 232 mi of track (137 mi underground, 72 mi elevated, and 23 mi open bed) and 4,560 buses.

- An estimated 3,500 acres of landfill and nine marine-transfer stations.

- Over 25,000 acres of parkland.

1.5   Life-Cycle Analysis in Planning and Design

The infrastructure construction boom of the 1950s, 1960s, and 1970s in the United States, Europe, and other parts of the world resulted in numerous projects and advances in planning and design practices [Hudson 97]. Since the 1990s more money is being invested in the United States for maintenance and rehabilitation of airports, roads, and bridges and transit. In response to the financial and economic crisis of 2008, $150 billion additional federal funds were allocated to infrastructure with most expenditures on roads and bridges. Many other world governments followed this with large infrastructure spending totaling billions of dollars for 2009 to 2015 [Gerritsen 09].

Availability and use of computers and mobile/telecommunication facilities have helped with infrastructure development and preservation. Operational management of the facilities and services has also advanced. However, until recently, much of the infrastructure development did not consider ongoing maintenance, future rehabilitation/renovation, and replacement activities in overall planning and costing. This points out that life-cycle analysis and planned monitoring over the life cycle is essential.

Figure 1-6 illustrates the life-cycle cost stream of a facility. The condition of the facility can be preserved over its service life if the condition-responsive maintenance and rehabilitation actions are properly timed. This requires predicting condition, deterioration, or performance over the service life as discussed in more detail in Chapter 8. Performance of an infrastructure element or facility is considered good if it performs as designed and provides an acceptable level of service over its intended life. Poor performance indicates that the facility. (1) may have deteriorated faster than predicted; (2) and/or provides an inadequate level of service; and/or (3) is aged beyond its design life without any major rehabilitation, renovation, or replacement. Moreover, sudden failures caused by catastrophic events or natural disasters or other incidents may require emergency response.

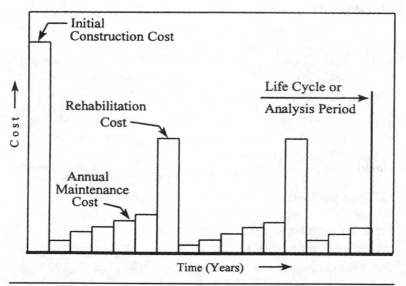

FIGURE 1-6   Life-cycle cost streams for infrastructure analysis.

One pioneering concept of performance in terms of the Present Serviceability Index (PSI) was developed at the AASHO Road Test [HRB 62] involving highway pavements. Appraisal methods and condition-rating procedures for bridges were developed in the early 1970s after the collapse of several structures resulted in loss of life. The concept of maintenance and rehabilitation during the service life of pavement assets was developed in the mid-1970s [Haas 94].

The development of good performance models depends on condition-assessment methods, load and demand data estimates, material behavior prediction, and understanding of climatic and environmental conditions.

The Building Research Board (BRB) reorganized the concept of life-cycle costs in its report *Pay Now or Pay Later* [BRB 91], which stated, "Decisions about a building's design, construction, operation and maintenance can, in principle, be made such that the building performs well over its entire life cycle and the total costs incurred over this life cycle are minimized." The GASB 34 standards also outline the importance of inventory databases, asset valuation, condition monitoring, and funding expenditure's reporting for all critical lifeline infrastructure assets, including roads, bridges, and other public works infrastructure [DOT 00, NCHRP 08].

In summary, the overall process of infrastructure management goes beyond planning and design; it includes construction and acceptance testing, periodic condition assessment, and maintenance and improvement programs during the service life.

1.6   Magnitude of Infrastructure "Crisis"

Starting in about the late 1970s, infrastructure caught the attention of the media and the public, largely initiated by the book *America in Ruins: The Decaying Infrastructure* [Choate 81]. Its condition, decay, and aging, and sometimes disastrous effects (including loss of lives and property), made for news headlines. Citizens, media, investors, and public officials became more concerned after hearing the evidence of and the publicity surrounding some critical incidents of sudden collapse and failure of various infrastructure components. The public awareness of these incidents and identification of potential failure areas has led to a perception of reality of an infrastructure crisis in the United States. The decline has continued to date.

1.6.1   Examples of Extreme Incidents in the Infrastructure Crisis

The following examples are taken from published literature and Internet sources.

1967: Bridge Collapses in West Virginia. The Silver Bridge, on the Ohio River between Point Pleasant, West Virginia, and Gallipolis, Ohio, collapsed during rush-hour traffic in 1967. Many vehicles were stopped on the structure for a traffic signal when the instantaneous fracture of an eyebar led to the loss of 46 lives. This disaster was highly publicized and drew attention to the aging condition of the nation's bridges. The U.S. Congress added provisions to the Federal Aid Highway Act of 1968, which required the Secretary of Transportation to establish a national bridge-inspection (NBI) system and to develop a bridge-inspection program [Hudson 87]. The standard was issued in April 1971. Since then, bridge inspection has been continuously improved. This was the most influential milestone in the recognition of the infrastructure crisis.

1974: The West Side Highway Collapses in New York. A major section of the West Side Highway, an important major arterial in New York City, collapsed, resulting in the closing of the highway; no life was lost [Wagner 84].

1982: An 80-Year-Old Aqueduct Fails in New Jersey. After the failure of an 80-year-old aqueduct in July 1982, 300,000 residents of Jersey City, New Jersey, had no drinking water for three days [Kwiatkowski 86].

1983: Mianus River Bridge Collapses in Connecticut. A section of the Mianus River Bridge, on Interstate 95 in Connecticut, collapsed in 1983, killing three persons and seriously injuring three others [Kwiatkowski 86].

1983: Water Main Breaks in New York City. Water main breaks in midtown Manhattan closed streets and subways, caused a blackout, and seriously disrupted business [Wagner 84].

1988: A Woman Falls through a Sidewalk Grate. In December 1988, New York City tabloids ran wild with the story of a 28-year-old Brooklyn woman who was critically injured when she fell through a weakened sidewalk grate and landed on subway tracks 50 ft below [Perry 89].

1992: Chicago River Floods Ravage the City. A hole in a 100-year-old freight-tunnel system allowed 250 million gallons of Chicago River water to flood the basements of 200 buildings in downtown Chicago, causing $2 billion in damages, evacuation of a quarter-million people, and taking six weeks to pump the tunnel dry [DHS 12].

2003: North America's Blackout. A massive power blackout struck on August 14, 2003, and affected Ontario, New York, and many other U.S. northern states. New York State faced the largest blackout in the history of America for about 30 hours before the electric power was restored. This showed the vulnerability of the electric-power-grid network infrastructure.

2007: Mississippi River Bridge Disaster in Minnesota. On August 1, 2007, the I-35W bridge collapsed during a busy hour in Minnesota killing 13 people. It was rated "structurally deficient" in 2005, but no action was taken to rehabilitate the bridge. The bridge was finally replaced at a cost of $234 million and was opened for traffic on September 18, 2010.

2011: Japan's Nuclear Disaster. The 9.0 magnitude earthquake and tsunami that hit Japan on March 11, 2011, caused the meltdown of the Fukushima Daiichi Nuclear Power Station and massive flooding, destruction of infrastructure, and evacuation of people. This highlights the need for disaster planning and crisis management in such situations.

2012: India's Blackout Affects 620 million People. The July 2012 India blackout was the largest power outage in history, affecting over 620 million people across north India, twice the size of the U.S. population.

1.6.2   Effects of Natural Disasters, Manmade Disasters, and Other Incidents on Infrastructure

Natural disasters (earthquakes, floods, hurricanes, tornados, volcanic eruptions, ice storms) and serious incidents (fires, riots, terrorist attacks) can cause extensive damage to infrastructure and disrupt essential services and business activities, as shown in the following examples:

1992—Hurricane Andrew in South Florida.

1993—Bomb explosion in World Trade Center, New York City.

1994—Earthquake in Los Angeles.

1995—Earthquake in Kobe, Japan.

1995—Bomb explosion in a federal building, Oklahoma City.

2001—Terrorist attacks in New York, Washington, DC, and Pennsylvania.

> The unforgettable terrorist attacks on the World Trade Center buildings in New York and the Pentagon in Washington, DC, and the Pennsylvania crash, on September 11 (known as 9/11), killed over 3,000 people, destroyed the World Trade Center towers, and changed the way security is managed since that time to safeguard transportation and other critical lifeline infrastructure facilities. This tragic event led, in large part, to a vastly increasing demand for security, and, for example, the creation of the Department of Homeland Security by President George W. Bush in the United States.

2004—Asian Tsunami disaster, Indian Ocean, December 26, 2004, affected many countries, destroyed infrastructure, and killed hundreds of thousands of people.

2005—Hurricane Katrina disaster on the Gulf Coast, August 29, 2005, flooded New Orleans, destroyed roads and bridges, caused over $100 billion in damages, and affected large areas of Mississippi and Louisiana on the Gulf Coast.

2005—Earthquake disaster, Northern Pakistan, caused more than 75,000 deaths and affected large mountainous areas and cities.

2007—Gas pipeline disaster in Mississippi.

2007—Fiery crash disaster on Interstate 5 tunnel in Southern California.

2010—Massive earthquakes hit Chile, Haiti, and Indonesia; super-flood of Pakistan.

2011—Costliest year of worldwide natural disasters. The Christchurch, New Zealand earthquake, magnitude 6.3, on February 22, 2011; record magnitude 9.0, Tohoku, Japan, earthquake on March 11, 2011, tsunami destruction, and nuclear disaster; floods in China, Pakistan, and Thailand leaving a total of over 36 million people affected, killing 355 and with direct economic losses of $380 billion; earthquake of magnitude 7.2 in Turkey on October 23; Mississippi River floods in May; tornado disaster and medium-severity 5.8-magnitude earthquake in DC/southern Virginia, and Category-1 Hurricane Irene disaster in August 2011 in the United States (Irene washed away more than 200 bridges and many roads).

2012—Tornados in Midwest and Southeast U.S.; Hurricane Sandy (October 27 to 30) in Northeast turning into Category 1 superstorm Sandy, it made landfall on the evening of October 29 on New Jersey shores with 90 mph winds and caused a total of 94 deaths including 26 in the United States, 14,000 flights were cancelled, New York and LaGuardia airports were closed, and it left 6.5 million customers without power across 13 states and Washington, DC, for several weeks.

1.7 Maintenance, Preservation, and Innovation Challenges

A nation's infrastructure represents a sizable asset. The value of this asset in the United States was estimated in 1995 at $20 trillion in civil infrastructure systems, including all installations that house, transport, transmit, and distribute people, goods, energy, resources, services, and information [NSF 95]. Because of aging, overuse, exposure, misuse, mismanagement, and neglect, many of these systems are deteriorating and becoming more vulnerable to catastrophic failure, particularly when earthquakes, hurricanes, and other natural hazards strike. It would be prohibitively costly and disruptive to replace these vast networks.

They must instead be renewed in an intelligent manner, which includes the prudent and effective use of our economic, material, and human resources, and which focuses on optimizing the performance of both individual subsystems and of the civil infrastructure systems complex as a whole.

In the public works arena, these systems have evolved in a piecemeal fashion, with new extensions grafted onto existing systems and designs often governed by expediency and low construction costs rather than true life-cycle costs. We have inherited a complex network of systems comprised of subsystems with wide variations in age and functionality. How systems interact is often poorly assessed, and maintenance has often been inadequate [NSF 95].

1.7.1 Preservation and Repair Estimates

Expenditures on infrastructure preservation and major repair are about 20 to 40 percent of total new construction in the United States, more than $120 billion annually. The ways in which we allocate and manage these resources influence the results. Inefficiencies of only a few percentage points will represent substantial losses.

The Congressional Budget Office estimated that for the 756 urban water-supply systems in the United States, between $63 and $100 billion was needed by the year 2000 to replace all water mains over 90 years of age and to replace other mains as necessary. Extrapolations to all community water systems (adjusted for population variations) suggested that total replacement and rehabilitation needs for all communities could run as high as $160 billion [O'Day 84].

As reported in *Fortune,* the National Council on Public Works Improvement, which was created in 1984 to assess the state of America's public works (the report card), concluded in its final report to Congress in 1988 [Perry 89]:

> While America's infrastructure is not in ruins, it is inadequate to sustain future economic growth. America has to face its needs even as the federal government is moving out of the infrastructure business. Its last heroic public works project, the interstate highway system, will be completed in 1992. Preoccupied with an unyielding budget deficit, Uncle Sam is doling out less and less money for infrastructure. Heywood Sanders estimates that of the $40.5 billion spent to expand highways, airports, sewers, mass transit, and waterworks last year, 46% came from federal funds, down from 54% in 1985. State and municipal governments are stuck with the rest.

The 2009 infrastructure report card by the American Society of Civil Engineers (ASCE), compiled from the opinion survey of public-works agencies and other stakeholders, shows both the average overall condition at grade D and an estimated repair budget of

$2.2 trillion over 5 years (http://www.infrastructurereportcard.org/, accessed May 28, 2012). The 2009 average condition grade of each infrastructure category follows: Aviation, D; Bridges, C; Dams, D; Drinking Water, D−; Energy, D+; Hazardous Waste, D; Inland Waterways, D−; Levees, D−; Public Parks and Recreation, C−; Rail, C−; Roads, D−; Schools, D; Solid Waste, C+; Transit, D; Wastewater, D−.

1.7.2   Infrastructure Maintenance Problems—National Perspective

A report on the problems facing the preservation of the infrastructure [Perry 89] asks, "What needs fixing? Many of the sewers, bridges, and water systems in America's older cities, built around the turn of the century, are now in disrepair. Poor roads cost American motorists billions of dollars in wasted fuel, added tire wear, and extra vehicle repair. Breakage does not always imply decay. The financing of highway pavement preservation has been improving since 1982, when Congress raised the federal gas tax by five cents per gallon but is now grossly inadequate." The Federal Highway Administration's 2007 report classifies 23 percent of the bridges included in the national bridge inventory in the United States as structurally deficient [DOT 07]. They are either closed or restricted to light traffic.

The following excerpt from *Fortune* gives more examples of maintenance programs needed to improve the infrastructure and related services [Perry 89]: "Still, travelers feel that the roads and airports they use are overwhelmed by congestion. By 1993 up to one-half of the country's landfills, which collect 95 percent of the 450,000 tons of solid waste Americans generate each day, will fill up and be closed. Partly in an effort to promote alternatives, such as recycling and incineration, cities in some parts of the U.S. are doubling or tripling the price of waste disposal."

The recent state of public infrastructure condition and its funding needs are discussed in *Digging into the Infrastructure Debate* [Kliesen 09], which are summarized as follows:

- The public infrastructure in the United States is crumbling and in dire need of repair. This view seems to have become more strident after the Minneapolis bridge collapse in 2007.

- The ASCE's infrastructure report card gave poor grades and estimated that the United States needs to more than double planned infrastructure spending over the next five years, or by about $1.1 trillion, to put the nation's infrastructure in "good condition." About half of this infrastructure gap is due to deteriorating roads and bridges.

- The U.S. National Governors Association (NGA) published *An Infrastructure Vision for the 21st Century* in 2009. According to the NGA, "The nation's infrastructure system is no longer adequately meeting the nation's needs and faces several long-term challenges that affect our ability to maintain and enhance our competitiveness, quality of life and environmental sustainability."

- The U.S. Congressional Budget Office estimated in 2008 that spending on the U.S. transportation infrastructure was roughly $16 billion below the spending needed to maintain current levels of service.

- The American Recovery and Reinvestment Act of 2009 (ARRA) addressed this concern by providing $111 billion for infrastructure and science projects. Of this amount, about a quarter ($27.5 billion) was set aside for spending on highway construction.

- The Organisation for Economic Cooperation and Development (OECD) said in 2007 that advanced countries besides the United States face similar problems: "A gap is opening up in OECD countries between the infrastructure investments required for the future, and the capacity of the public sector to meet those requirements from traditional sources."

1.7.3 Airport Capacity Problems

Airports in Europe and other regions of the world risk becoming at or near capacity in terms of runways, taxis, and terminals. Even with capacity improvements, growth in traffic is placing high demand on these facilities. In 2006, 2.1 billion passengers departed on scheduled airline flights worldwide, and the 2011 top-ranked freight airline was FedEx at 15,939 million ton-km total domestic and international freight [IATA 12]. The development of modern large jet airliners and increased use of satellite navigation in the 21st century has propelled significant capital investments in landside and airside infrastructures at public airports in the United States and abroad.

1.7.4 Maintenance of Urban Infrastructure

The extensive and extraordinary diversity and size of the urban infrastructure of New York City has been described earlier. In parallel there are clear indications of neglect to maintain and preserve the condition of these facilities [Wagner 84]:

In 1978, the Koch administration found a pattern of neglect almost frightening in its extent. The desirable rate for repaving streets is once every 25 to 50 years, depending on usage. In 1978 the city was repaving streets at the rate of once every 200 years. Engineers say a water main will require replacement probably every 100 years. In 1978 we were replacing water mains at the rate of once every 296 years, in 1978 the state found 135 waterway bridges and highway structures to be in poor condition and requiring major reconstruction. The same pattern would apply to all other parts of the city's physical plant.

1.7.5   Infrastructure Research and Development Efforts: Funding Crisis

In 1987, the U.S. Congressional Office of Technology Assessment reported that Japan outspent the Unite States on research and development for public works by a factor of 30. Europe outspent the United States in this area by a factor of 8. Furthermore, our civil infrastructure has not benefited much from either U.S. or foreign advances in technology. A combination of the highly competitive nature of the infrastructure construction industry, the difficulty of disseminating new technologies, negative incentives for adopting new technologies, and excessive litigation has impeded technological advances in the industry [NSF 95].

1.8   Infrastructure Asset Management— An Integrated Approach

1.8.1   Reasons for Infrastructure Deterioration

Infrastructure problems have compounded over recent decades for several reasons: (1) the underinvestment in public works programs; (2) the lack of good management systems for infrastructure; (3) failure to recognize the importance to the future economy of maintaining a sound physical infrastructure; (4) cutbacks that have slashed public-works budgets; (5) failure to replace the infrastructure as fast as it wears out; (6) failure to realize that lack of physical infrastructure seriously impacts the level and types of services government can provide to their citizens; (7) tendency by national, state, and local officials to defer the maintenance of public infrastructure; and (8) increased costs to taxpayers to repair and rebuild the obsolescent public infrastructure.

Aschauer has shown that the productivity (i.e., output per unit of private capital and labor) is positively related to government spending on infrastructure, including roads [Aschauer 89]. Analyzing data from the United States for the period 1949 to 1985, he observed that underinvestment in infrastructure started in about 1968, and the effects of deterioration became evident half a decade later, when a productivity slump began in the United States [Queiroz 92].

1.8.2   Overall Framework for Infrastructure Asset Management

The past underinvestment in infrastructure maintenance and the lack of overall system management principles point out the need for better management and financing approaches. In turn, it is essential that available funds be spent in a cost-effective and timely way. The thrust should be on the preservation of the condition and value of assets. An example of leadership in this regard is the Intermodal Surface Transportation Efficiency Act (ISTEA) of 1991,

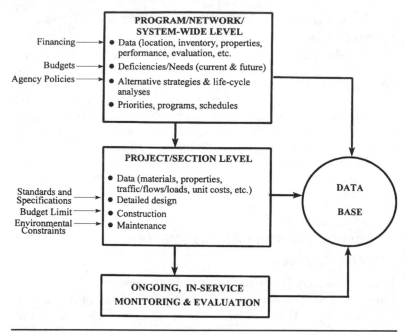

Financing ⟶
Budgets ⟶
Agency Policies ⟶

**PROGRAM/NETWORK/
SYSTEM-WIDE LEVEL**
- Data (location, inventory, properties, performance, evaluation, etc.
- Deficiencies/Needs (current & future)
- Alternative strategies & life-cycle analyses
- Priorities, programs, schedules

Standards and Specifications ⟶
Budget Limit ⟶
Environmental Constraints ⟶

PROJECT/SECTION LEVEL
- Data (materials, properties, traffic/flows/loads, unit costs, etc.)
- Detailed design
- Construction
- Maintenance

DATA BASE

**ONGOING, IN-SERVICE
MONITORING & EVALUATION**

FIGURE 1-7   Overall framework for infrastructure asset management.

which specified six management systems for state DOTs (pavement, bridge, safety, congestion, transit, and intermodal). As well, the Federal Highway Administration (FHWA) has been active in supporting efforts to improve bridge management system concepts and monitoring of all highway assets [Hudson 87, Golabi 92, NCHRP 08].

An overall framework for any IAMS is illustrated in Figure 1-7. The major point of the diagram is that management can be divided into two distinct but closely integrated levels: program/network/system-wide and project/section. Key elements of the overall framework for infrastructure management are ongoing, in-service monitoring and evaluation, and a database. Each of the two levels of management must consider exogenous elements over which little or no control may exist, such as financing, budgets, and agency policies for the network level, and standards and specifications, budget limits, and environmental constraints for the project level.

Comprehensive infrastructure assets can be found in large urban areas, as illustrated in Figure 1-8. Municipal infrastructure includes many facilities providing transportation services, utilities, health and education services, public buildings, parks, and other recreational areas. The exact scope of an IAMS for a specific community and jurisdiction will depend on the network size and extent of services to be covered.

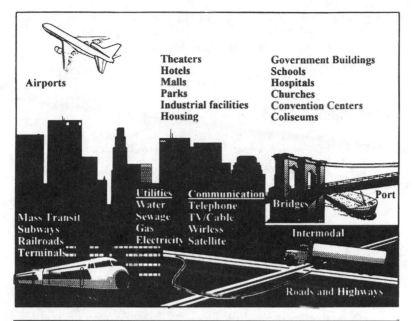

Airports

Theaters
Hotels
Malls
Parks
Industrial facilities
Housing

Government Buildings
Schools
Hospitals
Churches
Convention Centers
Coliseums

Utilities   Communication
Water   Telephone   Bridges
Sewage   TV/Cable
Gas   Wirless   Intermodal
Electricity Satellite

Mass Transit
Subways
Railroads
Terminals

Port

Roads and Highways

FIGURE 1-8   An illustration of infrastructure assets in a large metropolitan area.

1.8.3 Challenges of Environmental Sustainability and Funding

Modern infrastructure asset management in the 21st century is facing the following challenges while striving for improving infrastructure condition and capacity to meet the demand:

- New technologies and scientific developments and the impacts of natural and manmade disasters.

- Valuation of infrastructure assets, largely in response to government accounting standards requirements, such as GASB34 in the United States (published in June 1999), PSAB3150 in Canada, and similar standards in Australia.

- Quantifiable performance measures tied to realistic policy objectives and implementation targets.

- Environmental stewardship and the adverse environmental impacts of deteriorating air quality, excessive noise, and lack of available mitigation strategies.

- Consideration of greenhouse gas (GHG) emissions from transportation, construction, power generation, etc., and mitigation strategies.

- Comprehensive approaches to sustainability including "green" infrastructure methodologies, resource conservation, reuse of materials, renewable energy, etc.

- Innovative financing for publicly owned infrastructure and for Public-Private Partnerships (P3s) with emphasis on maintenance in industrialized countries and new infrastructure in developing countries.

1.9 References

[Abstract 91] Bureau of Census, *Statistical Abstracts of the United States 1991: The National Data Book,* U.S. Department of Commerce, Washington, DC, 1991.

[Abstract 94] Bureau of Census, *Statistical Abstracts of the United States 1994: The National Data Book,* U.S. Department of Commerce, Washington, DC, 1994.

[AGCA 82] Associated General Contractors of America, *Our Fractured Framework: Why America Must Rebuild,* Washington, DC, 1982.

[Aschauer 89] D. A. Aschauer, "Infrastructure Expenditures and Macro Trends," *Proceedings of the Africa Infrastructure Symposium,* The World Bank, Washington, DC, 1989.

[Bragdon 95] C. R. Bragdon, "Intermodal Transportation Planning for the 21st Century, a New Paradigm: An Urgent Call for Action," Presented at the *74th Annual Meeting of the Transportation Research Board,* National Research Council, Washington, DC, 1995.

[BRB 91] Building Research Board, *Pay Now or Pay Later,* National Research Council, Washington, DC, 1991.

[BTS 12] Bureau of Transportation Statistics, "National Transportation Statistics," U.S. Department of Transportation. http://www.bts.gov/publications /national_transportation_statistics/, accessed October 22, 2012.

[CBI 12] "Better Connected, Better Business," Report, Confederation of British Industry (CBI) and KPMG LLP, United Kingdom, September 2012. http://www .kpmg.com/UK/en/Issues AndInsights/ArticlesPublications/NewsReleases/ Pages/Positive-outlook-for-digital-infrastructure-according-to-CBI-KPMG-Infrastructure-Survey.aspx, accessed October 15, 2012.

[Choate 81] P. Choate, and S. Walter, *America in Ruins: The Decaying Infrastructure,* Duke Press, Durham, North Carolina, 1981.

[CIA 09] Central Intelligence Agency, *World Fact Book.* https://www.cia.gov, accessed April 5, 2009.

[Colorado 84] University of Colorado at Boulder, *Hard Choices: A Report on the Increasing Gap between America's Infrastructure Needs and Our Ability to Pay for Them,* Denver, 1984.

[DHS 12] Department of Homeland Security, "35,000 Gallons of Prevention," http://www.dhs.gov/files/programs/st-snapshots-35000-gallons-of -prevention.shtm, accessed April 7, 2012.

[DOT 94] Department of Transportation, *National Transportation Statistics,* U.S. Department of Transportation, Washington, DC, 1994.

[DOT 00] Department of Transportation, *Primer: GASB 34,* Federal Highway Administration, Office of Asset Management, Washington, DC, November 2000.

[DOT 07] U.S. Department of Transportation, *2006 Status of the Nation's Highways, Bridges, and Transit: Condition & Performance.* USDOT Report to Congress, Executive Summary, March 2007.

[DOT 12] U.S. Department of Transportation, Maritime Administration, "Vessel Calls at U.S. Ports by Vessel Type 2010," http://www.marad.dot.gov/library _landing_page/data_and_statistics/Data_and_Statistics.htm, accessed March 22, 2012.

[EIA 11] Energy Information Administration, "International Energy Statistics, Total Electricity Net Generation," U.S. Department of Energy, 2011, http://205.254.135.7 /cfapps/ipdbproject/iedindex3.cfm?tid=2&pid=2&aid=12&cid=all,&syid=2006 &eyid=2010&unit=BKWH, accessed May 20, 2012.

[FAA 08] Federal Aviation Administration (FAA), "National Plan of Integrated Airport Systems (NPIAS)," FAA Airports, U.S. Department of Transportation, Washington, DC, http://www.faa.gov/airports_airtraffic/airports/planning_capacity/npias/, accessed January 10, 2008.

[Faiz 87] A. Faiz, C. Harral, and F. Johansen, "State of the Road Networks in Developing Countries and a Country Typology of Response Measures," in *Transportation Research Record 1128*, Transportation Research Board, National Research Council, Washington, DC, 1987, pp. 1–17.

[FHWA 91] Federal Highway Administration, *Highway Statistics*, U.S. Department of Transportation, Washington, DC, 1991.

[FHWA 96] Federal Highway Administration, "The National Highway System: A Commitment to America's Future—Rodney Slater," *Public Roads, Journal of Highway Research and Development*, U.S. Department of Transportation, Washington, DC, Winter 1996, pp. 2–5.

[Gerritsen 09] E. J. Gerritsen, "White Paper: The Global Infrastructure Boom of 2009–2015," Commentary, *The Journal of Commerce Online*, May 19, 2009, http://www.joc.com/commentary, accessed August 29, 2011.

[Golabi 92] K. Golabi, P. Thompson, and W. A. Hyman, *Pontis Technical Manual*, prepared for the Federal Highway Administration, January 1992.

[Grigg 88] N. S. Grigg, *Infrastructure Engineering and Management*, John Wiley and Sons, New York, 1988.

[Haas 94] R. Haas, W. R. Hudson, and J. P. Zaniewski, *Modern Pavement Management*, Krieger Publishing Company, Malabar, Fla., 1994.

[Hamiel 92] J. Hamiel, "Changing Trends at U S. Airports as a Part of the International Scene," *Transportation Research Circular 393—Trends and Issues in International Aviation*, Transportation Research Board, Washington, DC, pp. 21–24.

[Horonjeff 94] R. Horonjeff, and F. X. McKelvey, *Planning and Design of Airports*, McGraw-Hill, New York, 1994.

[HRB 62] Highway Research Board, "The AASHO Road Test: Report 5-Pavement Research," *HRB Special Report 61-E*, National Research Council, Washington DC, 1962.

[Hudson 87] S. W. Hudson, R. F. Carmichael III, L. O. Moser, W. R. Hudson, and W. J. Wilkes, "Bridge Management Systems," *NCHRP Report 300*, National Cooperative Highway Research Program, Transportation Research Board, National Research Council, Washington, DC, 1987.

[Hudson 97] W. R. Hudson, R. Haas, and W. Uddin, *Infrastructure Management*, McGraw-Hill, New York, 1997.

[IATA 12] International Air Transport Association (IATA), "Publications & Interactive Tools," https://www.iata.org/ps/publications/Pages/wats-freight-km.aspx, accessed October 10, 2012.

[Kliesen 09] Kevin L. Kliesen, and Douglas C. Smith, "Digging into the Infrastructure Debate," *The Regional Economist*, Federal Reserve Bank of St. Louis, Missouri, July 2009. http://www.stlouisfed.org/publications/re/articles/?id=1309, accessed October 21, 2012.

[Kwiatkowski 86] V. F. Kwiatkowski, "Infrastructure Assets: An Assessment of User Needs and Recommendations for Financial Reporting," *Ph.D. thesis*, University of Kentucky, 1986.

[NCHRP 08] National Cooperative Highway Research Program, *GASB 34-Methods for Condition Assessment and Preservation*, Report 608, Transportation Research Board, Washington, DC, 2008.

[NSF 94] National Science Foundation, *Civil Infrastructure Systems Research*, Washington, DC, 1994.

[NSF 95] National Science Foundation, *Civil Infrastructure Systems—An Integrated Research Program*, Washington, DC, 1995.

[O'Day 84] D. K. O'Day, "Aging Water Supply Systems: Repair or Replace," *Infrastructure—Maintenance and Repair of Public Works, Annals of the New York Academy of Sciences*, Vol. 431, December 1984, pp. 241–258.

[Perry 89] N. J. Perry, "Good News about Infrastructure," *Fortune*, April 10, 1989, pp. 94–99.

[Queiroz 92] C. Queiroz, and S. Gautam, "Road Infrastructure and Economic Development—Some Economic Indicators," *Working Papers WPS 921,* Western African Department and Infrastructure and Urban Development, The World Bank, Washington, DC, June 1992.

[Queiroz 94] C. Queiroz, R. Haas, and Y. Cai, "National Economic Development and Prosperity Related to Paved Road Infrastructure," *Transportation Research Record 1455,* Transportation Research Board, National Research Council, Washington, DC, 1994, pp. 147–152.

[Stone 74] D. C. Stone, *Professional Education in Public Works/Environmental Engineering and Administration,* American Public Works Association, Chicago, 1974.

[Uddin 06] W. Uddin, "Air Quality Management Using Modern Remote Sensing and Spatial Technologies and Associated Societal Costs," *International Journal of Environmental Research and Public Health,* ISSN 1661-7827, MDPI, Vol. 3, No. 3, September 2006, pp. 235–243.

[Uddin 09] W. Uddin, "Light Detection and Ranging (LIDAR) Deployment for Airport Obstructions Survey," *Final Report,* Center for Advanced Infrastructure Technology, University of Mississippi, Prepared for Airport Cooperative Research Program, Transportation Research Board, Washington, DC, October 2009.

[Uddin 12a] W. Uddin, "Mobile and Area Sources of Greenhouse Gases and Abatement Strategies." Chapter 23, *Handbook of Climate Change Mitigation.* (Editors: Wei-Yin Chen, John M. Seiner, Toshio Suzuki, and Maximilian Lackner), Springer, 2012, pp. 775–840.

[Uddin 12b] W. Uddin, "Pavement Evaluation and Structural Strengthening Considering Surface Materials, Environmental Conditions and Natural Disaster Impacts," *Proceedings, MAIREPAV7—The Seventh International Conference on Maintenance and Rehabilitation of Pavements and Technological Control,* Auckland, New Zealand, August 28–30, 2012.

[Veldhuis 92] J. Veldhuis, "Impact of Liberalization on European Airports," *Transportation Research Circular 393—Trends and Issues in International Aviation,* Transportation Research Board, Washington, DC, pp. 25–30.

[Wagner 84] R. F. Wagner, Jr., "Infrastructure Issues Facing the City of New York," *Infrastructure—Maintenance and Repair of Public Works, Annals of the New York Academy of Sciences,* Vol. 431, December 1984, pp. 21–26.

[World 90] The World Bank, *World Development Report 1990,* Washington, DC, June 1990.

[World 09] The World Bank, *Research Guides,* http://researchguides.worldbanki.org, accessed April 5, 2009.

CHAPTER 2

Framework for Infrastructure Asset Management

2.1 Background

Infrastructure assets are physical facilities that provide essential public services as required by the economic and social needs of the public. As Sullivan stated at the 1983 New York Conference on Infrastructure: Maintenance and Repair of Public Works [Sullivan 84]: "There is nothing new about infrastructure. It is simply the fundamental basis on which this country was built—public works." Infrastructure assets and related services are provided by both public works agencies and private enterprises, which include energy, transportation, manufacturing, supply chain, utilities (water, gas, electric), solid waste disposal, parks, sports and athletic fields, recreational facilities, and housing. It is also reasonable to say that infrastructure asset management is not new. Management decisions must be made every day in public works agencies and private enterprises for project development, financing, construction, and maintenance. The exact style and type of management organization and activity often depends more on historical practice of the individual agency than on need or reality. Unfortunately, for a long time public agencies did not realize the importance of performance evaluation, maintenance programming, or other important keys to successful asset management [Hudson 97].

In the past, managing infrastructure has not always been systematic. There is a need to change the way we do business. The infrastructure asset management system (IAMS) concepts introduced in this book provide a framework and methods to integrate all phases involved with the provision of physical infrastructure facilities.

The terms infrastructure, asset, management, and system can mean different things to different people. Therefore, it is important to identify key terminology and provide technical definitions so that a uniform, common vocabulary exists for use in the infrastructure asset management field.

2.1.1   Terminology and Definitions

2.1.1.1   System

The word *system* has been appropriated for many purposes, such as circulatory system, drainage system, sprinkler system, highway system, parking system, wireless and online system, etc. The dictionary says that a system is a regularly interacting or interdependent group of items comprising a unified whole. This last definition will be used herein.

2.1.1.2   Infrastructure

The word *infrastructure*, as used in this book, refers to a group of physical systems or facilities that provide essential public services, such as:

- Transportation (mass transit, roads, airports, ports, railways, waterways, intermodal)
- Utilities (water, sewer, gas, electricity)
- Energy (grid network, traditional coal and other fossil-fuel power plants, hydroelectric, nuclear, geothermal, biofuel, and wind and solar renewable energy)
- Telecommunications and wireless communications
- Information technology and online social media
- Waste disposal and landfills
- Parklands, landmarks, and theme parks
- Sports complexes and recreational buildings
- Hospitals, public housing, rental housing, and education buildings
- Industrial, manufacturing, and commercial facilities
- Supply chain infrastructure assets (intermodal terminals, distribution centers)

Infrastructure can also include the management and human resources associated with providing a physical facility. Caution should be exercised referring to infrastructure as it is loosely used in recent online and social media, such as education infrastructure, health infrastructure, financial infrastructure, library IT infrastructure, mobile or cellphone infrastructure, banking infrastructure, social network infrastructure, etc. In this book, infrastructure only refers to physical systems and associated services essential for sustaining acceptable quality of life and social and economic needs of the public.

2.1.1.3 Assets

Assets in traditional literature and discussions usually refer to financial instruments. Infrastructure assets, as used in this book, is associated with a physical facility or an integral component of a particular physical system that is constructed and maintained to serve public needs. Infrastructure assets can be owned and managed by either or both public agencies and private enterprises.

2.1.1.4 Management

The word *management* has diverse meanings. To some it means to administer, to others it means control, and to still others it means coordinating the various elements of some unit or system. The dictionary definition of management is "the act or art of managing," or, less circularly, "the judicious use of means to accomplish an end." In this book, management means the coordination and judicious use of means and tools, such as funding and economic analysis to optimize output or accomplish a goal of infrastructure operation.

2.1.1.5 Infrastructure Asset Management

Infrastructure asset management includes the systematic, coordinated planning and programming of investments or expenditures, design, construction, maintenance, operation, and in-service evaluation of physical infrastructures and associated facilities. It is a broad process, covering those activities involved in providing and maintaining infrastructure assets at a level of service acceptable to the public, intended users, or owners. These activities range from initial information acquisition to the planning, programming, and execution of new construction, maintenance, rehabilitation, and renovation; from the details of individual project design and construction to periodic in-service monitoring and evaluation, and financial management.

2.1.1.6 Infrastructure Asset Management System

An IAMS framework consists of the operational package (methods, procedures, data, software, policies, decisions, budgets and funds, etc.) that links and enables the carrying out of all the activities involved in infrastructure asset management.

2.1.2 Ideal Infrastructure Asset Management System

An ideal IAMS would coordinate and enable the execution of all activities so that optimum use is made of the funds available while maximizing the performance and preservation of infrastructure assets and provision of services. It would serve all management levels in the organization (public or private) and would be structured to be adaptable to all of its infrastructure. In other words, it would be general in scope and incorporate particular models, methods, and procedures needed for specific types of infrastructure. For example, the concept of pavement management systems, initiated in the late 1960s [Haas 78], was later generalized to bridges [Hudson 87] and underground utilities in the 1980s and 1990s, and later to highway and municipal infrastructure management [Hudson 97, Hudson 11].

2.2 Key Issues for Infrastructure Asset Management

The infrastructure crisis was identified and discussed in Chapter 1. Public attention was certainly caught in the 1980s by such headlines as "America in Ruins" and "Crumbling Infrastructure" [Choate 81]. Huge needs and lack of funds to maintain and improve the U.S. infrastructure are often cited as the cause of this problem [ASCE 09], and this issue has been discussed vigorously in the post-2000 era within the U.S. government and Congress, as well as public forums worldwide. In recent years, public awareness of environmental sustainability has become important in infrastructure asset management practice by both public agencies and private entities [Uddin 12]. While cost is a factor, a major problem has been the lack of a comprehensive systematic approach to managing the infrastructure that involves the major issues explored in the following sections.

2.2.1 Infrastructure Decaying/Aging

The following key points are related to infrastructure decay and aging:

- The condition and level of service of infrastructure deteriorates through aging and usage.
- Catastrophic failure may happen because of design and/or construction deficiencies or overload/impact incidents.
- Some infrastructure components fail because of natural disasters, such as earthquakes and floods.
- Historically, design processes have not given adequate consideration to environmental effects and their interaction with loads and material variability.

2.2.2 Lack of Rational Maintenance, Preservation, and Renovation Programs

Lack of rational maintenance, preservation, and renovation programs is related to the following observations:

- Generally, past design practices were geared toward producing physical systems that would last a given design life with maintenance or future preservation treatments, such as renovation, given limited consideration.

- Routine maintenance was considered to be the responsibility of a maintenance group. The condition and level of service concepts suggested that facilities fail and are replaced.

- "Ad-hoc" maintenance practices in response to public complaints, emergency situations, and catastrophic failures are not adequate to sustain healthy infrastructure.

- Changes in use and inability to accurately predict future loads and service requirements have caused problems.

- Inadequate attention has been given to performance-prediction models.

These issues indicate a need for appropriate management of infrastructure assets.

2.2.3   Scarcity of Financing Resources

Traditionally, the federal government has financed most of the national public works infrastructure, while states and local agencies have financed infrastructure related to their jurisdictions. However, except for the 1998 to 2001 period with no deficit, the accumulated U.S. federal budget deficit has risen steadily since 2001 from $208 billion in 1983, $203 billion in 1994, $158 billion in 2002, $364 billion in 2005, to $1,293 billion (estimated 8.9 percent of GDP) in 2010 (http://www.usgovernmentspending.com/federal_deficit_chart.html, accessed November 1, 2012). There is strong pressure to cut federal spending and bring the deficit under control within 10 to 20 years. At the same time, competing demands of other societal needs make the federal budget a combination of solemn and deeply felt commitments to people, high-priority emergencies, and absolutely essential expenditures. The common justifications for public spending have extremely strong political appeal [Mathiasen 84], which has been used to revive economy and job creation in the United States and abroad after the 2008 global economic recession [Gerritsen 09].

Innovations are needed to identify financing resources. State and local governments need more flexibility in using available funds. Cost-effective solutions and better fund management are essential. Incidents like the bankruptcy of Orange County, California, in 1994 can shatter investor confidence in public works agencies. Better analytical tools are needed to improve priority programming. These considerations make it essential to adequately educate and train engineers and decision makers for cost-effective infrastructure management. This book is intended to provide the basic principles and techniques for improved infrastructure management, including innovative funding mechanisms.

2.2.4   Historic Inadequate Financial Reporting and Recent Developments

Infrastructure inventory and cost monitoring are important issues that have not been fully recognized in government accounting and financial reporting procedures. As a consequence, needed infrastructure information is not always available to decision makers. The U.S. Governmental Accounting Standards Board (GASB), established in 1984, initiated a first comprehensive look at accounting and reporting of infrastructure/fixed assets. Kwiatkowski's research assessed the needs of selected users of financial information about infrastructure in order to recommend procedures for financial reporting of infrastructure asset information [Kwiatkowski 86].

Valuation of public capital assets in the United States was promulgated in a GASB 34 statement on June 15, 1999. This standard was implemented by the USDOT for reporting of transportation infrastructure asset valuation and condition beginning after June 15, 2001 for agencies with revenues of $100 million or greater, after June 15, 2002 for agencies with revenues between $10 and $100 million, and after June 15, 2003 for agencies with revenues less than $100 million or greater [DOT 00]. The GASB 34 considers billions of dollars of government expenditures on long-lived public infrastructure assets built and owned by government agencies as long-term investments. Other similar public infrastructure asset valuation standards include the Public Sector Accounting Board Handbook Section 3150 (PSAB 3150) in Canada (http://www.municipalaffairs .alberta.ca/documents/ms/PSAB_3150_4_toolkit_full_document .pdf, accessed January 2, 2013), and similar standards in Australia.

GASB 34 requires all capital assets to be reported at historical costs over their estimated useful lives except infrastructure assets, which are reported using the modified approach [DOT 00]. The traditional financial reporting of capital assets includes: (1) beginning-of-year and end-of-year balances, (2) capital acquisitions, (3) sales or other dispositions, and (4) depreciation expenses. This financial-reporting approach is not new to most nongovernmental and commercial businesses. They have been following similar reporting format at the time of annual income-tax returns and take advantage of tax savings from straight-line depreciation of equipment and other physical assets with relatively short useful life. The modified approach of financial reporting (according to GASB 34) does not require government-owned infrastructure assets to be depreciated if the assets are:

- Managed using an IAMS
- Preserved at an established condition level

According to GASB 34, an IAMS should include an inventory of assets and condition assessment at least every three years.

The results of the three most recent assessments should show that the infrastructure assets are being preserved at or above the established condition level. Most highway agencies in the United States have been traditionally keeping computerized records of inventory and condition monitoring data for pavements and bridges, as mandated by the FHWA since the mid-1990s through implementation of pavement management systems and bridge management systems. These practices and data form the backbone of a highway agency's asset management system, considering that about 90 percent of the value of the assets of a highway agency are pavements, bridges, and maintenance treatments [Hudson 11]. Specifically, the following information and disclosures are required for reporting government-owned infrastructure assets using the modified approach [DOT 00, NCHRP 08]:

- Annual capital investment records
- Allocation of costs among capital investment, preservation (including replacement), and maintenance
- The basis of overall condition-rating scales and specific condition-level targets for each asset class as acceptable
- Linkage of budget levels to achieve condition targets
- Periodic condition assessment of assets, at least every three years
- Results of the assessed condition from the three most recent condition assessments
- Annual maintenance and preservation costs, both estimated and actual, for the past five years
- Factors that significantly affect the trends in the information of the preceding two items
- An account of infrastructure asset additions, retirements, and replacements
- Comprehensive annual financial reports (CAFR) and other pertinent information required annually for bond-rating agencies

The concepts and detailed guidelines for an effective asset management system presented in this book can help establish an effective working IAMS.

2.3   Application of Systems Methodology

A systems methodology can address various modeling and analytical issues related to infrastructure management. It has been successfully used in areas such as pavement management, bridge management, and infrastructure management of other assets [Haas 78, 94, Hudson 87, 97, 11].

Systems engineering comprises a body of knowledge developed for the efficient planning, design, and implementation of new systems, and for structuring the state of knowledge on an existing system or modeling its operation. There are two main, interrelated applications of systems methodology:

- The framing or structuring of a problem, body of knowledge, or process
- The use of analytical tools for actually modeling and solving the problem or for incorporation in the process

These uses are complementary and interrelated; one is insufficient without the other. The framing of a problem is usually too generalized by itself to achieve a useful operational solution, whereas applying analytical techniques to an inadequately structured problem may result in an inappropriate solution [Stark 72, 05].

The structure or framework of any problem-solving process should provide for systematic consideration of all the technical, economic, social, and political factors of interest. Moreover, it should be a logical simulation of the progression of activities involved in efficiently solving or dealing with a problem. A framework for infrastructure management is subsequently presented in this chapter. The following discussion concentrates on a summary of some analytical tools and some precautions. These are applicable within the general systems approach shown in Figure 2-1.

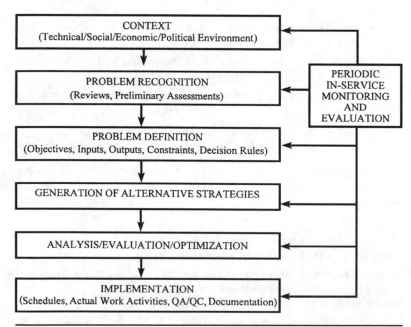

FIGURE 2-1   General systems methodology approach.

2.3.1   Some Analytical Tools

The structuring alone of a problem is usually too general to be useful in finding an operational solution. Moreover, applying analytical techniques, no matter how elegant or complex, to an inadequately structured problem will likely result in an inappropriate solution. In other words, analytical techniques for solving problems have maximum usefulness when the problems are well formulated or structured.

This section provides a "catalogue" of some of the more widely used techniques, tools, and models. Use of such techniques should facilitate reaching a decision on an objective basis. Their use depends largely on the available knowledge of the outputs of the system, which can be classified as follows:

- Certainty, where definite outputs are assumed for each alternative strategy (i.e., deterministic types of problem)

- Risk, where any one of several outputs, each with an identified or estimated probability, can occur for each alternative strategy

- Uncertainty, where the outputs are not known for the alternative strategies; thus probabilities cannot be assigned

A majority of engineering practice has treated problems in terms of certainty decisions because of convenience and the need to deal with many variables. However, there is considerable need to incorporate risk concepts into practice. Where practical problems are too complex for symbolic representation, Stark suggests that they may be modeled on an analogue or a scalar basis. Alternatively, it is possible to "force" a solution by experimentation, gaming, or simulation for some types of problems [Stark 05].

The optimization problems for finding cost-effective solutions can be solved by several approaches:

- Mathematical programming techniques, which produce an exact solution

- Heuristics methods, which are more common for large problems and produce suboptimal solutions

- Probabilistic approaches, which are based on random selection, biased sampling, and Monte Carlo simulations

- Graphical solutions

One of the most widely applied and useful classes of systems models involves linear programming. These techniques have been used in everything from construction to petroleum refinery operations because they are well suited to allocation-type problems [Gass 10]. A typical problem for a linear-programming application might involve determining how much of each type of material a materials supplier should produce, given production capacity, the number and capacities of trucks, available materials, and their costs, etc. There are several variations of linear-programming models and several methods of solutions, including parametric-linear programming, integer-linear programming, and piecewise-linear programming. The latter is used to reduce a nonlinear problem to approximate linear form in the area of interest.

Nonlinear methods range from the so-called classical use of differential calculus, Lagrange multipliers (and their extension to non-negativity conditions and inequality constraints), and geometric programming to iterative search techniques [Künzi 79]. These latter techniques often apply where more rigorous methods are impractical. There are some types of nonlinear problems not easily solved by analytical techniques that may lend themselves quite well to graphical solutions. For example, a simple graphical solution has been shown to be applicable to a construction problem involving a discontinuous-cost function [Haas 73].

Problems involving multistage decisions can be represented as a sequence of single-stage problems. These can be successively solved by dynamic programming methods [Dreyfus 65]. Each single-variable or single-stage problem that is involved can be handled by the particular optimization technique that is applicable to that problem. These techniques do not depend on each other from stage to stage and can range from differential calculus to linear programming. Combinatorial-type problems are often well suited to dynamic programming. A typical example is an aggregate producer with several mobile crushers and several sources of raw materials who wants to determine how many crushers should be assigned to each site for a given profit margin.

Random and queuing models have a wide range of applicability to systems problems, and a lot of literature is available. One class of models involves Monte Carlo methods, which are useful when adequate analytical models are not available. These probabilistic methods require distribution functions for the variables. There are also a large range of problems to which reliability, random walk, and Markov-chain techniques can be applied. The latter can be used to extend stochastic and chance-constrained programming models. Queuing models have been used extensively in engineering, including various air-terminal operations, traffic-facility operations, rail operations, and canal operations.

Many problems involve allocating and scheduling personnel, equipment, money, and materials. Several project-management techniques have found widespread use for these types of problems, including sequencing, routing, and scheduling.

This section has noted only a few of the many analytical tools that have potential applicability to various aspects of infrastructure management. Those desiring more in-depth information may consult some of the many references cited and others available in libraries.

2.3.2   Some Precautions in Application

The general systems approach of Figure 2-1 models the logical, systematic pattern that is used by efficient problem solvers. It must, however, be used with recognition that there are limitations.

- Successful application inherently depends on the capabilities of the people involved. The approach is not a substitute for poor judgment or poor engineering.

- The point of view of the individual or agency involved must be clearly recognized and identified. Otherwise, confusion and contradiction can result. For example, a materials-processing problem for a public works project might be viewed differently by the contractor than by the government agency involved. They may have competing objectives, and they will have different constraints.

- The components or extent of the system under consideration should be clearly identified. For example, to one person the term "parking system" might mean the actual physical parts of the parking lot, such as pavement, curbs, and gates; to another it might mean the method used to operate the facility; and to still another person it might mean a combination of the two.

- A fourth point concerns the inherent danger of generating precise solutions to an imperfectly understood problem. That is, the problem has been recognized but not yet rigorously defined. It is, of course, common to perceive some general solutions in the problem-recognition phase, but these may be inadequate or incomplete if the problem solver does not go further to define the problem.

2.4   Development of an Infrastructure Asset Management System

2.4.1   The Infrastructure Asset Management System Process

From the systems-engineering perspective, a system consists of a set of interacting components that are affected by certain exogenous factors or inputs. In a physical airport-facility system, the set of mutually interacting asset components usually includes:

- Ground-access facilities connecting the airport to the nearest city

- Parking structures and rental-car facilities

- An air-traffic control tower and other ground-based navigation equipment

- A terminal building (includes baggage handling, security areas, gates, concessions)
- Airside facilities (such as jetways, hangers, fire stations)
- Pavements: apron(s), taxiway(s), exitway(s), and runway(s)

Each of these components involves different types of construction. For example, all paved surfaces serving vehicular traffic and aircraft operations usually consist of pavement (with a surface layer, base and subbase layers, and subgrade), shoulder or sidewalk, and other appurtenances. The exogenous or external factors that affect the building structures and paved surfaces are age, traffic, environment, material degradation, disasters/accidents, and maintenance actions. Maintenance is carried out to preserve the functionality and structural integrity by reducing the rate of deterioration from the impacts of the traffic and environment inputs.

An IAMS, on the other hand, consists of such mutually interacting components as planning, programming, design, construction, maintenance and renovation, and evaluation. The overall framework for such a system is illustrated in Figure 1-7. Exogenous factors affecting an IAMS include budgets, decision criteria and maintenance policies, and nonquantifiable agency policies, including political climate, environmental stewardship, and sustainability considerations.

An ideal IAMS would help provide and maintain comfortable, safe, and economical physical infrastructure systems and related services at acceptable standards to the public, within the available funds. It would recognize the consequences of unwanted delays in the implementation of maintenance programs and assist decision makers in spending the available funds cost-effectively. The minimum requirements of such a system would include adaptability, efficient operation, practicality, quantitatively based decision-making support, and good information feedback. There is no ideal single IAMS that is best for all agencies. Every agency presents a unique situation with specific needs for various components of the physical infrastructure system. Any existing decision support or management system for infrastructure components should be carefully integrated if an overall infrastructure management system is to be developed. Each agency must establish objectives and goals for an infrastructure management system. These objectives and goals may vary considerably, depending on the jurisdiction of the agency at the national, state, and regional levels, and local levels of government or private enterprises.

The scope of an IAMS depends on the extent and size of the physical components of infrastructure systems for which an agency is responsible. In the case of a municipal IAMS, all public works infrastructure may be included in the scope. That generally implies city-street infrastructure, water-supply and sewer infrastructure, solid-waste facility, electricity and gas-supply infrastructure, mass-transit infrastructure, airport infrastructure, Intelligent Transportation System (ITS) infrastructure for video surveillance of traffic operations and incident management, and coliseum/convention hall/school and recreational facility infrastructure. Additionally, privately managed infrastructure may provide other essential services to communities, such as hospitals and medical infrastructure, malls and office complexes, manufacturing/industrial facilities, movie theaters and TV stations, fiber-optic-based landlines and wireless-communication infrastructure. There is a network or portfolio of each category of a physical-infrastructure system, which can be publicly, privately, or jointly owned and operated.

2.4.2 The Network and Project Levels of Infrastructure Management

Infrastructure management has two basic working or operational levels: network and project, as illustrated previously in Figure 1-7. Figure 2-2 expands the major activities occurring at each level for highway/street infrastructure. These activities are discussed in more detail in subsequent chapters.

Network level management has as its primary purpose the development of a priority program and schedule of work, within overall budget constraints. Project-level work thus comes on-stream at the

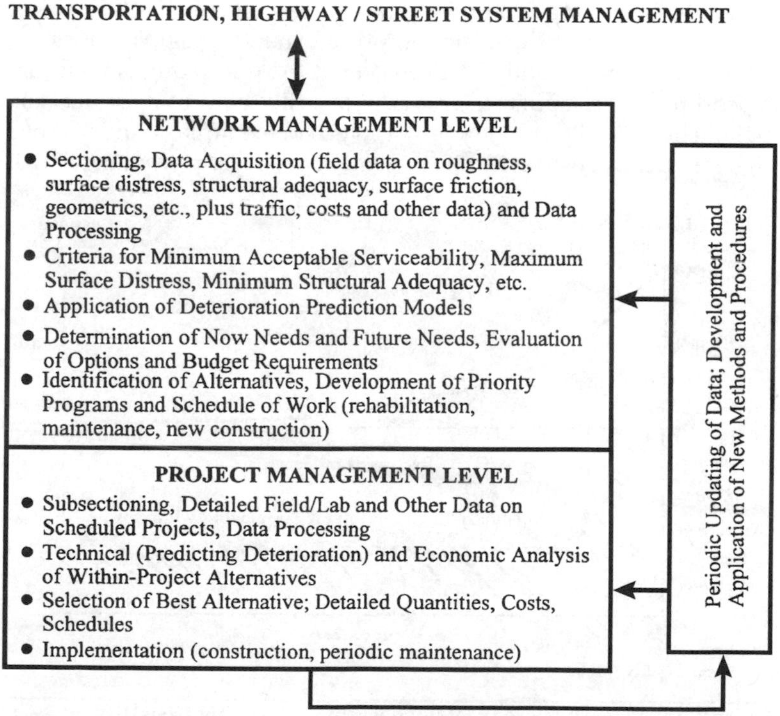

TRANSPORTATION, HIGHWAY / STREET SYSTEM MANAGEMENT

↑

NETWORK MANAGEMENT LEVEL

- Sectioning, Data Acquisition (field data on roughness, surface distress, structural adequacy, surface friction, geometrics, etc., plus traffic, costs and other data) and Data Processing
- Criteria for Minimum Acceptable Serviceability, Maximum Surface Distress, Minimum Structural Adequacy, etc.
- Application of Deterioration Prediction Models
- Determination of Now Needs and Future Needs, Evaluation of Options and Budget Requirements
- Identification of Alternatives, Development of Priority Programs and Schedule of Work (rehabilitation, maintenance, new construction)

PROJECT MANAGEMENT LEVEL

- Subsectioning, Detailed Field/Lab and Other Data on Scheduled Projects, Data Processing
- Technical (Predicting Deterioration) and Economic Analysis of Within-Project Alternatives
- Selection of Best Alternative; Detailed Quantities, Costs, Schedules
- Implementation (construction, periodic maintenance)

Periodic Updating of Data; Development and Application of New Methods and Procedures

FIGURE 2-2 Basic operating levels of pavement management and major activities [Haas 94].

appropriate time in the schedule, and represents the actual physical implementation of network decisions.

2.4.3   Influence Levels of IAMS Components

Four of the major components or subsystems (planning, design, construction, and maintenance) have important but changing impacts in terms of a "level-of-influence" concept. This concept (Figure 2-3), which has been used in sectors of industry, such as manufacturing and heavy industrial construction [Barrie 78], shows how the potential effect on the total life-cycle cost of a project decreases as the project evolves [Haas 94].

The lower portion of Figure 2-3 presents a simplified picture, in bar-chart form, of the length of time each major component acts over the life of the infrastructure facility. The upper portion shows plots of increasing expenditures and decreasing influence over the infrastructure life. Expenditures during the planning phase are relatively small compared with the total cost. Similarly, the capital costs for construction are a fraction of the operating and maintenance costs associated with service life. However, the decisions and commitments made during the early phases of a project have far greater relative influence on later-required expenditures than some of the later activities.

At the beginning of a project, the agency controls all (100 percent influence) factors in determining future expenditures. The question is, to build or not to build? A decision not to build requires no future expenditure for the project. A decision to build requires more decision making but initially at a very broad level. For example, in the case of a highway, should it be a flexible pavement or a rigid pavement, and, if

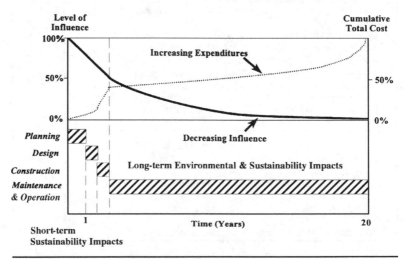

FIGURE 2-3   Influence levels of IAMS subsystems on the total costs [after Hudson 97].

rigid, with joints or continuously reinforced? How thick should it be and with what kind of materials? Once decisions are firm and commitments are made, the further level of influence of future actions on the future project costs will decrease.

In the same manner, decisions made during construction, even within the remaining level of influence, can greatly impact the costs of maintaining or rehabilitating the infrastructure. For example, lack of quality control or substitution of inferior materials may save a few dollars in construction costs, but the extra maintenance costs and user costs caused by more frequent maintenance activities may consume those "savings" several times over.

With construction completed, attention is now given to maintaining the existing infrastructure at a satisfactory level. The level-of-influence concept can also be applied to the subsystems of a maintenance management system (MMS). Expenditures during the planning phases of rehabilitation and renovation are relatively small compared to the total maintenance cost during the service life of the facility. However, the decisions and commitments made during early phases of a rehabilitation and renovation project have far greater relative influence on what other maintenance expenditures and user costs will be required later.

Some define a three-level concept of IAMS that represents the situation that exists in a number of agencies. Also, the terminology sometimes overlaps; for example, in some papers when "project level" is mentioned, the "project-selection level" actually is meant, and in other cases when the "network level" is mentioned, it is the "program level" that is involved.

This three-level concept is illustrated in Figure 2-4. The lower-left triangle represents an area of unreliability because too little

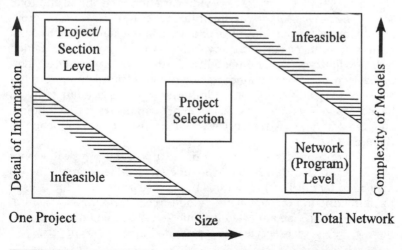

FIGURE 2-4 Information detail and complexity of models for a three-level IAMS.

information is available for models at the project level, and the upper-right triangle is an area infeasible for modeling because of the size and complexity of the required models.

2.5   Life-Cycle Analysis Concept

2.5.1   Service Life or Whole Life

The physical infrastructure facilities discussed in Chapter 1 are generally fixed assets. From the design and analysis point of view, some finite number of years of the design life/analysis period is associated with each component of infrastructure. In reality, the public and users expect the infrastructure to provide a particular service forever, unless a catastrophic failure occurs or the area is uninhabited. However, the responsible agency managers and decision makers know that there comes a time when the infrastructure facility cannot provide adequate service because of one or more reasons, such as:

- It is structurally unsafe.
- It is functionally obsolete.
- It causes delay and inconvenience to the users because of overuse and overdemand.
- It is costly to maintain and preserve.
- There is catastrophic failure caused by natural disasters or other incidents.

This leads to the concept of the "service life" or "whole life" of an infrastructure within a life cycle. Unlike the design/analysis period, service life is typically not a single number. The same type of facility (e.g., a steel bridge) may have a wide variation in its initial and total service life because of the varying influence of traffic history, environmental inputs, and maintenance practices. Maintenance history has a significant influence on total service life. An adequately maintained facility will have a better probability of extended service life, as compared to a poorly maintained facility. A good infrastructure management system recognizes the importance of service-life analysis, including agency costs (for construction, maintenance, rehabilitation, and renovation/replacement) as well as user costs and benefits (Figure 1-6).

The concept of service life is based upon the physical service life, as contrasted to a social/economic service-life estimate, which may be different. Throughout this book, physical service life is used for infrastructure management. Some typical expectations of infrastructure service life are listed in Table 2-1. A detailed discussion on methods to estimate service life is provided in the next chapter.

Infrastructure Facility and Components		Expected Service Life
Airports	Buildings/structures	Up to 150 years
	runways/taxiways/ aprons	Up to 50 years
Bridges	Decks	Up to 50 years
	Substructure/ superstructure	Up to 125 years
Tunnels	(For auto traffic, water)	Up to 200 years
Ports, rail, and intermodal facilities	(Concrete/steel/stone construction)	Up to 300 years
Public buildings and sports complexes	(Concrete/steel/brick construction)	Up to 300 years
Electricity Transmission/ Telephone lines	(Concrete/steel construction)	Up to 400 years
Nuclear power plants	(Concrete/steel construction)	500 years or more
Hydraulic dams	(Concrete/steel construction)	300 years or more

TABLE 2-1 Typical Expectations of Infrastructure Service Life

2.5.2 Environmental Stewardship and Sustainability Dimensions

This book presents the framework and elements of life-cycle management of civil-infrastructure assets in the context of current practice, as well as addressing the adverse environmental impacts of deteriorating air quality, water-quality degradation, and sustainability concerns related to carbon footprint and resource conservation.

2.5.2.1 Environmental Stewardship Concerns

The environment and energy resources are greatly impacted by growth of cities and land-use practices, traffic congestion, and transportation-system management. The built-infrastructure of densely populated cities and intercity travel by passenger and freight traffic lead to significant adverse impacts on traffic congestion, air quality, energy consumption, toxic vehicle emissions, and carbon dioxide (CO_2). A metropolitan area includes both the urban area(s) and rural area(s) that are socially and economically integrated with a particular city. Frequent congestions and transportation-choke points in most urban areas and cities adversely affect travel time, business operating costs, and air quality [Uddin 12]. Increased air pollution in cities and areas with industries, particularly and increase in particulate matter (PM) and tropospheric ozone (O_3), is affecting public health because of the increased risk of respiratory diseases and higher mortality rates [Uddin 06]. Associated medical costs related to these public-health hazards and lost productivity can be enormous.

Despite the progress made in the United States over the last 40 years in cleaner fuel technology, fuel-efficient automobiles, vehicle-inspection regulations, over 160 million tons of pollution are emitted into the air each year, and about 146 million people live in counties where monitored air in 2002 was unhealthy at times because of high levels of at least one of the six principal air pollutants. This has resulted because the transportation infrastructure has expanded; economic prosperity and vehicle ownership has increased; longer trips are made in a typical day because of urban sprawl; and congestion hours increase travel time, wastage of fuel, and CO_2 and other toxic emissions of nitrogen oxide (NO_2), carbon monoxide (CO), and PM [Uddin 05, 12].

Many emerging megacities of the developing world with a population of 10 million or more (São Paulo in Brazil, Beijing and Shanghai in China, Delhi in India, and Mexico City) are creating air pollution at an alarming rate, which has exceeded some of the most polluted megacities of the industrialized nations (Moscow, Tokyo, New York City, and Los Angeles). Urban growth and sprawl in both metro and non-metro cities destroys parks, agricultural land, forestland, and open spaces. Biodiversity and the ecosystem are adversely impacted. Additionally, urban growth leads lots of city population at risks of flood and other natural hazards.

2.5.2.2 Sustainability Dimensions

Sustainability is broadly defined as a preservation, development, and management measures that meet the needs of the present without compromising the ability of future generations to meet their own needs [Uddin 12]. Sustainability dimensions and goals with respect to built infrastructure, passenger and freight traffic, urban growth, energy demand, and the environment include:

- Enhancing health, safety, and security
- Conserving energy and enhancing the environment
- Creating equitable and livable communities
- Reducing costs and finding innovative financing solutions
- Promoting economic prosperity

Current mobility and life-style dependence on fossil-fuel consumption is definitely not sustainable, as indicated by global demand on natural sources and its adverse impacts:

- Higher GDP is associated with higher energy consumption and CO_2 emissions.

- Travel demand (number of vehicles and vehicle-km traveled) is at its highest level.

- Migration from rural areas to urban areas and mobility needs are all accelerating adverse impacts on biodiversity and the ecosystem.

Table 2-2 shows 2009 energy generation by source for the United States and the world. The share of non-fossil-fuel electricity generation has been increasing for the last decade. Although nuclear-power infrastructure is still a sizable source of energy in the United States, Japan, France, and other countries, the 2011 tsunami disaster of the Fukushima nuclear power-plant meltdown in Japan has fueled public opinion in Japan and many other countries against building more nuclear power plants. On the energy front, efforts have been made in the United States and abroad to reduce dependence on fossil-fuel–derived electricity generation and increase innovative renewable-energy infrastructure (wind and solar) in addition to traditional renewable hydraulic-power generation. Renewable-energy infrastructure assets are relatively new areas of asset management considering their rapid expansion because of increased levels of capital investment. In 2011, total new investment in clean energy increased five times more than 2004 levels to $260 billion, despite the sluggish global economy and

Sources	2001 U.S.	2009 U.S.	2009 World
Fossil Fuel	70% (total)	69% (total)	67% (total)
Coal	51%	45%	41%
Petroleum	3%	1%	5%
Natural Gas	16%	23%	21%
Nuclear	20%	20%	13%
Hydroelectric	7%	7%	16%
Other renewable energy (geothermal, solar, wind, waste)	3%	4%	3%

Data sources: U.S. Energy Information Administration (EIA); International Energy Agency (IEA)

TABLE 2-2   Sources of Energy Generation, United States and World [after EIA 09]

Sectors of Economy	U.S.	World
Energy	34%	46%
Transportation	28%	25%
Industry (manufacturing)	20%	23%
Others (residential, agriculture, etc.)	18%	6%

Data sources: U.S. Energy Information Administration (EIA); International Energy Agency (IEA)

TABLE 2-3   Distribution of GHG Emission from Fuel Combustion by Sectors, 2007 [after EIA 09]

funding constraints faced by manufacturers, with investment in solar far outstripping that in wind (http://www.bnef.com/PressReleases/view/180, accessed October 30, 2012).

Anthropogenic CO_2 emission is a major component of greenhouse gas (GHG) emissions. Between 1990 and 2007, the U.S. inventory of anthropogenic GHG emissions included electricity, transportation, industrial, agricultural, residential, and commercial sectors. The transportation sector accounted for 28 percent of all anthropogenic GHG emissions in the United States trailing the energy sector at 34 percent [EIA 09]. Table 2-3 compares these numbers with global GHG distribution by sectors of economy. These two sectors are the top producers of GHG emissions. Several international accords have been made since the 1997 Kyoto protocol to establish country targets for reducing GHG emissions [Uddin 12]. On-road vehicles contribute 85.5 percent of all 2008 transportation-related GHG emissions in the United States, 86 percent in Canada, and 91 percent in Mexico; these three countries are all partners in the North American Free Trade Agreement (NAFTA) corridor [CEC 11].

Figure 2-5 shows that road–traffic-related CO_2 emission is higher per capita for several rural and smaller cities compared to large metropolitan areas in the United States and other megacities in Latin American and Asian countries. It is estimated that 75 percent of all GHG emissions worldwide are produced by cities, and one-third of this is from transportation sources [Uddin 12]. Anthropogenic GHG emissions are generally considered a primary source of global warming.

Climate effects, whether in terms of global warming caused by GHG emissions or climate change (as addressed, for example, in the Kyoto Protocol of 1997 and subsequent international agreements) are reportedly having major impacts on some types of civil infrastructure because of increased weather-related natural disasters worldwide. It is important to have adaptation strategies where needed and/or feasible. Some examples of international efforts to reduce GHG emissions include the 2010 Cancun agreement targeting reduction of deforestation and the 2012 Rio de Janeiro

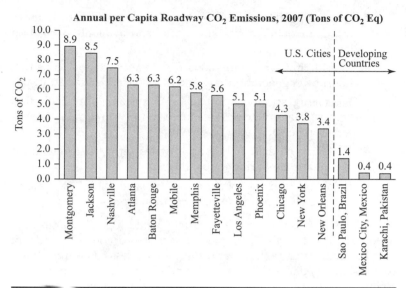

FIGURE 2-5   Annual per capita CO_2 emissions from road traffic in selected cities, 2007 [after Uddin 12].

accord emphasizing clean mass transit and energy-efficient transportation infrastructure.

2.5.3   Concluding Remarks

A very important contribution to the subject of this chapter is an ASCE Monograph [Antelman 08], which documents effective infrastructure reporting practices, demonstrates the linkage between asset management and good infrastructure, and identifies opportunities for improving current practices.

Key concerns in developing and managing infrastructure assets, sustainability issues, preservation of the environment, and technological solutions are highlighted as follows:

- Adding new and/or updated methodologies on performance measures related to sustainability, security, level-of-service provisions, safety, institutional effectiveness, preservation of assets and investments, technologies for in-service monitoring, data analysis, prioritization, and decision support and knowledge management

- Providing a foundation for improved leadership in infrastructure management, more effective and long-term life-cycle maintenance, explicit recognition of stakeholder needs and issues, promoting sustainability, and developing long-range strategic plans that maximize human, financial, technical, and environmental resources

- Addressing the adverse environmental impacts of deteriorating air quality, excessive noise, and available mitigation strategies
- Considering GHG emissions from transportation, construction, power generation, etc., and available mitigation strategies
- Examining specific approaches to sustainability including "green" methodologies, resource conservation, reuse of materials, renewable energy, etc.
- Financing models for publicly owned infrastructure and for Public-Private-Partnerships (P3s)
- Using quantifiable performance measures tied to realistic policy objectives and implementation targets

This book implicitly recognizes that while we will continue to need most if not all the infrastructure services of today, the advent of cyber infrastructure (Internet resources, wireless communications, online social media, mobile apps, and network and cloud-computing technologies that control and operate most critical lifeline infrastructure and essential services, and others) can profoundly impact the way these services are assessed, protected from cyber attacks, and in some cases even the way they are provided over the life cycle or whole life.

2.6 References

[Antelman 08] Albert Antelman, James J. Demprey, and Bill Brodt, "Mission Dependency Index—A Metric for Determining Infrastructure Criticality," *ASCE Monograph on Infrastructure Reporting and Asset Management*, American Society of Civil Engineers, 2008.

[ASCE 09] American Society of Civil Engineers (ASCE), "Report Card for America's Infrastructure: 2009 Grades," http://www.infrastructurereportcard.org/, accessed May 28, 2012.

[Barrie 78] D. Barrie, and B. Paulson, *Professional Construction Management*, McGraw-Hill, New York, 1978.

[CEC 11] Commission for Environmental Cooperation (CEC), "Destination Sustainability," CEC Secretariat Report Pursuant to Article 13th of the North American Agreement on Environmental Cooperation, Montreal, Canada, 2011.

[Choate 81] P. Choate, and S. Walter, *America in Ruins: The Decaying Infrastructure*, Duke Press, Durham, North Carolina, 1981.

[DOT 00] Department of Transportation, "Primer: GASB 34," Federal Highway Administration, Office of Asset Management, Washington, DC, November 2000.

[Dreyfus 65] F. E. Dreyfus, *Dynamic Programming and the Calculus of Variations*, Academic Press, San Diego, Calif., 1965.

[EIA 09] Energy Information Administration, "International Energy Annual Outlook," U.S. Department of Energy, Washington, DC, 2009.

[Gass 10] S. I. Gass, *Linear Programming Methods and Applications*, Fifth Edition McGraw-Hill, New York, 2010.

[Gerritsen 09] E. J. Gerritsen, "White Paper: The Global Infrastructure Boom of 2009–2015," Commentary, *The Journal of Commerce Online*, May 19, 2009, http://www.joc.com/commentary, accessed August 29, 2011.

[Haas 73] R. C. G. Haas, W. A. McLaughlin, and V. K. Handa, "Systems Methodology Applied to Construction," *Proceedings, National Conference on Urban Engineering Terrain Problems*, Montreal, May 1973.

[Haas 78] R. C. G. Haas, and W. R. Hudson, *Pavement Management Systems*, McGraw-Hill, New York, 1978.

[Haas 94] R. Haas, W. R. Hudson, and J. P. Zaniewski, *Modern Pavement Management*, Krieger Publishing Company, Malabar, Fla., 1994.

[Hudson 87] S. W. Hudson, R. F. Carmichael III, L. O. Moser, W. R. Hudson, and W. J. Wilkes, "Bridge Management Systems," NCHRP Report 300, National Cooperative Highway Research Program, Transportation Research Board, National Research Council, Washington, DC, 1987.

[Hudson 97] W. R. Hudson, R. Haas, and W. Uddin, *Infrastructure Management*, McGraw-Hill, New York, 1997.

[Hudson 11] W. R. Hudson, and R. Haas, "Keynote Presentation: Progress Assessment of PMS," *8th International Conference on Managing Pavement Assets*, Santiago, Chile, November 15–19, 2011.

[Kwiatkowski 86] V. F. Kwiatkowski, "Infrastructure Assets: An Assessment of User Needs and Recommendations for Financial Reporting," *Ph.D. thesis*, University of Kentucky, 1986.

[Künzi 79] H. B. Künzi, W. Krelle, and R. von Randow, and W. Oettli, *Nonlinear Programming*, Blaisdell, New York, 1979.

[Mathiasen 84] D. G. Mathiasen, "Federal Perspectives on Public Works Infrastructure: One Person's View from the Executive Branch," *Infrastructure—Maintenance and Repair of Public Works, Annals of the New York Academy of Sciences*, Vol. 431, December 1984, pp. 5–11.

[NCHRP 08] National Cooperative Highway Research Program, GASB 34—Methods for Condition Assessment and Preservation, Report 608, Transportation Research Board, Washington, DC, 2008.

[Stark 72] R. M. Stark, and R. L. Nicholls, *Mathematical Foundations for Design: Civil Engineering Systems*, McGraw-Hill, New York, 1972.

[Stark 05] R. M. Stark, and R. L. Nicholls, *Mathematical Foundations for Design: Civil Engineering Systems*, McGraw-Hill, New York, 2005.

[Sullivan 84] R. J. Sullivan, "Look Ahead, Never Back," *Infrastructure—Maintenance and Repair of Public Works, Annals of the New York Academy of Sciences*, Vol. 431, December 1984, pp. 15–17.

[Uddin 05] W. Uddin and K. Boriboonsomsin. "Air Quality Management Using Vehicle Emission Modeling and Spatial Technologies," *Proceedings, Environment 2005 International Conference: Sustainable Transportation in Developing Countries*, Abu Dhabi, UAE, January 30–February 2, 2005.

[Uddin 06] W. Uddin, "Air Quality Management Using Modern Remote Sensing and Spatial Technologies and Associated Societal Costs," *International Journal of Environmental Research and Public Health*, ISSN 1661–7827, MDPI, Vol. 3, No. 3, September 2006, pp. 235–243.

[Uddin 12] W. Uddin, "Mobile and Area Sources of Greenhouse Gases and Abatement Strategies." Chapter 23, *Handbook of Climate Change Mitigation*. (Editors: Wei-Yin Chen, John M. Seiner, Toshio Suzuki and Maximilian Lackner), Springer, New York, 2012, pp. 775–840.

CHAPTER 3

Planning, Needs Assessment, and Performance Indicators

3.1 Infrastructure Planning

Planning involves future-related activities that are concerned with achieving desired goals. As shown in Figure 1-7, planning functions are primarily concentrated at the system-wide or network level and deal with demand forecasting, site/corridor selection, financing, budgeting, and policy issues. Modern concepts of planning recognize two distinct strategic and tactical approaches [Binder 92]; additionally, sustainability planning must be incorporated.

- *Strategic planning* is generally long-range and reflects financial and business aspects of planning, involving senior administration and/or corporate managers.

- *Tactical planning* usually reflects technical aspects of facilities and involves technical managers who are responsible for facility management and future expansion within the bounds of the strategic plans in consultation with senior administration and/or corporate managers.

- *Sustainability planning* is related to both short-term and long-term impacts of any infrastructure planning, construction, and service phases.

The major emphasis of tactical planning is on network-level needs, including such activities as preparing and updating master plans, assessing needs and budgets, predicting future demands and developing strategies to preserve and upgrade facilities, and annual and multiyear work programming. Planning for sustainability has become an important focus of the public perception in the 21st century, which requires decisions makers at both strategic and tactical levels to embrace environmental stewardship and develop/use measures for sustainability performance. More on these sustainability issues is discussed in Part 5.

A practical and effective infrastructure asset management system (IAMS) integrates planning, design, and construction with the service-life activities of maintenance, rehabilitation and renovation, replacement and reconstruction (M,R&R). This concept is illustrated in Figures 1-6 and 2-3. Additionally, in the 21st-century world all these IAMS functions need to be executed with:

- Maximum security and safety of critical lifeline-infrastructure assets to safeguard against natural disasters and manmade disaster concerns

- Minimum adverse impacts on the environment (including biodiversity and ecology)

- Reduction in energy consumption where feasible and use of new raw materials; more recycling and reuse

- Sustainability performance by minimizing carbon footprint during construction and service life

- Consideration of other sustainability dimensions related to financial and technical operations

3.1.1   Modeling and Simulation

Physical facilities are subjected to many conditions, depending upon their location and type of loading or usage. The complexity of behavior and performance of a facility is impacted by construction quality, material degradation and aging, and their interrelationships. Because of these factors, a coordinated framework of systems-analysis concepts is appropriate to infrastructure management. This idea was first conceptualized by Hudson et al. for pavement design [Hudson 68]. A simplified diagram of this modeling process, which is also generically applicable to other infrastructure, is shown in Figure 3-1.

The scientific and engineering aspects of a systems problem span a spectrum of activities [Haas 94]:

- Use of physical observation or measurements to characterize behavior

- Statements of mathematical models that describe or approximate the physical phenomena

- Development of a system for prescribed behavior using the mathematical models

- Physical realization of the system

Therefore, it is essential to formulate a system in terms of physical or mathematical models or to use computer simulation to develop

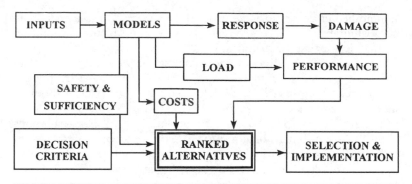

Figure 3-1   A simplified diagram of the major steps of modeling and system analysis to improve project-level pavement activities.

the necessary outputs. As an example, consider the bridge infrastructure in California. The project-level IAMS, in this case a bridge management system (BMS), is concerned with improved bridge design that resists seismic loading. An ideal way to look at selected alternate designs is with computer simulation.

The computer models involve earthquake-like motions based on previously recorded earthquakes in the region. Subsequently, the model outputs are acceleration, stresses, and deformation, which are compared with the measured observations, wherever possible. Once a design and the simulation results are verified, then further simulations can be done under different design inputs and magnitudes of seismic motion to optimize a specific design. This modeling process also allows studying of the use of new, innovative, and high-performance materials of construction.

Network-level IAMS (BMS in this example) is concerned with setting priorities for M,R&R work on substandard bridges in the network that are posted for load restrictions, near failure, approaching a critical condition threshold, and/or needing strengthening in the active seismic areas in California.

Computer modeling and simulation applications are commonly used in optimizing electricity-grid networks, cell towers for wireless networks, and transit routing. Transport modeling and simulation applications in a supply-chain network are targeted to evaluate both multimodal-freight flow and logistics optimization to reduce travel time and overall costs of freight inventory and distribution centers.

3.1.2   Space-Use Planning
For planning new facilities and M,R&R actions, it is essential to examine the space use and forecast the usage demand on the facility for the selected performance period during the service life of the facility. This will assist in selecting the most appropriate combination of

materials and design to achieve satisfactory performance. Adequate consideration should be given to space use and future demand on the facility during the planning stage.

The concept of a three-dimensional space for planning of infrastructure is useful for maximizing the available space in modern urban areas. Land-use planning has been traditionally carried out by architects, engineers, and planners. This is a two-dimensional view of planning needs. Most of the master plans and zoning regulations using this two-dimensional approach are prepared for cities, counties, and regional metropolitan agencies. The concept of "air space" (e.g. the vertical dimension) has been largely restricted to aviation facilities and airways. Similarly, underground and subsurface space has been solely used for subways, sewer and water lines, utilities, and shipping channels. "Space-use" planning [Bragdon 95] is a more accurate approach for today's crowded urban regions, as illustrated in Figure 3-2.

Citing many examples of space-use planning in Japan, Europe, and Washington State, Bragdon (1995) outlines a strategic master-planning process considering and utilizing vertical and horizontal dimensions [Bragdon 95]. Examples include multilevel golf courses in Tokyo, commercial/office complexes in Seattle Freeway Park over Interstate 5 in Washington, and the historic bridge on the Po River in Florence, Italy, with mixed recreational, transportation, and commercial uses. The three-dimensional space concept and

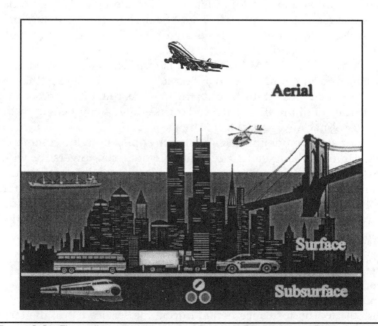

FIGURE 3-2 Three-dimensional space planning [after Bragdon 95].

other innovative aviation and intermodal transportation concepts are being pursued at the National Aviation and Transportation (NAT) Center on Long Island, New York, and at the NAFTA (North American Free Trade Agreement) Intermodal Transportation Institute at the University of Texas at El Paso [Bragdon 96]. This discussion points out that future construction and M,R&R work should consider the air and underground spaces in addition to the existing ground structures, to generate cost-effective alternatives. Efficient use of all three spaces is imperative in cities and urban-infrastructure planning as urban population grows, demand of energy and transportation increases, and public awareness leads to environmental-degradation concerns and sustainability aspirations to reduce carbon footprint and demand on fossil fuel and other natural resources.

3.1.3 Demand Forecasting

The accurate prediction of future demand on an infrastructure facility is crucial to the selection of an appropriate alternative for new construction or M,R&R actions. Many statistical and analytical tools are available for developing forecasting models. These tools include cross-correlation analysis, regional market-share methods, regression analysis, time-series models, and neural network methods.

All of these techniques and models require historical data of possible explanatory variables and demand (response) variables. If the historical data is not available, then either the data for a similar situation/location or simulated data must be used to develop preliminary demand-forecasting models. Plotting of historical data can show the possible form of models if either age or usage indicator (e.g., cumulative traffic on a bridge) is used as a single explanatory variable (independent variable). Figure 3-3 illustrates some examples of possible linear and nonlinear predictive models.

Model development is a three-step process:

1. Preliminary model based on historical data
2. Model verification using another set of data
3. Model calibration using alternate data collected under different conditions

Note that some nonlinear models can be linearized by a transformation of variables, such as using the logarithm of the variable, as shown in the bottom-right plot. These modeling techniques are further discussed in Chapters 4 and 8.

Once a predictive equation is verified and calibrated, it can be used to predict the demand in a future year by entering the known or estimated values of the independent variable(s) for that year.

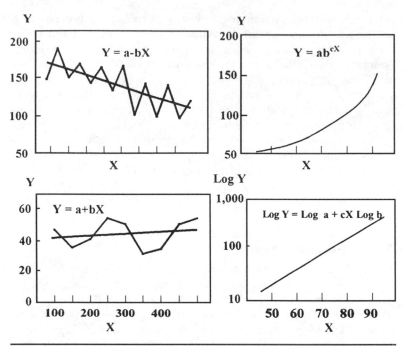

FIGURE 3-3   Typical trend variation of data and possible model forms.

3.1.4   Environmental Impact Studies and Sustainability Concerns

For capital investment projects, environmental-impact studies are often required to evaluate alternatives during the planning stage to comply with state and federal regulations. The effect of planned development must be considered on surrounding communities, water bodies, wetlands, ecological systems, air quality, surface and subsurface contaminations, noise pollution, and other areas of community concern. The National Environmental Policy Act of 1969 in the United States (Public Law 91-190) requires a detailed statement on the environmental impact of the proposed action and effects on the quality of the human environment [FAA 86]. There are four broad groups that must be considered to evaluate the impact of any infrastructure development: (1) pollution factors, (2) ecological and biodiversity factors, (3) social factors, and (4) engineering factors. Some examples of these factors are:

- *Pollution factors:* Air quality, water quality, noise, construction impacts on surface and subsurface soil contamination, urban-runoff pollution, wastewater treatment, appropriate waste disposal
- *Ecological and biodiversity factors:* Wetlands, coastal zones, wildlife and waterfowl, endangered species, animal and bird habitats, flora and fauna, landscape and drainage, ecosystem disturbance

- *Sustainability factors:* Energy consumption, carbon emissions from coal-fired electricity generation, industrial and vehicle emissions from burning of fossil fuel, other greenhouse gas (GHG) emissions

- *Social factors:* Displacement and relocation of residences and businesses, parkland and recreational areas, historical and archeological sites, cultural and religious places, natural and scenic beauty, land development, and social justice for vulnerable communities

- *Engineering factors:* Stormwater drainage, flood hazards, use of energy and natural resources, costs and benefits of alternatives

Demand forecasts are also used to analyze environmental impacts and evaluate effects of the development on the above factors and related areas of community concerns. For example, noise pollution and their effects on surrounding communities are studied whenever an airport or highway expansion and construction is planned.

3.1.6   Safety and Security

Safety and security are important considerations for planning, designing, and operating infrastructure facilities. Security is needed to: (1) prevent losses caused by theft, vandalism, and arson; (2) minimize the risk of possible safety hazards to the occupants and users of the facility; and (3) enforce measures for complying with the applicable laws to avoid liability claims.

Planning for safety and security will depend upon risk assessment (e.g. a low-risk or a high-risk facility). The most important federal laws pertaining to safe and healthful working and operating conditions was passed in 1970 and led to the creation of the U.S. Occupational Safety and Health Administration (OSHA).

The Department of Homeland Security was created in the United States after the terrorist attacks on September 11, 2001 on the World Trade Center (WTC) buildings in New York, the Pentagon in Washington, DC, and the Pennsylvania crash, which led in large part to a vastly increased demand for transportation security initially. A post-2001 review of the collapse of the targeted WTC towers following the initial impact of the jet-fuel filled aircraft and the collapse of the 47-story WTC 7 showed the structural vulnerability of these modern steel-and-concrete skyscrapers, mainly in terms of the damage from very high temperatures due to the fire in the WTC 7 building [NIST 03]. In view of this tragic event, security became an important function of infrastructure-asset management of airports and other mass-transit stations, as well as ports, electricity-power stations, nuclear-power plants, dams, and sports stadiums where large crowds could become targets. This increased the overall cost of operations of these facilities because of the erection of physicals barriers and electronic video-surveillance equipment and staff.

Increased investment in facility and supply-chain security by manufacturing companies and commercial businesses has generally resulted in economic benefits because of improved condition of facilities, such as: reduction in excess inventory, less lost cargo and better product quality, improved on-time delivery, and increased customer satisfaction. The electronic-security technology demand led to creation of new businesses and jobs, which are additional benefits of the post-2001 era.

3.1.6 ADA Concerns

The Americans with Disabilities Act (ADA), passed in 1990, is one of the most significant federal regulations to guarantee access and physical accommodations to people with disabilities at work and public places. The law is applied to all organizations with 15 or more employees. The implementation of ADA provisions requires awareness and sensitivity to the psychological and physical environment and removal of architectural and communications barriers. All existing facilities and new facilities must accommodate the ADA requirements. Future planning must, therefore, take ADA seriously, as suggested by Prior in the following key points of influence of the ADA [after Prior 94]:

- ADA has changed how people think about themselves in relationship to the workplace.
- ADA is a reflection of heightened human concerns.
- ADA is making us aware that we can do a better job when it comes to workplace accommodations.
- ADA should be viewed as a blueprint for the workplace of the future to help standardize workplace and worker expectations.

3.2   Examples of Planning Studies

Planning studies involve population projections, land-use and space-use predictions, usage (traffic) demand projections, and economic studies. The following case studies are presented to illustrate the application of analytical tools for facility planning.

3.2.1   An Airport-Planning Study Example

The air-travel market is sensitive to prevailing business cycles, and it requires frequent updating of travel-demand forecasts. Airline-passenger data collected at the Robert Mueller Municipal Airport in Austin, Texas, were used for a planning study in 1983. The following regression equations developed in the study suggest sales tax revenue as a strong predictor of annual airline-passenger data, as shown in Figure 3-4 [Uddin 84].

$$PAX = 2071959.8 + 0.1809 \ (STR) - 6.2428 \ (POP) \quad R^2 = 0.987 \quad (3.1)$$

$$PAX = 0.1081 \ (STR) \qquad\qquad\qquad\qquad\qquad R^2 = 0.991 \quad (3.2)$$

$$\log_n(PAX) = 249.79123 - 466193.63 \ (1/YEAR) \qquad R^2 = 0.960 \quad (3.3)$$

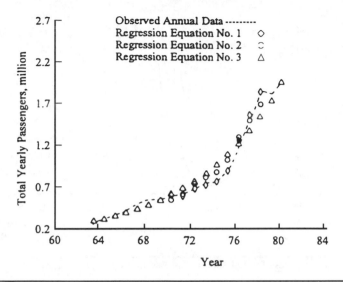

FIGURE 3-4 Observed and optimated annual-airline passengers at Austin Municipal Airport [Uddin 84].

where *PAX* = total yearly airline passengers
    STR = annual sales tax revenue in dollars
    POP = population of metropolitan area (thousands)

To project passenger forecasts using Eqs. (3.1) and (3.2), the independent-variable sales tax revenue (*STR*) also should be projected for the future year of choice. This is not always possible, and it needs to be estimated first. However, the last equation is easy to use, because it contains only *YEAR* as an independent variable. In this study, the equations are extrapolated to predict travel demands in future years. This has some possible dangers and should be used with discretion.

The annual arriving and departing passengers were 2.5 million in 1983, as compared to the Eq. (3.3) forecast of 2.4 and 2.39 million predicted by a Box-Jenkins time-series model. Monthly time-series data were used to develop the Box-Jenkins Auto Regressive Integrated Moving Average (ARIMA) model that can predict reliable seasonal variation, as shown in Figure 3-5.

The study also investigated the impact of these projections on aviation (runway, taxiway, and gates, etc.) and ground (terminal building, parking areas, gates, etc.) facilities, noise pollution caused by projected increases in flight operations, and air-space problems because of the proximity of Bergstrom Air Force Base [Uddin 84]. Planning of alternative sites was actively initiated to meet the future needs because the existing site had limited available space at the time of that study.

The airport has witnessed a steady growth in recent decades. Bergstrom Air Force Base was closed in 1991, selected by the city of Austin as the prime choice for the site of a new airport, and approved

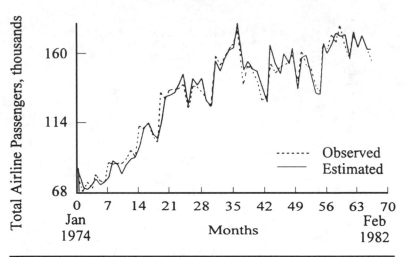

FIGURE 3-5 Plots of the observed and estimated series of ARIMA model [Uddin 84].

in 1991 by the Federal Aviation Administration (FAA) as Austin-Bergstrom International Airport [Amick 96].

3.2.2 A Transportation-Planning Study Example

Montgomery County, Maryland, a large municipality north of Washington, DC, has experienced rapid growth in employment and housing during the past several decades. The county's economic base is largely focused on information and communications technologies, biotechnologies, activities of the U.S. government, and support services. In the late 1980s, the county had 350,000 jobs; 700,000 people living in 270,000 households; and one-fourth of the workforce employed in the District of Columbia. The county completed a comprehensive growth-policy study in 1989 for the future 30 years to assess choices for transportation [Replogle 90].

The study considered the following four land-use scenarios: *fast* but balanced growth, *slow* but balanced growth, *jobs* favoring employment growth, and *housing* favoring housing growth. These were tested against the following mobility choices:

- Auto—Continue current policies and build out the highway master plan.
- Van—Add a network of high-occupancy-vehicle (HOV) lanes.
- Rail—Add to auto a light-rail network with certain assumptions to increase the parking fee and road pricing to effectively double the cost of automobile operation.

Figure 3-6 shows the mode shares for selected land-use and mobility choices based on analysis using the logit-mode choice model and future population and travel-demand projections. The rail choice

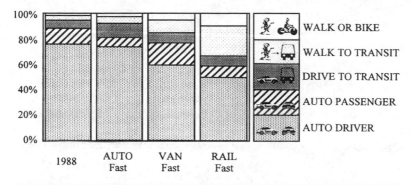

Figure 3-6   Montgomery County origin mode share for home-to-work trips [Replogle 90].

allows the county to meet its traffic-congestion standards, depending on the land-use balance between housing and jobs [Replogle 90]. The study findings point to the need for significant changes in the direction of master-plan development and infrastructure planning for the future.

3.3   Life-Cycle Management

In the postconstruction stage of infrastructure management, in-service evaluation of the infrastructure should have high priority. Appropriate rules and checklists should be established regarding the use of the facility and maintenance/repair. Unfortunately, deferred maintenance has been the general rule in most public infrastructure including public buildings. For example, the average age of buildings at the University of Mississippi campus in Oxford, Mississippi, was 40 years in 1994, with four buildings over 100 years old [Miss 94]. Many old buildings needed extensive repair and renovation to comply with current building codes and ADA requirements, which has been accomplished since then as a result of high priority given by the university leadership (http://infrastructureglobal .com/?p=3997, accessed January 20, 2013). According to an article in *Building Operating Management* magazine (February 1990), by 1991 the amount of deferred maintenance costs in public school facilities was $14 billion, and college and universities had over $60 billion in deferred maintenance, renovation, and new construction [Binder 92].

In reality, satisfactory performance and service quality cannot be preserved very long unless a life-cycle management plan is put into operation, preferably when the facility is opened for use. The management plan should consider the following activities:

- Rules for appropriate use of the facility
- Regulations for routine/minor maintenance to keep up with normal use and aging

- Plans for emergency management of fire, accident, natural disaster (tornado, floods, earthquake, etc.), or sabotage
- A program of scheduled maintenance of equipment and structures
- A framework and methodology for planning condition-based and demand-responsive maintenance, rehabilitation, and renovation, as well as replacement or reconstruction (M,R&R) actions; the framework should also include analysis of "do-nothing" and deferred maintenance actions
- A financial management plan to pay for the operation and life-cycle M,R&R requirements

3.4   Infrastructure Service Life

The most important component of life-cycle analysis is the estimate of service life of a facility. Infrastructure service life depends on design and construction methods, usage and environment, and in-service maintenance and operation practices. Service life for specific examples of any class of infrastructure may still vary greatly. Service life is not the same as design life or economic life. The terminology used in this book is described in the following sections.

3.4.1   Terminology Related to Service-Life Needs

The following terminology is based on the Building Research Board publication *Pay Now or Pay Later* [BRB 91]:

Service Life: "The period in years over which a building, component, or subsystem provides adequate performance; a technical parameter that depends on design, construction quality, operations and maintenance practices, use, and environmental factors; different from economic life." This definition is equally applicable to all categories of infrastructure assets.

Performance: "The degree to which a building or other facility serves its users and fulfills the purpose for which it was built or acquired." In other words, performance is history of serviceability that shows the quality and length of service that a facility provides to its users.

3.4.2   Evaluation of Infrastructure Service Life

Service life is the period in years from the time of completion of the facility to the time when the complete facility or its components are expected to reach a state where it cannot provide acceptable service because of physical deterioration, poor performance, functional obsolescence, or unacceptably high-operating costs. Evaluation of the service life of infrastructure assets is quite complex, because different components of a facility can have varying ranges of possible service life. Service life of critical structural components should be taken as a representative estimate to plan new construction, major repairs, and reconstruction. At best, only an estimate of an average service life can be made based on: (1) an acceptable level of performance; and (2) average life of the facilities in each group or similar infrastructure facilities after falling below an acceptable limit or performance or at failure.

In general, service life of a public infrastructure or privately owned building is not normally less than 40 years. Service life can be estimated from the historical infrastructure database using survivor techniques. A systems-engineering approach is particularly useful to improve the performance model and predict service life based on some minimum-condition acceptance criteria.

3.4.2.1 Survivor Curve Method

The survivor curve shows the number of units of a property (miles or kilometers of water pipelines, bridges, original cost, or percentage of units, etc.) that survive in service at given ages. Property surviving is generally expressed in percentage of the base cost at zero age. Figure 3-7 shows a survivor curve developed for the box-culvert data [Winfrey 69]. The area under the survivor curve is a direct measure of the average service life of the property units. The probable life of the surviving units at any age can be calculated from the remaining area by dividing the remaining area by the amount surviving of that age. The top of the probable life curve is the maximum number of years expected for failure in the database. A line driven vertically down from the top of the probable life curve intersects the survivor curve

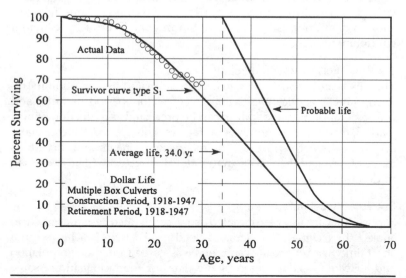

FIGURE 3-7 A survivor curve for box culverts based on dollar units [Winfrey 69].

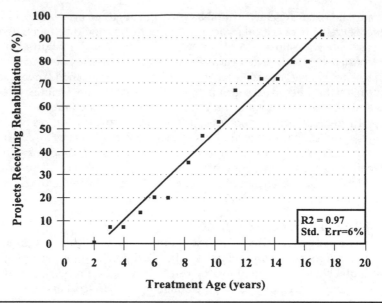

FIGURE 3-8   A survivor curve for SAMs applied on state routes in Arizona [Flintsch 94].

and gives the expected average life of the structure. Figure 3-8 shows a survivor curve for pavement-rehabilitation treatment using a stress-absorbing membrane (SAM) for state routes in Arizona [Flintsch 94].

3.4.2.2   Reference to Previous Experience
Service life can be estimated based on previous experience with similar facilities. This approach is particularly useful for large facilities such as dams, nuclear plants, etc., but may be in error for different environments.

3.4.2.3   Performance Modeling
The physical-deterioration rate can be estimated by condition monitoring and in-service evaluation over a short period of time, and future deterioration and failure can be predicted as a function of age, load/demand, and environmental factors. This approach is especially useful to predict service life of major components and structures made of new and innovative materials. This subject will be discussed in-depth in Chapter 8.

3.4.2.4   Accelerated Testing
Service life can also be estimated from accelerated tests designed to subject the facility to the demand/load until failure is achieved in a short time. The service life can then be assessed from the interpolation and/or careful extrapolation of the data for the facility in the same region or other environmental and general conditions. Durability and

service life for new construction materials is often assessed by accelerated laboratory testing. The AASHO (American Association of State Highway Officials) Road Test, constructed and tested to failure within a few years during 1957 to 1962, has been a valuable source of performance data and prediction of service life of asphalt and concrete roads [HRB 62].

3.4.3   Example Estimates of Service Life

Service life is generally expressed in number of years. In cases where service life is difficult to estimate accurately, it is still useful when items are ranked in order of durability of the primary construction materials and assigned expected service life based on experience. Service life can also be expressed in terms of load repetitions (e.g., total number of aircraft coverage for an airport runway or total number of equivalent single-axle truckloads in the life cycle of a road pavement), or in other cases, as periods/cycles of use (e.g., total kilowatt hours or total cumulative running time in hours). The prevailing load, environmental, ground, and operating conditions should always be considered to predict the service life. These factors may also interact with the material of construction and the constructed facility as a whole. These interactions may lead to a shorter service life for the facility than its individual components. For example, a building wall or a bridge structure made of bricks may have a lower service life than the predicted service life of bricks.

Table 3-1 lists general ranges of service lives for highway components [Winfrey 69]. Table 3-2 shows a comparison of building lives

Highway Components	Years
Right-of-way land	75 to 100
Right-of-way damages (suggested write-off period)	10 to 30
Right-of-way buildings to be moved or destroyed (suggested write-off period)	10 to 30
Earthwork	60 to 100
Culvert and small drainage facilities	25 to 50
Retaining walls and general concrete work	40 to 75
Riprap and other bank protection	20 to 50
Bridge and other major structures	50 to 75
Granular roadway surfaces	3 to 10
Low-type bituminous surfaces	12 to 20
Rigid and flexible high-type pavements	18 to 30
Signs and traffic-control devices	5 to 20

TABLE 3-1   General Service Lives for Highway Components [Winfrey 69]

Building Facility	U.S. Practice	Canada [CSA 94]	UK [BS 92]	Japan [AIJ 93]	
Industrial buildings	50–300	25–49	Minimum 30	25–40 or more	Minimum 25
Commercial, health, education, residential	30–300	50–99	Minimum 60	60–100 or more	Minimum 60
Civic, monumental, national heritage, landmark	70–500 or more	Minimum 100	Minimum 120	60–100 or more	Minimum 60

TABLE 3-2   General Service Lives of Buildings, and Comparison of Different Countries

based on general practice in the United States and codes and recommendations extracted from relevant document sources in Japan, Canada, and England. Overall service life of some infrastructure facilities is recommended in Table 8-1.

3.5   Infrastructure Needs Assessment

One key to effective infrastructure management is the establishment of network-needs assessment on a regular basis and timely planning of M,R&R actions. Historically, this has been done using the past experience and judgment of the facility and maintenance engineers. This ad-hoc approach is less effective because of the expansion of infrastructure assets, changing demands on systems, and the growing shortage of funds for construction and maintenance actions at all levels of government, as well as by nongovernmental organizations and private owners.

The Urban Institute's six-volume report series *Guides to Managing Urban Capital* was an early comprehensive study of capital planning and budgeting practices of 40 local-government agencies. The study recommends three basic strategies that a government can employ to reduce capital investment and facility-maintenance problems [Urban 84]:

Strategy 1   Better identify capital needs and priorities, to screen out marginal needs and to make the best use of available funds.

Strategy 2   Build community support for facility maintenance and repair and reinvestment.

Strategy 3   Find new revenue sources, or reorganize the local revenue system so that it provides a stable source of revenues to maintain and replace basic facilities.

Most agencies operate under constrained budget conditions with competition from other important items of public spending;

therefore, a rational methodology of infrastructure-needs assessment is required to convince management and legislators to allocate available funds fairly. The establishment of needs does not by itself indicate which M,R&R alternative should be considered and which is most cost-effective. The M,R&R selection and priority-programming methods will be discussed in Chapters 11, 13, and 15.

User demand reaching system capacity and extreme incidents involving long disruptions of essential services to the public and businesses indicate system vulnerability to the peak-demand levels and the need for future planning and expansion or conservation strategies. The following examples of extreme disruptive incidences, extracted from an Internet search, illustrate the lack of infrastructure system capacity to meet user demands and urgency for timely needs assessment:

- On August 1, 2007, the I-35W bridge collapsed during rush hour in Minnesota, killing 13 people. This fifth-busiest bridge in Minnesota was rated "structurally deficient" in 2005 during bridge inspection, but no action was planned or taken to rehabilitate the bridge. Post-failure investigation showed design flaws leading to cracks in steel-gusset plates. During the August 23, 2011, seismic tremors in the Virginia and DC areas in the United States, most cell-phone services were disrupted due to an exceptionally heavy use of the wireless network. State-emergency agencies discourage the general public calling emergency cell numbers unless it is real emergency, such as Superstorm Sandy that struck the U.S. East Coast states on October 30 to 31, 2012.

  On July 30 and 31, 2012, India's blackout affected 620 million people, which was the largest power outage in world history. A massive blackout struck on August 14, 2003 and affected Ontario, New York, and many other U.S. northern states for about 30 hours before the power was restored.

3.6   Infrastructure Performance

Performance indicators of infrastructure must evaluate the quality of service provided by the facility for needs assessment. Functional performance of a facility is usually assessed from the user perspective. Hatry lists a number of performance measures including the following: quality of service and system effectiveness to meet the expectations of the users, productivity and efficiency, and resource utilization and cost-effectiveness [Urban 84].

Efficient mobility for the highly developed transport infrastructure in Germany is an example considered economically and socially; however, it also gives due importance to the consequences of consuming resources for transport, effects on health and climate change, and associated costs to the economy. The following highlights are extracted from the German Federal Ministry of Economics and Technology (BMWi) report on the third transport-research program [BMWi 08]:

- Number of transport users in Germany with reduced mobility will rise; traffic in transit will be more than doubled (136 percent by 2030; the amount of freight traffic (tonne-kilometer) is expected to increase by 69 percent between 2005 and 2030, with 70 percent weight by heavy trucks and 27 percent by rail and inland waterways. German federal spending of 19.5 billion euros in 2004 was mostly for rail (48 percent) and trunk roads (30 percent) followed by 9 percent on communities and cities, 8 percent on waterways, and 5 percent other; most federal funds spent on the German transport network are devoted to maintenance and preservation.

- Focus is given on reducing harmful exhaust emission and CO_2 emissions from transport vehicles, as well as reducing traffic-noise pollution (felt by 60 percent of the population); the federal-research agenda includes new ways of land consumption for transportation infrastructure involving cultural landscapes, natural balance of flora and fauna biodiversity, and climate changes.

On July 6, 2012, the Moving Ahead for Progress in the 21st Century Act (MAP-21) was signed into law by U.S. President Barack Obama, which implements the National Highway Performance Program (NHPP) and performance standards for highway asset management as a part of highway funding mechanisms [FHWA 12]. MAP-21 investments guided by the policy and programmatic frameworks are designed to create a performance-based surface-transportation program that builds on many of the existing highway, transit, bike, and pedestrian programs and policies [FHWA 12]. The bulk of MAP-21 funding is for NHPP to (1) support the monitoring of condition and performance of the National Highway System (NHS), (2) support the construction of new facilities on the NHS, and (3) ensure that investments of federal funds in highway construction are directed to support progress toward the achievement of performance targets established in a State's asset management plan. MAP-21 establishes the following national performance goals for federal-aid highway programs in the United States [FHWA 12]:

- *"Safety*—To achieve a significant reduction in traffic fatalities and serious injuries on all public roads.

- *Infrastructure condition*—To maintain the highway infrastructure asset system in a state of good repair.

- *Congestion reduction*—To achieve a significant reduction in congestion on the NHS.

- *System reliability*—To improve the efficiency of the surface transportation system.

- *Freight movement and economic vitality*—To improve the national freight network, strengthen the ability of rural communities to access national and international trade markets, and support regional economic development.

- *Environmental sustainability*—To enhance the performance of the transportation system while protecting and enhancing the natural environment.

- *Reduced project delivery delays*—To reduce project costs, promote jobs and the economy, and expedite the movement of people and goods by accelerating project completion through eliminating delays in the project development and delivery process, including reducing regulatory burdens and improving agencies' work practices."

Many of the highway-infrastructure goals and methodologies supported by the MAP-21 policies and guidelines are similar to the overall IAMS framework functions of inventory, performance monitoring, acceptable performance targets, life-cycle assessment of benefits and costs, and maintenance and preservation of assets originally defined in the first edition of this book [Hudson 97] and now expanded in this second edition.

3.6.1   Performance Indicators

Performance indicators have traditionally been grouped into four broad categories. A fifth category is added on sustainability performance:

1. Service and user perception

2. Safety and sufficiency

3. Physical condition

4. Structural integrity/load-carrying capacity

5. Sustainability related to the "triple bottom line" of environmental, social, and economic responsibility

3.6.2   Examples of Performance Indicators

All infrastructure assets include several structural and nonstructural components. Performance of the individual components may differ with the material and usage of the component. It is also desirable to establish an overall performance indicator, with weight given to the structural integrity and service to users. Additionally, sustainability performance measures are being developed and implemented by cities and agencies/companies involved in energy, manufacturing, supply chain, and aviation industries. Examples of performance indicators for various different infrastructure groups are presented in the following sections.

3.6.2.1   Transportation Infrastructure

Highways, roads, and streets constitute the backbone of transportation infrastructure. Pavement structural-thickness and surface-smoothness requirements vary according to intended use; however, there are common performance indicators based on pavement-condition assessment; for example, a structural-capacity index based on deflection response and a distress index based on distress type, severity, and extent [Haas 94]. Such indexes are generally graduated from 100 to 0, where 100 represents the best-possible condition and 0 represents the worst or failed condition. A third performance indicator is based on surface-riding quality. The objective measurement of surface-ride quality is performed by measuring surface smoothness or roughness. Smoothness and roughness represent the opposite ends of the same scale. A composite performance indicator is based on a combination of two or more of the pavement-condition attributes, using appropriate weights. Other components of paved facilities (excluding bridges) are markings, signs, and roadside appurtenances or furniture items. These are subjected to environmental forces and occasional failures because of accidents. Their performance can be conveniently evaluated using an overall condition rating on a scale of 10 or 100 (the best) and 0 (the worst or failed) that is linked with the maintenance requirement.

In the case of a railroad facility, the track provides the primary load-carrying component and riding surface. Condition deterioration of a track segment will depend upon the condition of ballasts, track gradient and alignment, ties and joints, wheel distribution of load and annual traffic, train speed, and environmental parameters. An overall condition index from 100 (the best) to 0 (failed and unusable) can be established based on the critical geometry, ballast depth, load, and environmental parameters. A track-quality index (TQI) has been formulated that considers different track structural parameters (number of bad ties, deflection under load), load and speed, environmental factors, and maintenance quality [Fazio 80].

Bridges are important components of all types of ground-transportation facilities. Bridges may be a part of an interchange, a grade separation, or a crossing over a water body or a river. Each major component of a bridge (deck, superstructure, substructure, and foundation) will perform differently and, therefore, requires a different condition rating. The national bridge-inventory program combines these different ratings into a composite overall-sufficiency rating (SR) on a scale of 100 (the best condition) to 0 (the worst condition) and uses it to identify deficient bridges [Hudson 87].

3.6.2.2   Water and Wastewater Infrastructure

The deterioration of a water supply and its distribution system primarily depends on the material of the main-pipeline network, joints, subsurface moisture and soil, demand on the system (e.g. gallons of water pumped annually), infiltration of groundwater, blockage to flow, and environmental effects on service life [Grigg 12]. The same is true for main sewer lines. However, from the users' and from the owner agency's points of view, the number of breaks per year, number of failures per 1,000 km, and leakage of water per year are the most important performance indicators.

3.6.2.3 Waste Management Infrastructure

Wastewater from pipelines goes to treatment plants. Other types of waste collection and treatment sites are landfills. Facility-specific performance indicators are needed in such cases. These facilities do not have direct user impacts until there is a serious breakdown or fire incident that causes pollutants to spread and threaten the health and safety of public in the vicinity. Leakage quantity per year and number of breakdowns per year are possible performance indicators.

3.6.2.4 Electric-Power Infrastructure

Electric-power infrastructure is vital in modern society and for the national economy of every country. The utilities in some countries and regions are publicly owned and operated but are private in others. Vulnerability of the infrastructure was revealed in long blackouts in July 2012 in vast areas of northern India and parts of New York and other U.S. states. Electric power-capacity limitations are affected by population growth and industrial development. Some performance measures are frequency and duration of service interruptions. The sustainability performance measure of CO_2 per capita can also be applied to evaluate energy-efficient transport policies [Uddin 12].

Emissions of CO_2 from electricity generation is one-third in the United States and even a higher 46 percent worldwide (Figure 3-9). Coal still remains the largest source of electric power in most countries

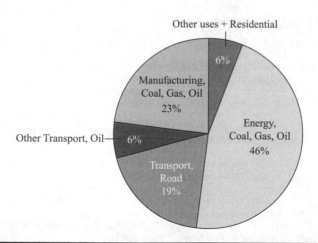

FIGURE 3-9 Global CO_2 emissions from fuel combustion by sectors, 2007 [EIA 09].

with a 45 percent share in 2009 in the United States, however, a cleaner natural-gas share of electric production has risen from 16 percent in 2001 to 23 percent in 2009 (Table 2-2). Hydroelectric power generation is at 7 percent in the United States and 13 percent in the world. Recent years have seen large investments and government initiatives in renewable electric-generation sources, such as wind and solar technologies. These renewable-energy resources should be integrated in the national power grid, such as was done in New Zealand for wind- and geothermal-electric generators.

3.7   References

[AIJ 93] Architectural Institute of Japan, *The English Edition of Principal Guide for Service Life Planning of Buildings*, Tokyo, Japan, 1993.

[Amick 96] J. P. Amick, and J. Almond, "Reconstruction of Bergstrom Air Force Base to Austin-Bergstrom International Airport," *Rebuilding Inner City Airports, Proceedings of the 24th International Air Transportation Conference*, American Society of Civil Engineers, Louisville, Kentucky, June 5–7, 1996, pp. 1–11.

[Binder 92] S. Binder, *Strategic Corporate Facilities Management*, McGraw-Hill, Inc., New York, 1992.

[BMWi 08] Federal Ministry of Economics and Technology (BMWi), "Mobility and Transport Technologies: The Third Transport Research Programme of the German Federal Government," April 2008. www.bmwi.de, accessed on October 20, 2011.

[Bragdon 95] C. R. Bragdon, "Intermodal Transportation Planning for the 21st Century, a New Paradigm: An Urgent Call for Action," *Proceedings of the 74th Annual Meeting of the Transportation Research Board*, National Research Council, Washington, DC, 1995.

[Bragdon 96] C. R. Bragdon and C. Berkowitz, "Solving Aviation and Intermodal Transportation Related Issue, a New Prototype for the 21st Century," *Rebuilding Inner City Airports, Proceedings of the 24th International Air Transportation Conference*, American Society of Civil Engineers, Louisville, KY, June 5–7, 1996, pp. 212–222.

[BRB 91] Building Research Board, *Pay Now or Pay Later*, National Research Council, Washington, DC, 1991.

[BS 92] "Guide to Durability of Buildings and Building Elements, Products and Components, BS 7543:1992," British Standards Institution, London, 1992.

[CSA 94] "Guideline on Durability in Buildings, CSA S478, Draft 9," Canadian Standards Association, Toronto, Ontario, September 1994.

[EIA 09] Energy Information Administration, "International Energy Annual Outlook," U.S. Department of Energy, Washington, DC, 2009.

[FAA 86] Federal Aviation Administration, "Policies and Procedures for Considering Environmental Impacts," Order 1050.1, Washington, DC, 1986.

[Fazio 80] A. E. Fazio, and R. Prybella, "Development of an Analytical Approach to Track Maintenance Planning," in *TRR 744*, Transportation Research Board, National Research Council, Washington, DC, 1980.

[FHWA 12] Federal Highway Administration, "MAP-21—Moving Ahead for Progress in the 21st Century: National Highway Performance Program (NHPP) Implementation Guidance," U.S. Department of Transportation. http://www.fhwa.dot.gov/map21/, accessed November 9, 2012.

[Flintsch 94] G. W. Flintsch, L. A. Scofield, and J. P. Zaniewski, "Network-Level Performance Evaluation of Asphalt-Rubber Pavement Treatments in Arizona," in *TRR 1435*, Transportation Research Board, National Research Council, Washington, DC, 1994.

[Grigg 12] N. S. Grigg, *Water, Wastewater and Stormwater Infrastructure Management*, 2nd ed., CRC Press, Boca Raton, Fla., 2012.

[Haas 94] R. Haas, W. R. Hudson, and J. P. Zaniewski, *Modern Pavement Management*, Krieger Publishing Company, Malabar, Fla., 1994.

[HRB 62] Highway Research Board, "The AASHO Road Test, Report 5—Pavement Research," *Special Report 61-E*, National Research Council, Washington, DC, 1962.

[Hudson 68] W. R. Hudson, F. N. Finn, B. F. McCullough, K. Nair, and B. A. Vallerga, "Systems Approach to Pavement Systems Formulation, Performance Definition, and Materials Characterization," *Final Report NCHRP Project 1-10*, Materials Research and Development Inc., March 1968.

[Hudson 87] S. W. Hudson, R. F. Carmichael III, L. O. Moser, W. R. Hudson, and W. J. Wilkes, "Bridge Management Systems," *NCHRP Report 300*, National Cooperative Highway Research Program, Transportation Research Board, National Research Council, Washington, DC, 1987.

[Hudson 97] W. R. Hudson, R. Haas, and W. Uddin, *Infrastructure Management*, McGraw-Hill, New York, 1997.

[Miss 94] "University Buildings Continue to Deteriorate," *The Daily Mississippian*, Oxford, Mississippi, Vol. 87, No. 28, September 29, 1994.

[NIST 03] National Institute of Standards and Technology, "NIST Response to the World Trade Center Disaster: World Trade Center Investigation Status," U.S. Department of Commerce, December 2003.

[Prior 94] A. P. Prior, "ADA and the New Work Place Culture," *Civil Engineering News*, March 1994, pp. 25–26.

[Replogle 90] M. Replogle, "Computer Transportation Models for Land Use Regulations and Master Planning in Montgomery County, Maryland," in *TRR 1262*, Transportation Research Board, National Research Council, Washington, DC, 1990, pp. 91–100.

[Uddin 84] W. Uddin, B. F. McCullough, and M. B. Crawford, "Methodology for Forecasting Air Travel and Airport Expansion Needs," in *TRR 1025*, Transportation Research Board, National Research Council, Washington, DC, 1984.

[Uddin 12] W. Uddin, "Mobile and Area Sources of Greenhouse Gases and Abatement Strategies." Chapter 23, *Handbook of Climate Change Mitigation*. (Editors: Wei-Yin Chen, John M. Seiner, Toshio Suzuki and Maximilian Lackner), Springer, New York, 2012, pp. 775–840.

[Urban 84] The Urban Institute, *Guides to Managing Urban Capital*, Vol. I, H. P. Harty, and G. E. Peterson (eds.), Washington, DC, 1984.

[Winfrey 69] R. Winfrey, *Economic Analysis for Highways*, International Textbook Company, Scranton, Pennsylvania, 1969.

PART 2

Information Management and Decision Support Systems

CHAPTER 4
Database Management, Data Needs, Analysis

4.1 Overview of Information Management

Information support and management is a critical step for the effective and successful operation of any infrastructure-management system. Access to information that is both correct and timely is essential for the satisfactory conduct of the management process. This applies equally to all key players in the organization, including planning, design, construction, and maintenance departments.

4.1.1 Information Technologies

During the last decade, computer-based information technologies have become broadly and deeply integrated with management. The rapid development of electronic database storage and retrieval capabilities is the primary force behind this development. The increased popularity of personal computers and the evolution of software have also played an important role. In the early 1970s, database management system (DBMS) software, for example, was considered exotic and unreliable, whereas it is now regarded as stable and mundane [Begley 95]. Begley and Sturrock identify key information technologies for material science and engineering [Begley 95], such as relational and object-oriented database systems, expert systems, and multimedia, that either have significantly impacted or very likely will significantly impact the development of many applications in the areas of IAMS. Other information technologies like video logging, neural networks, case-based reasoning, and virtual reality also offer potential for future infrastructure-management applications. Since the late 1990s the increased use of geospatial technologies, airborne and spaceborne remote sensing, and Internet sources have greatly impacted data-collection and information-management applications in all sectors of society including infrastructure management practice [Uddin 02].

4.1.2   Decision-Support Systems

Generally, the term *decision-support system* (DSS) refers to the use of computers to store, analyze, and display information that is used to support decision making. DSS implies, however, more than just data-processing business as usual; it includes analysis models that generate results useful for making rational decisions. In other words, DSS organizes the processing, analysis, and delivery of information necessary for decision making. The use of an information-support system, database management, and analytical studies can help engineers make better decisions through: (1) improved identification and information of the infrastructure assets; (2) access to condition data, usage, and history; (3) delineation of problem areas; (4) methodologies for needs assessment; (5) evaluation of alternative solutions; (6) projection of work programs and budgets; and (7) priority setting of projects and programs of inspection and evaluation schedules. Therefore, DSS is an integral component of IAMS.

DSS is not drastically different from the traditional staff activity of giving recommendations to the boss for making appropriate decisions. Before computers, databases existed as paper copies and were often poorly organized. Every department had its own paper files, and both information management and decision making were, if used at all, customized and more expensive than necessary, and relied heavily on personal preferences and judgment. These past practices have been replaced by computerized procedures to meet the challenge of today's decision-making management, and consequently provide better service to the public.

Once a DSS is established, the preparation of annual and multiyear work programs for maintenance, rehabilitation/renovation, and replacement/reconstruction (M,R&R) is streamlined as an automatic DSS output. Aside from the increased productivity, plans can be continually updated by providing feedback to the database. DSS should be designed to suit the needs and resources available to an agency, and to function within the primary organizational structure. DSS is also increasingly used for operations; for example, in utility companies, plant production, transport services, and water-resource management. Figure 4-1 shows an example of a DSS framework of a pavement management system (PMS) for roadway infrastructure [Uddin 95], and a DSS example for a bridge management system (BMS) is illustrated in Figure 4-2 [Hudson 87].

Successful implementation of a DSS must consider the flow of requests for decision support from the decision maker to the technical-support staff. The two main activities in a DSS are data management and study of alternatives. These activities generate meaningful results or "knowledge" from data or "information" to support decision making. The primary role of a DSS is thus to use data, together with the necessary analytical models, to produce the decision-support rationales.

A central database is the heart of a DSS for managing infrastructure assets. The database typically consists of many database files.

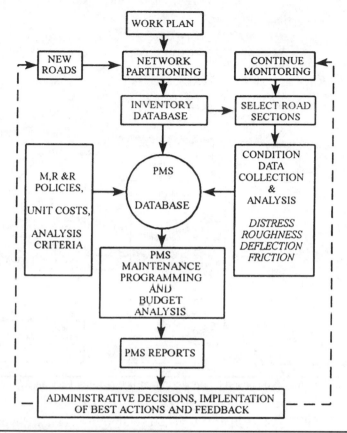

FIGURE 4-1 A DSS framework for road infrastructure [Uddin 95].

A computerized database system has several advantages over paper-based records keeping [FHWA 90], including:

- Data are stored in a compact space and shared by all users.
- Storage and retrieval of the data are much faster than a manual method, permitting the data to be updated on a regular basis and facilitating the use of the information.
- Use and processing of data are centrally controlled.

However, the advantages of centralization cannot be realized without a properly designed and maintained database system. Advantages of a centralized system are [FHWA 90, Haas 94]:

- Redundancy is reduced by having each piece of data stored in only one place. This also avoids inconsistencies when updating the data files.
- Data can be shared for various applications. For example, frequently the data needed for pavement management is

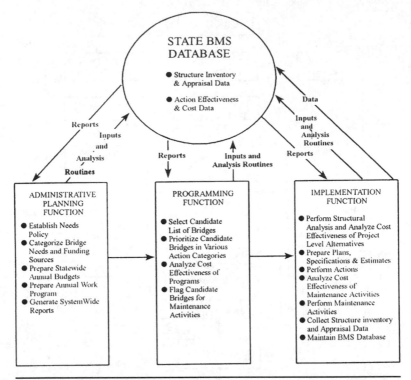

STATE BMS
DATABASE

● Structure Inventory
& Appraisal Data

● Action Effectiveness
& Cost Data

Reports

Inputs
and
Analysis
Routines

Data

Inputs
and
Analysis
Routines

Reports   Inputs and   Reports
                 Analysis Routines

ADMINISTRATIVE PLANNING FUNCTION	PROGRAMMING FUNCTION	IMPLEMENTATION FUNCTION
● Establish Needs Policy ● Categorize Bridge Needs and Funding Sources ● Prepare Statewide Annual Budgets ● Prepare Annual Work Program ● Generate SystemWide Reports	● Select Candidate List of Bridges ● Prioritize Candidate Bridges in Various Action Categories ● Analyze Cost Effectiveness of Programs ● Flag Candidate Bridges for Maintenance Activities	● Perform Structural Analysis and Analyze Cost Effectiveness of Project Level Alternatives ● Prepare Plans, Specifications & Estimates ● Perform Actions ● Analyze Cost Effectiveness of Maintenance Activities ● Perform Maintenance Activities ● Collect Structure inventory and Appraisal Data ● Maintain BMS Database

FIGURE 4-2   A DSS proposed for bridge infrastructure [after Hudson 87].

collected by different divisions within the agency. Having a centralized database ensures that all divisions will have access to the needed information.

- Standards can be enforced in terms of data formats, naming, and documentation.

- Security restrictions can be applied to control the flow of the data and the updating of the data.

- Data integrity can be maintained by controlling the database updating and using integrity checking whenever an update is made.

- Conflicting requirements of the individual users can be balanced to optimize the database for the agency. This is particularly important when sharing data between divisions.

An important consideration in the development of a database-management approach is the temporal and spatial identification of the data. Temporal identification is accomplished by storing the data by reference of time and/or date. It is important to establish historical records of construction, maintenance, and evaluation data. Spatial identification requires being able to physically relate the data to the location of a facility in the infrastructure network. Spatial referencing is accomplished through the section-definition process, in which homogeneous sections or components with similar characteristics are identified and their physical boundaries and descriptions are established. Geocoding (geographical-coordinate description) and the use of a Geographical Information System (GIS) can enhance this process.

Public works agencies have been using traditional decision practices since the evolution of urban society in the early 20th century in the United States. Planners in most cities are now either trying to improve the existing systems or are seeking a new approach for managing their infrastructure assets and associated services. GIS technology has emerged as a useful tool for developing a comprehensive management system so that all of the municipal infrastructures, such as pavement, bridges, water supply, wastewater, sewer, gas, and electricity, can be integrated on a common platform to improve managerial decisions. This concept of an integrated infrastructure-management system is illustrated in Figure 4-3, in which GIS is the common location-reference system [Zhang 94].

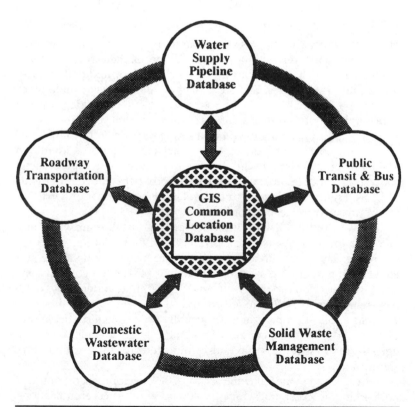

FIGURE 4-3   A concept of an integrated overall infrastructure-management system [Zhang 94].

GIS technology was originally developed using traditional low-resolution maps and 30-m resolution Landsat imagery. Geospatial analysis and geospatial mapping are more commonly used by the industry to replace GIS in the 21st century [Uddin 11a]. Geospatial/GIS applications are being used in many sectors of society from news media to social sciences; it is not just limited to traditional applications in geography, geology, landuse planning, transportation, and environmental areas. Geospatial technologies have evolved rapidly since the availability of modern 1-m high-resolution commercial imagery and public use of GPS (global-positioning system) technology for satellite-based location referencing and topographical surveys. *GoogleEarth* is an example of an online Internet geospatial/GIS tool that provides many basic spatial-analysis functions of GIS software (such as linear measurements) and lets the user mark and save several nongeographical attributes on the satellite-imagery layer [Uddin 11a].

4.2   Database Development and Management

4.2.1   System Design

The term *data base* written as two separate words, or *database* written as a single word, refers to a large collection of data in a computer, organized in such a way that it can be expanded, updated, and retrieved rapidly for uses. The term *database*, written as a single word and adopted throughout this book, also means a specific group of data within the structure of a database-management software system. The database may be organized as a single file or as multiple files or sets. The need for database-software packages should be identified at the outset of system design, based on the information processing, analysis methodologies, and the scope of other specific requirements of DSSs. Figure 4-4 shows the interrelationships of computer hardware and software needed for DSS.

Software serves as the interface among users, hardware, and data. It allows the user to access data by giving instructions to the hardware. There are three important components for database systems: (1) the operating system; (2) database-management software; and (3) application programs. Generally, operating systems and database-management software are commercially available off-the-shelf, whereas application programs are usually unique to a particular application. The Internet, computer networking, and wireless technologies in the 21st century have also impacted the software industry significantly.

As discussed in the following sections, the development of computer networking, Internet, fiber-optic-based and wireless communications, and desktop- and handheld-computer hardware in the 21st century have revolutionized software, visualization, and mobile apps.

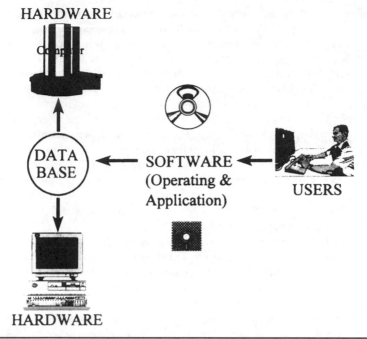

HARDWARE

DATA
BASE

SOFTWARE
(Operating &
Application)

USERS

HARDWARE

Figure 4-4 Interrelationships of computer software and hardware [after FHWA 90].

These strides in information technology (IT) have revolutionized networked-data collection, processing, access, and other primary functions of database-management systems.

4.2.2 Operating Systems

The operating system is the first-level software required by the computer hardware. No hardware will work without an operating system. Therefore, the operating system will typically be preselected by the data-processing section when the computer is purchased, rather than being a choice to make when setting up a database. Since the publication of the 1997 edition of this book [Hudson 97], innovations in computer hardware, software, and networking, including Internet resources, have made most of the older computer technologies from 1980s and 1990s obsolete.

4.2.2.1 1980s–1990s Computer Technologies

Microsoft DOS was a typical operating system for microcomputers, UNIX for minis and workstations, and MVS-XA for mainframes. Because of the ease of graphical user interfaces (GUIs), Windows-based operating systems, such as Microsoft's Windows 3.1, Windows 95, and Windows NT, became popular. Apple's Macintosh operating system is another popular Windows operating system. Desktop microcomputers that rely on 16-bit processors and operating system, such as DOS and Windows 3.1, were less expensive and easy to maintain. They became common platforms for many mapping and database-management software packages. Many users, however, demand capabilities that exceed the limits of microcomputers. Despite the development of faster microcomputer processors, such as Intel's Pentium processors, larger random-access memory (RAM), larger mass-storage devices and larger hard-disk space, and better video-graphics cards and graphics resolution, microcomputers are somewhat limited for handling large volumes of data and complex analysis requirements. DOS- and old Windows-based microcomputers with 16-bit processors and bulky video monitors became obsolete by the end of the 1990s due to strides in hardware technologies in the 2000s, such as 32-bit and 64-bit operating systems and high-resolution video display units.

Thirty-two-bit processor workstations, employing UNIX or other operating systems optimized for 32-bit processors, deliver great processing speed and can support sophisticated software, high-resolution graphics, large RAM, and mass storage. These were better suited for the multiuser market among the local, state, regional, and national agencies responsible for managing infrastructure assets. Workstations also permitted the establishment of a better local area network (LAN) for multiuser environments to share data and output devices, as compared to the limited LAN capability offered by microcomputers.

4.2.2.2   The 21st-Century Computer Technologies

"The successive waves of innovation over the last century has taken us from electricity to telecommunications to microchips all the way into the ubiquitous nature of the Internet and has changed the way humans live and advance into the future" [Uddin 11b]. The 21st-century computer-hardware and computer-network technologies have led to enhanced visualization interfaces and large data-processing and graphical capabilities. The following sections provide milestones of modern IT and Internet technologies:

- *Advancements in microcomputer hardware:* Modern microcomputers feature faster 64-bit central-processing units (CPUs), 1- to 2-gigabyte (GB) video cards or graphics processing units (GPUs), up to 16-GB RAM, hard-disk memory space approaching 1 or more terabyte (TB), optical drives, multimedia- and digital-card readers, and Ethernet landlines and wireless-network cards. Very-high-storage external hard-disk drives and high-definition flat light-emitting-diode (LED) video monitors are available at fairly cheap prices. (These capabilities are available in both desktop and notebook computers.) Following the current trend of computer-hardware technology, it seems apparent that continually decreasing prices and increasing performance can be expected in future years.

- *Computer networking and Internet:* Fiber-optic landlines starting in the early 1990s and development of Internet protocol
  (IP) for networking and worldwide web (www) services
  have made it possible to connect computers anywhere in the
  world.

- *Internet-based visualization and GIS technologies:* Internet-based
  database analytics, data visualization, and data access and
  processing have connected computers and people better than
  ever before. For example, *Mind Map* is a useful visualization
  tool for organizing thought processes preceding a project
  plan with interconnected activities. *GoogleEarth,* created using
  NASA's Blue Marble, is a good example of an Internet GIS
  and many new software applications generate files that can
  be viewed on *GoogleEarth.*

- *Internet application programming using XML and Java languages:*
  Starting from hypertext links, introduced first by Apple/
  McIntosh in the mid-1980s, Internet program coders now use
  hypertext markup language (HTML) for webpage development, extensible-markup (XML), and *Java* languages. For
  georeferenced data manipulation, geography-markup language (GML) is used, which refers to XML components for
  encoding geospatial data.

- *Wireless mobile communication technologies and apps:* The development of wireless communication is based on cell (wireless
  transmission) towers, which provide wireless coverage to
  most regions on the globe. The widespread use of cell or
  mobile phones worldwide in the last decade led to the development of thousands of specially coded *apps* for mobile
  phones and tablets. These apps for handheld devices bring
  the same functionality and access to the Internet, which are
  available on desktop computers using traditional IT network
  infrastructure.

- *Cloud-computing infrastructure:* The need to reach remote databases in real time and the availability of Internet Protocol (IP)
  for online data access led to the cloud-computing infrastructure, which allows an Internet-application program to
  retrieve data from remote-networked computer servers, process and analyze data, save the results on selected server(s),
  and show the desired end results on the user-interface screen
  [Uddin 11b].

Currently most of the large-scale databases (in excess of 500 GB)
are established on mainframes and supercomputers. These large
computers are preferred over other computer workstations because
of their massive storage capacity. The architecture of these large computers allows multiple mass-storage disk packs. The disk packs can
be easily daisy-chained together. Each of these disk packs can have

500 GB or more. Workstations can retrieve data from mainframe databases by connecting mainframes on the wide area network (WAN) connections. Following the current trend of computer-hardware technology, it seems apparent that decreasing prices in future years will make these high-performance and larger-storage workstation systems more attractive to organizations that formerly would have purchased a relatively cheaper microcomputer system with limited RAM and hard-disk storage space. Internet-networking and cloud-computing technologies will be integral to the future networked enterprise IT and cyber infrastructure.

4.2.3 Database-Management Software

Database-management software is the next level of software above the operating system (OS) software. This database software platform allows the user to define data structures and models without worrying how the data are physically stored on the hardware. Database-management software provides a mechanism for an application programmer to write programs that can perform various data-access, retrieve, and manipulation functions. They also provide the capability for certain queries, such as ad hoc queries, to be made on the database. Three kinds of database-management software are available: (1) relational, (2) hierarchical, and (3) network. The relational database-management software (RDBMS) is the most recent and by far the most used. The other two types are used primarily in the mainframe environment. Hierarchical databases are structured by users as a tree in which users must start at the root and follow specific branches to get to a particular "leaf" they want. Network databases are similar to hierarchical ones, except any particular "leaf" can be attached to more than one branch. Relational databases (e.g., DB2 for IBM mainframes and OS/2 workstations, Oracle, or Access for microcomputers) allow users to organize the database as a collection of tables that can be linked together. Object-oriented database software with easy-to-use GUIs are commonly used in the 21st century. Some database-software packages now include GIS functionality as well.

4.2.4 Application Programs

Application programs are software specifically written to accomplish data-manipulation and analysis functions, and to provide input interfaces and design output generators, as required by the end-users. Inventory and condition-assessment computer programs for a railroad company or for the water department of a city public works agency are examples of application programs. High-level programming languages have been adopted by the data-processing community as a means to develop and modify applications more quickly. Fourth-generation languages, usually associated with a particular database-management package, have greatly increased the efficiency and productivity of programmers in application-development work. Some GIS software packages also provide an integrated high-level programming language. In the 21st century, decade application programs are being developed using object-oriented languages (Visual Basic, C++, Visual Pascal, Delphi XE3, Python, etc.). Internet-based application programs use the XML and Java languages.

Manufacturers of geospatial analysis and GIS packages rely to a greater degree on third-party relational-database packages. Many vendors have started offering the option of integrating or linking their software to a variety of relational packages.

4.2.5 Database Management and Data Manipulation

Since the early 1980s, the data-processing community as a whole has embraced the relational model and relational techniques. Recent advances in relational-database software have greatly enhanced the user interface, and the architecture of computer processors has become more suitable for relational data structures. The current database-management software packages combine the flexibility of relational models with an ability to perform interactive transaction processing on large databases. Advances in high-level-query and programming languages have prompted the popularity of relational-database systems. Data-query languages have adapted English-like commands to perform complex retrievals from the database. A recently accepted standard is the structured-query language (SQL). Initially developed by IBM, SQL is used by many relational-database software and GIS software developers. For example, the Intergraph-enterprise MGE suite of GIS application program was implemented on workstations in the early 1990s. Intergraph's GeoMedia Pro software (www.intergraph.com http://geospatial .intergraph.com), introduced in 2000, can read data from a variety of database sources (SQL, Oracle, Access, etc.), but it writes in specific database types (such as Access and Oracle). The software also facilitated linking to external databases from remote servers available through the Internet [Uddin 11a], which is an early example of cloud computing.

In relational-database software, the size of infrastructure networks and data activities influences the memory and hard-disk storage-space requirements. Typically, nongraphic-relational databases can be managed using relatively small hard-disk space. Memory and storage-space requirements for graphic databases can be many times higher than for the nongraphic databases depending upon the type of geographic data (vector or raster) and extent of coverage area. For large-size agencies the use of large mainframes, supercomputers, and high-performance workstations on a network becomes a necessity, particularly when the database is expected to fill from 100 GB to terabytes of storage space.

4.2.6   Geographic and Nongeographic Data

GIS software (available for workstations and microcomputers) provides the user with mapping capabilities and links attribute databases to several map features. This is useful for the end-user to visualize the physical layout and specific attributes of an infrastructure facility. Figure 4-5 illustrates this concept. A GIS database usually consists of two basic types of data: geographic and nongeographic. Each type has specific characteristics and requirements for efficient data storage, processing, and display.

4.2.6.1   Nongeographic Data

Nongeographic data are representations of the characteristics or attributes of map features. They are stored in conventional numerical formats, although data such as graphic images can be linked to nongeographic data with geospatial technology. The term *nongeographic* is used here to differentiate those data that do not represent the geographic features. Examples include name of landmark building, parcel address, population or income attributes associated with a city or state, daily traffic volume by road, and name of a power plant. They are related to geographic locations through common data fields or identifiers. In geospatial or GIS software, nongeographic data are managed separately from the geographic data because of their different characteristics.

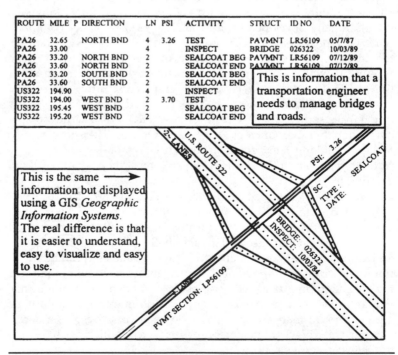

FIGURE 4-5   An example of a visual display of data [after Keystone 93].

4.2.6.2   Geographic Data

Geographic data represent map features in a computer-readable form. Geographic data use six types of graphic elements (i.e., points, lines, areas, grids, pixels, and symbols) to depict map features and annotation. There are two groups of geographic data: *vector* and *raster*. Vector data define geometrical features by a combination of points, lines, and areas. Raster data represent the pattern of pixels that define a scanned image or a digital aerial and satellite imagery. Any scanned image is raster data, where each location on the map directly corresponds to a location in the raster-data storage grid. Raster data of scanned images require large amounts of storage memory [Antenucci 91].

Just like the AUTOCAD software for computer-aided graphics, a series of layers is often used to describe the graphic component of the GIS database, each of which contains map features that are related functionally. Figure 4-6 shows the layering concept. Each layer is a set of homogeneous features that is registered positional to the other database layers through the common coordinate system. The separation into layers is based on logical relationships. One major purpose of the layering is to simplify the combination of features for display. This electronic-layering scheme is comparable to a series of overlays in a manual-mapping system.

Not all geospatial programs have specific layering functions. For example, layering is integrated within legend features in Intergraph's GeoMedia Pro geospatial software [Uddin 11a].

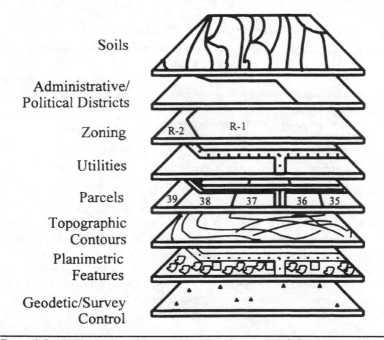

Soils

Administrative/
Political Districts

Zoning

Utilities

Parcels

Topographic
Contours

Planimetric
Features

Geodetic/Survey
Control

Figure 4-6   An illustration of the database layering concept [after Antenucci 91].

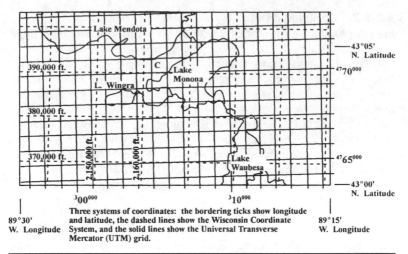

Three systems of coordinates: the bordering ticks show longitude and latitude, the dashed lines show the Wisconsin Coordinate System, and the solid lines show the Universal Transverse Mercator (UTM) grid.

FIGURE 4-7   An illustration of commonly used coordinate systems [after Robinson 69].

4.2.7   Geocoding

Geocoding (geographic coordinates) data are important for georeferenced graphic displays. Figure 4-7 shows commonly used coordinate systems. Typically, data can be obtained by using a digitizer on a paper map, a differential GPS receiver, satellite-imagery data, Topological Integrated Geographic Encoding and Referencing (TIGER) files, and digitized-data files produced from computer-aided drafting graphics software, such as AUTOCAD. The TIGER system was developed by the U.S. Geological Survey (USGS) and the Bureau of the Census for the 1990 census of the United States. The TIGER system is the first comprehensive digital map of the nation's roads and streets at a scale of 1:100,000, street addresses in urban areas, railroads, and all significant hydrographic features [Antenucci 91]. Higher-resolution geospatial databases are in use today [Uddin 02].

4.3   Data Needs

4.3.1   Data Requirements for Infrastructure Management

Databases and data-management software are needed in any department of an agency that is responsible for infrastructure management. Some of these databases may be used by one or more departments. For example, financial databases are usually used by accounting departments; contract and project-information data are used by contract and construction departments; and maintenance-work databases are used by accounting and maintenance departments. The key to effective database management in IAMS is the linkage among various databases so that the necessary information is easily accessible. However, the following data are essential to a generic IAMS:

- Inventory data for physical descriptions of infrastructure facilities, including construction and metrical data, preferably geographically based
- Usage-history data (e.g., water usage for water-distribution systems, aircraft coverage for airport pavement)
- Condition monitoring and evaluation data
- Real-time monitoring data
- Maintenance history and operation data
- Maintenance-intervention criteria, decision criteria, maintenance policy, and unit cost and budget data
- Design and analysis data
- Maintenance and construction priority list data
- Sustainability-performance data such as an emission inventory of both harmful emissions and GHG from transportation, energy, and industrial sources

The basic framework of a typical asset-management system (AMS) for transportation, recommended by the U.S. DOT based on the GASB 34 guide, is discussed in Chapter 5. The required data types and associated details are provided in Chapters 5 and 6 with examples from transportation, as well as other public infrastructure.

4.3.2 Data Details

Classification of data needs and levels of detail are very important issues because data collection can be a very expensive effort. Haas et al. proposed a three-level concept of information detail and complexity of models, as shown in Figure 2-4 [after Haas 94]. The lower-left and upper-right triangles are infeasible areas because while detailed data is required for analysis at the project-level, detail and complexity of models restrict the amount of detail of data at the network. Similarly, performance modeling for project-level design applications requires a detailed material-property database, as described by Uddin for road-pavement applications [Uddin 95].

A study by the World Bank recognizes five functional levels of data needs [Paterson 90], which offer a framework to establish infrastructure-management data needs for specific agency objectives:

Sectoral Level: Aggregation of data from the road-infrastructure system; e.g., annual statistics and road-user charges for comparison with the education sector, for example.

Network Level: Planning, programming, and budgeting: strategic planning (three to five years, medium-term needs); strategic regional-transportation planning; tactical network-level work programming for M,R&R.

Project Level: Project-level programming for selection and extent of treatment and exact location; may require intensive data collection and analysis.

Operational Level: Facility and operation management related to construction, maintenance, traffic, and safety.

Research and development: Data needs are more detailed and precise than that for project level and operational level. Data are usually study-specific.

Element-specific data detail is then best defined for groups of similar information, since these groups have features in common for more than one application and form a natural basis for collection and storage in a database. For each category, the specific data needs relevant to a functional application can be identified. The amount of detail required for these various applications increases progressively from overall summary statistics through planning, programming, design, and research. In accordance with Figure 2-4, as the system becomes more complex the data-collection efforts increase per unit from network-level analysis to project-level analysis. More sophisticated methods are required to gather the more detailed data required for each group of information. This leads to the definition of four information-quality levels that represent the ranges of both the data requirements and the methodologies for collecting the data.

The road-inventory and pavement condition data needs can be identified for four information-quality levels (IQLs) suggested in the World Bank's report [Paterson 90]. These IQLs are listed in Table 4-1.

IQL	Amount of Detail
I	Most comprehensive level of detail; normally used at high-class project-level design and research studies; requires high level of institutional resources to collect and process the data
II	Level of detail sufficient for programming models and standard design methods; would usually require sampling and automated acquisition equipment for network-level programming; requires reliable institutional sources
III	Level of detail sufficient for planning and standard programming models for full-network coverage; network data collection can be combined using automated and manual methods
IV	Level suitable for the simplest planning and programming models; suitable for standardized road-design catalogs; not adequate for high-class project-level design; the simplest data-collection methods, either entirely manual or partly semi-automated; requires the least resources

TABLE 4-1   Classification of Information-Quality Level (IQL) and Detail, with Example Application to Road Infrastructure [after Paterson 90]

Level 1 is the greatest level of detail and is typically benchmark-type information that will be collected only on a project-level basis. Level II is the greatest amount of detail that would be gathered at a network level and usually requires sampling. Level III detail is adequate for planning purposes, and data may be collected using a combination of manual and automated methods. Level IV is the minimum detail needed for generating the basic statistics and other information on the network for planning models. The concept of IQL from roads also can be applied to other infrastructure-management applications.

The general classification of detail in terms of the number of parameters is complicated by the question of reliability. It is possible, for example, to compensate for some lack of detail by increasing the reliability of each data point by raising the accuracy of measurement and/or intensifying the measurement sample rate.

4.3.3   Data Terminology and Data Dictionary

The agency should compile a comprehensive data file of all pertinent local terminology and definitions for various data elements (inventory, condition, equipment, data-summary statistics, and indices) and appropriate units of measurement. The adoption of international standards for new data elements would be useful. This is crucial for consistency among agency staff and other users and should be considered as a step to ensure quality.

4.4   Analysis and Modeling Techniques

Data analysis is an integral part of IAMS that distinguishes IAMS from a traditional information-management system. Analysis requires the use of models and analytical techniques. Models serve various purposes, including: (1) a performance model to predict life; (2) a model to backcalculate material properties; (3) a failure-analysis model, such as fatigue models; or (4) cost models, such as maintenance-cost or user-cost models.

Nishijima outlines several steps in material-property modeling [Nishijima 95]:

- Catch general trend and scatter of the data.
- Evaluate quality of the data set.
- Deduce characteristic values of the property.
- Compare the data with other data (statistical tests).
- Predict values by interpolation and extrapolation.
- Determine confidence limits of the property.
- Obtain mathematical expressions for further tests.

These steps are also generally valid for other types of data and model development. A brief overview of available modeling techniques is presented next. These techniques require a reliable database as the first step.

4.4.1 Regression Analysis

Traditionally, data have been analyzed using statistical regression analysis. This method is commonly used to develop empirical models by estimating parameters and coefficients of independent or explanatory variables in mathematical relationships that can explain most of the variations in the dependent or predictor variable [Box 78]. A scattergram of the data should be reviewed first, to estimate the model shape. Analysis of variance (ANOVA) should precede regression analysis to identify statistically significant independent variables. The regression model may be linear or nonlinear. Regression modeling has its limitations, especially if the scattergram does not show a known model shape, if many different independent variables influence the dependent variable, or if the goodness of fit is low (low R-square values). Many new analysis and modeling techniques have emerged during the last decade.

4.4.2 Expert Systems

The definition of an expert system is "interactive software that uses simplified methods to drastically limit search time in large problems for certain tasks" [Begley 95]. The simplest architectural model of an expert system, shown in Figure 4-8, indicates that the main components of expert systems are a knowledge base and an inference engine.

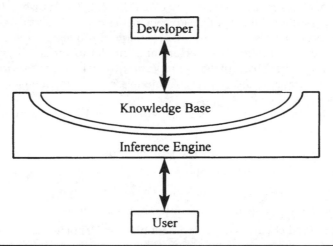

FIGURE 4-8 Architecture of an expert system [after Begley 95].

Expert systems differ from conventional computer programs in two ways:

1. Expert systems work with domain-specific knowledge that is symbolic as well as numerical.

2. Expert systems use domain-specific methods that are heuristic (simplified) as well as algorithmic.

4.4.3   Object-Oriented Database-Management System

Object-oriented database-management systems (OODBMS) employ data structures and programming concepts originally developed in object-oriented programming (OOP) languages. OOP languages were developed to address cost and efficiency issues associated with the implementation of large software applications. Object-oriented software uses cell-like components called *objects* that communicate with one another via messages, much as cells communicate through chemical messages in the human body. An object, or abstract data type, is a block of code that combines data with operations to manipulate that data. The development of OODBMS solved the database-query problems associated with CAD applications in automotive and aerospace industries. Object-oriented databases can be viewed as database-management systems based on relationships and functional characteristics of data where data-access paths are built into the object types [Begley 95].

4.4.4   Artificial Neural Networks

Artificial neural networks (ANNs) are computing systems made up of several simple, highly interconnected elements that process information by dynamic responses to external inputs. Neural-network architecture was inspired by studies of brains [Begley 95, Ghaboussi 92]. The basic model of an individual neuron is shown in Figure 4-9 for

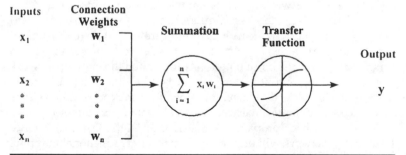

Figure 4-9   The processing elements in an artificial neural network [Begley 95].

a simple artificial neural network. It does not execute a series of fixed instructions like a traditional computer program or statistical analysis; rather, it responds, in parallel, to the inputs presented to it during a training period. Each neuron computes and maintains a dynamic variable called activation, S_j, which is a function of the sum of inputs that arrive via weighted pathways. The input from a particular pathway is an incoming signal, X_i, multiplied by the weight w_{ij} of that pathway. The outgoing signal from neuron i is the activity of the unit, computed as $O_j = f(S_j)$, where $f(S_j)$ is called the activation or transfer function, usually a binary threshold or a sigmoid function. The neuron can also have a bias term, b_j, which acts as a form of threshold. The relationship between incoming signals and the outgoing signal can be expressed as:

$$S_j = \Sigma W_{ij} X_i + b_j \tag{4.1a}$$

$$O_j = f(S_j) \tag{4.1b}$$

$$f(S_j) = 1/[1 + e^{-Sj}] \tag{4.1c}$$

The major variables of a neural network are network topology (the number of nodes and their connectivity), rules of computation of the activations of the processing units, rules of propagation, and the rules of self-organization and learning [Ghaboussi 92]. One of the common types of rules involves back-propagation networks, also referred to as *feed-forward networks*. The processing units in back-propagation neural networks are arranged in layers. Each neural network has an input layer, an output layer, and a number of hidden layers. A neural network performs "computations" by propagating changes in activation between the processors. Propagation takes place in a feed-forward manner, from the input layer to the output layer. Processing elements and interconnections from a sample neural network are shown in Figure 4-10. Each unit receives input from all the units in the processing layer and sends its output to all the units in the next layer.

The neural network gains its knowledge through training. A supervised training method is commonly used to train feed-forward networks. In supervised learning, a set of training data (consisting of input/output patterns) is presented to the network, one example at a time. For each set, the input pattern is propagated through the network and the resulting output pattern is compared to the target. A learning algorithm is employed to incrementally adjust the connection weights to reduce the difference between the calculated and target outputs. This is done by back-propagating the error through the network. Once trained, the network will provide an approximate functional mapping of any input pattern onto its corresponding output pattern. Once a neural network has processed all of its training data and achieved a state of equilibrium, new input data can be presented for evaluation.

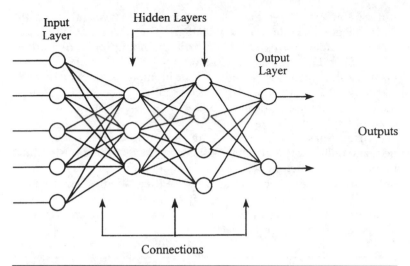

FIGURE 4-10   ANN-processing elements and interconnections [Begley 95].

4.5   Database Security

When working with high-volume IAMS databases, it is imperative to develop and maintain a comprehensive data backup and security system [Hudson 97]. Furthermore, Internet-based data access and connectivity in the 21st century requires safeguards against computer viruses, spam, and hacking.

Considerable time and money is invested in the development of databases, and without the implementation of appropriate security protocols the agency could face serious consequences. These protocols should be documented, the responsible staff trained, and frequent surveillance undertaken to ensure continued implementation. Off-site storage of the latest permanent backup is essential to recover the lost data and other information in case of a catastrophe like fire, flood, earthquake, or arson. The protocols should include standard formats for backup identification, storage, and retrieval.

A variety of backup media are available, depending on the computer hardware and operating software. These range from backup diskettes, 8-mm tapes, and spools to removable high-memory optical disks. In all options some type of data-compression utility is used to compress, or zip, the data files to maximize the capacity of the storage medium. Removable flash drives, hard disks, and cloud servers are commonly used nowadays.

An excellent example of the ultimate benefit of offsite backup storage is the digital GIS database of the road infrastructure of Kuwait. During the 1990 to 1991 Persian Gulf War, all computerized digital road-network databases were reportedly destroyed or corrupted; however, a backup copy available outside the country was used by Intergraph to print a detailed road map of Kuwait

[Intergraph 91]. Other examples of manmade devastations in the United States include: (1) the Oklahoma City federal building that was completely destroyed by a terrorist bombing on April 19, 1995 [Time 95] and (2) destruction of New York's World Trade Center and partial destruction of the Pentagon in Washington, DC on September 11, 2001 (www.911commission.gov/, accessed December 9, 2012). These incidences emphasize the need to keep an off-site permanent backup copy of IAMS databases.

Internet-based information management, wireless communication, and online data access have enhanced efficiency of infrastructure-management functions. This global cyber infrastructure includes online social media (such as Facebook, Twitter, Google+), which have enabled not just social groups, but also infrastructure agencies/ entities massive opportunities in sharing significant accomplishments and getting online-user feedback.

4.6   Data-Quality Control and Quality-Assurance Issues

Consistency of data collection and processing procedures and database accuracy should be given top priority. Principles of quality management should be followed to achieve quality outputs/products [Hayden 89, Uddin 91a]. Computer-database integrity and online security-breach concerns are important in the cyber world of the 21st century, as discussed in the previous section.

4.6.1   Quality Control (QC)

- *Definition:* The operational techniques and activities that are used to fulfill requirements for quality. In other words, it is the process of doing and checking the work before releasing it.

- Quality control is exercised through detailed step-by-step office and field procedures, instructions, and operational guidelines contained in the appropriate manuals. Examples: road-pavement inventory manuals [Uddin 91b]; pavement-distress manuals [Shahin 90, SHRP 93, Uddin 91c].

- Productivity level for all IAMS data-handling and database operations should be checked, and a system outlining the minimum desired level of productivity should be based upon assigned equipment and staff resources.

- QC principles and tailored procedures should be used continually for all future IAMS operations.

4.6.2   Quality Assurance (QA)

- *Definition:* All planned and systematic actions necessary to provide reasonable confidence that a product or service will satisfy quality requirements.

- For effectiveness, QA usually requires continuing verification, inspection, and evaluation of the operations and outputs. This can be accomplished for IAMS operations through on-the-job training, manual instructions, dedicated data collection, and processing forms and procedures.

- IAMS activities must be appropriately documented for records and follow-up.

- Work supervisor(s) should ensure data consistency and implement quality assurance through a system of routing files, follow-up, and random checks.

- QA principles and specially developed protocols should be continually used for all future IAMS operations.

4.7 References

[Antenucci 91] J. C. Antenucci, K. Brown, P. L. Croswell, M. J. Kevany, and H. Archer, *Geographic Information Systems*, Chapman & Hall, New York, 1991.

[Begley 95] E. F. Begley, and C. P. Sturrock, "Matching Information Technologies with the Objectives of Materials Data Users," *Computerization and Networking of Materials Databases*, Vol. 4: *ASTM STP 1257*, American Society for Testing and Materials, Philadelphia, Pa., 1995, pp. 253–280.

[Box 78] G. E. P. Box, W. G. Hunter, and J. S. Hunter, *Statistics for Experimenters: An Introduction to Design, Data Analysis, and Model Building*, John Wiley and Sons, New York, 1978.

[FHWA 90] "An Advanced Course in Pavement Management," Course Text, Federal Highway Administration, U.S. Department of Transportation, Washington, DC, 1990.

[Ghaboussi 92] J. Ghaboussi, "Potential Application of Neuro-Biological Computational Models in Geotechnical Engineering," *Numerical Models in Geomechanics*, Pande and Pietruszczak (editors), Balkema, Rotterdam, 1992, pp. 543–555.

[Haas 94] R. Haas, W. R. Hudson, and J. P. Zaniewski, *Modern Pavement Management*, Krieger Publishing Company, Malabar, Fla., 1994.

[Hayden 89] W. M. Hayden, "The Effective A/E Quality Management Program: How to Do It," short course notebook, American Society of Civil Engineers, New York, 1989.

[Hudson 87] S. W. Hudson, R. F. Carmichael III, L. O. Moser, W. R. Hudson, and W. J. Wilkes, "Bridge Management Systems," *NCHRP Report 300*, Transportation Research Board, National Research Council, Washington, DC, 1987.

[Hudson 97] W. R. Hudson, R. Haas, and W. Uddin, *Infrastructure Management*, McGraw-Hill, New York, 1997.

[Intergraph 91] W. Uddin, Personal Communications to Intergraph, Reston, Va., October 1991.

[Keystone 93] *Geographic Information System*, product brochure, Keystone Management Systems Inc., 1993.

[Nishijima 95] S. Nishijima, "Common Data Processing Needs for Material Databases," *Computerization and Networking of Materials Databases: Fourth Volume, ASTM STP 1257*, American Society for Testing and Materials, Philadelphia, 1995, pp. 9–19.

[Paterson 90] W. D. O. Paterson, and T. Scullion, "Information Systems for Road Management: Draft Guidelines on System Design and Data Issues," *Technical Paper INU77*, Infrastructure and Urban Development Department, The World Bank, Washington, DC, 1990.

[Robinson 69] A. N. Robinson and R. D. Sale, *Elements of Cartography*, John Wiley & Sons, Hoboken, New Jersey, 1969.

[Shahin 90] M. Y. Shahin, and J. A. Walter, "Pavement Maintenance Management for Roads and Streets Using the PAVER System," *USACERL Technical Report M-90/05*, Champaign, Ill., 1990.

[SHRP 93] "Distress Identification Manual for the Long-Term Pavement Performance Project," *Report SHRP-P-338*, Strategic Highway Research Program, National Research Council, Washington, DC, 1993.

[Time 95] "Oklahoma Bombing," *Time Magazine*, Vol. 145, May 1, 1995, pp. 1–4.

[Uddin 91a] W. Uddin, "Dubai Road Pavement Management System (DRPMS)— DRPMS Manual of Data Processing, Analysis & Generation of DRPMS Reports," *U.N. Expert's Report No. PM-14, Dubai Municipality, U.N. Project UAE/85/012*, United Nations Centre for Human Settlements (Habitat), Dubai, United Arab Emirates, July 1991a.

[Uddin 91b] W. Uddin, "Dubai Road Pavement Management System (DRPMS)— DRPMS Manual of Road Network Partitioning and Inventory Data Collection," *U.N. Expert's Report No. PM-12, Dubai Municipality, U.N. Project UAE/85/012*, United Nations Centre for Human Settlements (Habitat), Dubai, United Arab Emirates, July 1991b.

[Uddin 91c] W. Uddin, "Dubai Road Pavement Management System (DRPMS)— DRPMS Manual of Pavement Inspection and Distress Data Collection," *U.N. Expert's Report No. PM-9, Dubai Municipality, U.N. Project UAE/85/012*, United Nations Centre for Human Settlements (Habitat), Dubai, United Arab Emirates, July 1991c.

[Uddin 95] W. Uddin, "Pavement Material Property Databases for Pavement Management Applications," *Computerization and Networking of Materials Databases*, Vol. 4: *ASTM STP 1257*, American Society for Testing and Materials, Philadelphia, Pa., 1995, pp. 96–109.

[Uddin 02] W. Uddin, "Evaluation of Airborne Lidar Digital Terrain Mapping for Highway Corridor Planning and Design," *CD Proceedings, International Conference Pecora 15/Land Satellite Information IV ISPRS*, American Society for Photogrammetry & Remote Sensing, Denver, November 10–15, 2002.

[Uddin 11a] W. Uddin, "Geospatial Analysis for Visualization Applications," Lecture Notebook, (3-credit-hour course for seniors and graduate students), University of Mississippi, May 2008, updated 2009–2011, © W. Uddin, 2008–2011, p. 258.

[Uddin 11b] O. W. Uddin, "Mind Map: Cloud Computing for Non-Profits," February 23, 2011. http://uvisionconsulting.com/technology, accessed on December 15, 2012.

[Zhang 94] Z. Zhang, T. Dossey, J. Weissmann, and W. R. Hudson, "GIS Integrated Pavement and Infrastructure Management in Urban Areas," in *Transportation Research Record 1429*, Transportation Research Board, National Research Council, Washington, DC, 1994, pp. 84–89.

CHAPTER 5

Inventory, Historical, and Environmental Data

5.1 Infrastructure Asset Management Data Needs

5.1.1 Background

The implementation of infrastructure management in the form of pavement management systems (PMSs) since 1970 [Haas 94, Hudson 94], and in bridge management systems (BMSs) since the late 1980s [Hudson 87, Golabi 92] clearly demonstrate the requirement for valid inventory or identification databases. The national bridge inspection standard database serves for BMSs, and each state individually has covered their needs in PMSs. These systems demonstrate the importance of good data and a well-balanced database in managing the design, construction, and M,R&R (maintenance, rehabilitation/renovation, and replacement/reconstruction) activities for infrastructure facilities. The U.S. DOT has taken the GASB 34 initiative of asset management beyond pavements and bridges to safety and other transportation assets [DOT 00, 11], which require more data collection. Figure 5-1 shows the GASB asset-valuation framework, which is the primary guide adopted by the U.S. DOT, AASHTO, and individual states [after DOT 00]. It requires that an IAMS include an inventory of assets and condition assessment at least every three years. The results of the three most recent assessments should show that the infrastructure assets are being preserved at or above the established condition level.

This asset-management framework is generic and does not specify types of asset data and levels of their details for inventory and monitoring. All this is left to agency discretion. This book helps fill this gap in knowledge of effective and detailed guidelines to establish a working IAMS. The database for any IAMS involves the following:

- Good identification and location-referencing data
- Inventory of physical description, construction, and historical data of past actions

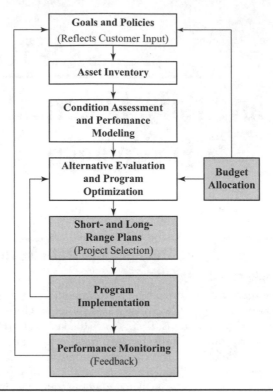

Figure 5-1   A general asset-management framework [after DOT 00].

- Condition monitoring and evaluation data
- Usage data and maintenance history
- Sustainability performance-measure data
- Feedback system to update and improve IAMS activities

These system components are essential for any IAMS, as well as for an integrated IAMS that ties together several types of infrastructure. In some agencies, the location-referencing and physical-inventory data attributes for infrastructure are available but are not systematically coordinated. Some agencies have made progress in linking their primary databases, as reported for New York [Cohn 95] and California [Robison 96]. These initial efforts may be time consuming but lead to the advantages of common computer-system operations and quick access to the primary inventory database for emergency management. In the case of New York City, a $1.7 million integrated system of computer-aided drafting and design (CADD) and GIS provides a single electronic information channel for everything from improving pothole repair and road design to developing tree-planting plans and mapping the city's bridges. The city of Oakland, California, collected over $6 million in compensation

Transportation Infrastructure	Bridge	Pavement	Safety	Public Transportation	Congestion	Intermodal
Bridge	—	Major	Major	Major	Minor	Major
Pavement	Major	—	Minor	Major	Major	Minor
Safety	Major	Minor	—	Major	Major	Major
Public transportation	Major	Major	Minor	—	Major	Minor
Congestion	Minor	Major	Minor	Major	—	Minor
Intermodal	Major	Minor	Minor	Minor	Minor	—
Emissions	Minor	Major	Minor	Minor	Major	Minor

TABLE 5-1   Model Interactions Among the Transportation Information Systems [after Hudson 94]

from the Federal Emergency Management Agency (FEMA) for post *facto* damage to city streets after a devastating fire in October 1991 on the basis of a detailed pavement management report that quantified the damage to streets resulting from heavy reconstruction traffic.

Typical GIS software accesses database files, and a GIS database file can be easily linked with large IAMS databases. Figure 4-3 illustrates an integrated database based on GIS. Such an integrated database is cost-effective for facilities that share common information needs. This is illustrated by Table 5-1, which models interactions among the transportation subsystems [after Hudson 94].

The issues of sustainability and environment have added to the need for inventory data collection. For the last 20 years the U.S. Environmental Protection Agency (EPA) has been requiring states to make annual emission inventory and monitor air pollutants harmful to human lives. Additionally, many cities including New York City [NYC 07] have embarked upon a collection of CO_2 emission inventories and long-range sustainability data.

5.1.2   Data Details and Data Collection Frequency

A proper database will contain details ranging from a few mandatory elements to a long list of data elements, depending on the objectives of the IAMS. The use of specific data will vary with the functional level of IAMS (i.e., network-level planning function or project-level design). As discussed in Chapter 2 (Figure 2-4), the data details and system complexity required increase from the network-level to project-level IAMS functions. An IAMS decision support system (DSS) database consists of a number of files that depend on the data-collection frequency and updating. The database files, listed in Table 5-2, are

| DSS Component | Database Update | | Database File |
	Mandatory	Desirable	
1. Inventory (location reference, physical description, construction type and data, functional class, M,R&R history)	Once (original construction)	After every major M,R&R and capital improvement	Section-specific
2. Usage* (historical usage record and future estimate)	For a new IAMS, back cast for best estimates, then annually	Monthly and seasonal in addition to yearly total	Section-specific
3. Condition evaluation* (condition monitoring and evaluation data)	Once every 1–3 years (on sampling basis)	More than once every 1–5 years (on 100% network)	Section-specific
4. M,R&R Policies and costs* (short-term, annual, long-term)	Once (used for analysis)	Current data (annual update of costs)	Facility-specific
5. Performance prediction (condition deterioration, effect of M,R&R on condition)	For life-cycle analysis	Every year	Facility-specific
6. Economic analysis (analysis period, economic evaluation parameters)	Once (used for analysis)	Multiple interest and inflation rates	Network-specific (global)
7. M,R&R Program reports* (economic evaluation, short-term and long-term M,R&R work program reports)	Every year annual M,R&R program	3-year, 5-year, 10-year, multi-year M,R&R program	Network-specific with facility-specific outputs
8. Environment* (rainfall, temperature, etc.)	Annually	Annually	Use NOAA or statewide access
9. Environmental impacts* (air pollution, CO_2 emission, deforestation, etc.)	Annually	Annually	Use EPA/DOT/ DOE (EIA) city and statewide access

*Not inventory data. See Chapter 6 for detailed discussion of item 3 (condition evaluation data). See Chapter 15 for discussion of item 7 (M,R&R program reports).

TABLE 5-2   The Interrelationship of IAMS Data Collection Frequency and Database Update

section-specific, facility-specific, and applicable to all sections in the network. The files include the following data types:

- Section-specific data (inventory, usage, and condition evaluation)
- Facility-specific data (M,R&R policies and unit costs, performance models)
- Network-specific data (economic analysis parameters, M,R&R work programs)

Data updating is a key activity that applies not only to usage data and condition evaluation data, but also to changes in the inventory resulting from major M,R&R projects. Items 2, 3, 4, 7, 8, and 9 in Table 5-2 are not asset inventory but are shown here for completeness.

5.2   Network Partitioning and Inventory Data

An inventory of infrastructure facilities generally refers to the data related to the physical features, including structural components, physical dimensions, material properties, and construction details (including date of original construction). For existing facilities, it is essential to determine and record in the database a history of all construction, rehabilitation, capital improvement, and other physical changes in the structure, drainage, and so forth. The M,R&R work and cost history, as well as the historical record of usage, are also important data attributes. Most characteristics do not change from year to year, but any items that change because of M,R&R should be updated immediately. The database design requires that all data be defined with respect to specific section and location-referencing codes. Some of the inventory data attributes are mandatory for M,R&R analysis to generate network-level M,R&R programs. Other data elements may be desirable for a broad and detailed M,R&R analysis and for project-level analysis of alternative design strategies. The following major classes of inventory and historical data are important to the IAMS databases:

- Section identification and location
- Functional class
- Geometrical data
- Structural data
- Material type and property data
- Appurtenance data
- Construction and M,R&R history
- Costs data
- Environmental data
- Usage history
- Baseline emission data

5.2.1   Network Partitioning

One of the important initial steps for building an inventory database is the partitioning of the network into homogeneous management units or sections. Each of these management units is given a unique identification number that is the backbone organization of the IAMS inventory, historical, and condition databases and all related analysis and management reports.

Generally, an IAMS unit or section should have the same material type, geometric parameters, construction history, functional classification, operating volume, and characteristics along its entire length. Other practical aspects include easy identification on the network map and in the field, and manageable length for monitoring. For example, a road network section length will be generally 2 km (1.2 mi) or less. Local streets may have one section per block. On the other hand, an expressway would be divided into a number of sections. Some facilities stand alone like a bridge or a dedicated building structure (e.g., a coliseum, monument, water tower, or air-traffic control tower) and should be treated as a unique unit.

Partitioning criteria should be established after a thorough study of the infrastructure network and appropriate numbering methodologies. The following common parameters can be used as partitioning criteria:

- Agency's jurisdiction and administrative zone boundaries
- Functional and use classifications (e.g., storm sewer or sanitary sewer)
- Primary construction material and geometry/structure characteristics
- Construction or major rehabilitation/renovation/replacement project number and completion date
- Boundaries based on major operating component/equipment locations (e.g., the junction of two sewers can be the beginning point for the first trunk sewer)
- Land parcel and adjacent road references (with geographical coordinates)

5.2.2   Dynamic Segmentation

Network partitioning based on location identifiers and physical descriptions establishes homogeneous sections. This approach is necessary to calculate age, area, and volume of M,R&R work, and future-condition predictions; however, the uniformity of units based on behavioral attributes, such as deflection and hydraulic characteristics, may be different. A dynamic segmentation procedure can be used to establish new boundaries of units without losing other physical inventory data in the original sections [Uddin 92]. Even a large building, such as the Pentagon in metropolitan Washington, DC, can be subdivided into manageable units based on wing, floor, and so forth.

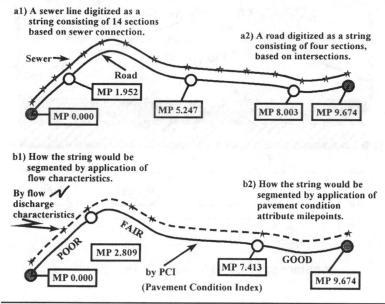

a1) A sewer line digitized as a string consisting of 14 sections based on sewer connection.

a2) A road digitized as a string consisting of four sections, based on intersections.

Sewer ➤

Road

MP 1.952

MP 0.000

MP 5.247

MP 8.003   MP 9.674

b1) How the string would be segmented by application of flow characteristics.

b2) How the string would be segmented by application of pavement condition attribute milepoints.

By flow
discharge
characteristics

FAIR

POOR

MP 2.809

MP 0.000

by PCI
(Pavement Condition Index)

MP 7.413

GOOD

MP 9.674

Figure 5-2   Dynamic segmentation based on evaluation data.

Figure 5-2 shows an example in which several main storm-sewer sections are included in larger segments based on discharge flow characteristics, and the corresponding street sections are segmented by attribute mile points using the pavement-condition rating.

Dynamic segmentation can be implemented in GIS software because of its powerful query operations. For example, the IAMS database can be queried to display all sections with similar attributes, or homogeneous segments can be characterized based on selected threshold values of evaluation data.

5.2.3 Location Referencing Methodology

Location references should be based on well-defined points at the start and end of the section, along with the geographical coordinates at both ends. The location should be easy to identify in the field as facilitated by GPS. Postal zip code and street address must be included for any building or fixed plant facility. Location references used for a road section can also be adapted for other spatially distributed facilities that are present along the road, as shown in Figure 5-3 [after Hudson 94]. Location-referencing codes should not be confused with record sequence numbers generated by software or input by a user. A unique number for each IAMS section or management unit is important for two reasons: (1) the IAMS database and analysis software packages use the information to track each section; and (2) the user identifies the section by its code, both in the field and on the reports generated. The properly designed location-referencing code can be used to link numerous other infrastructure components.

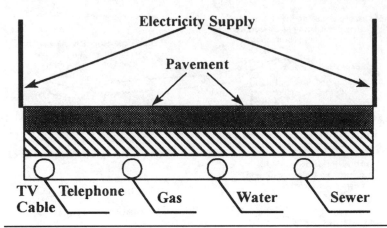

Figure 5-3   Example of common location reference for spatially distributed infrastructure facilities [after Hudson 94].

5.2.4   Inventory of Physical Assets

Considering the quantity of possible physical inventory data (functional class, geometry, structure, material type and properties, and appurtenances), it is essential to prioritize the data and/or collect the primary data elements that are required for M,R&R analyses first. Suggested mandatory inventory data elements for different infrastructure groups are shown in Table 5-3. Functional class is an important data item for identification of IAMS sections.

Facility	Identification/ Location	Construction/ Geometry/Material/ Structure*	Cost and Usage History*
(1) Highways, roads, streets, parking areas	Agency, state, county, city, name and number, section number, reference longitude and latitude, use, functional class	Construction number, type and date, lanes, pavement width and length, surface type, material types and thickness, shoulder, sidewalk, drainage, safety appurtenance, traffic control and lighting	Total/annual construction and M,R&R costs, unit costs, annual traffic and percent vehicle type
(2) Airports: pavements, buildings, aviation facilities	Agency, state, county, city, name and number, section number, reference longitude and latitude, use, functional class	Construction number, type and date, airport lighting, navigation facilities, etc.; see (1) for pavement data; see (11) for building data and air-traffic-control facilities	Aircraft mix, annual operations and passenger; see (1) for cost data and vehicular traffic data

Table 5-3   Examples of Infrastructure Inventory Data Elements

Facility	Identification/ Location	Construction/ Geometry/Material/ Structure*	Cost and Usage History*
(3) Bridges: interchange, over/ underpass, flyover, rail/ river crossing	Agency, state, county, city, road name and number, NBIS number, reference longitude and latitude, use, functional class	Construction number, type and date, spans, material and size of substructure and superstructure; see (1) for approach pavement and deck surface data	See (1) for cost data and vehicular traffic data
(4) Railroad: terminals, track, bridge	Agency, state, yard and terminal address, track sections and location, use, coordinates	Construction number, type and date, track geometry and material, track bed, traffic control; see (3) for bridge data; see (11) for terminal building data	Traffic mix, annual operations of trains/goods/ passengers; see (1) and (11) for cost data
(5) Waterborne transport: ports, inland terminals, intermodal facilities	Agency, state, port and terminal address, reference longitude and latitude, sections and location, use	Construction number, type and date, port components, water front protection material, navigation facilities, traffic control; see (1) for road data; see (3) for bridge data; see (11) for building data	Traffic mix, annual operations of container/ passenger ships; see (1) and (11) for cost data
(6) Mass transit: subway and metros, tracks, stations, vehicle stocks, maintenance facilities	Agency, state, control facility address, reference longitude and latitude, track sections and station locations, use	Construction number, type and date, pavement data, track geometry and material, track bed, traffic control; see (1) for road data; see (3) for bridge data; see (11) for building data	Traffic mix, annual operations of trains/goods/ passengers; see (1) and (11) for cost data
(7) Supply chain: container operations, manufacturing sites, trucks and intermodal fleet, distribution centers, retail stores	Owner and operator, state, central facility address, reference longitude and latitude, locations and use of all physical civil assets, use, goods inventory,	Construction number, type and date, pavement data, track geometry and material, track bed, traffic control; see (1) for road/parking data; see (3) for bridge data; see (5) for port/ shipping data; see (11) for building data	Traffic mix, annual operations of trains/goods/ passengers; see (1) and (11) for cost data

TABLE 5-3   Examples of Infrastructure Inventory Data Elements (*Continued*)

Facility	Identification/ Location	Construction/ Geometry/Material/ Structure*	Cost and Usage History*
(8) Water: treatment plants, storage, supply	Agency, state, county, city, plant number, address, tower, hydrants and pump stations, main pipeline section number, reference longitude and latitude	Construction number, type and date, tower and hydrant data, soil, pipe diameter and material, pipeline lent, pipe lining, pump and valve data; see (1) for road and parking data; see (11) for plant building data	Total/annual construction and M,R&R costs, unit costs, annual water consumed, annual consumer consumption data
(9) Wastewater and sewer	Agency, state, county, city, treatment plant number, address, pump stations, main pipeline section number, reference longitude & latitude	Construction number, type and date, manholes, soil, pipe diameter and material, pipeline length, pipe lining, joint and pump data; see (1) for road and parking data; see (11) for plant building data	Total/annual construction and M,R&R costs, unit costs, annual sewer pumped and treated, storm sewer and consumer sewer data
(10) Solid-waste facilities	Agency, state, county, city, treatment plant number, address, waste disposal section number, reference longitude and latitude	Construction number, type and date, length and width and size of plant and fill site, equipment, soil and treatment data; see (1) for road and parking data; see (11) for plant building data	Total/annual construction and M,R&R costs, unit costs, annual waste collected and treated, street trash and consumer data
(11) Buildings: public, commercial, industrial, multipurpose complexes	Agency/owner, state, county, city, name, street address, code number, reference longitude and latitude, use, functional class	Construction number, type and date, primary material type and data, number of floors, floor data (room number, use, length, width, height, door, window, light, air conditioning), soil and foundation data; see (1) for road and parking data	Total/annual construction and M,R&R costs, unit cost data, operation cost, yearly persons served and capacity used per event, yearly revenue

TABLE 5-3 Examples of Infrastructure Inventory Data Elements (*Continued*)

Facility	Identification/ Location	Construction/ Geometry/Material/ Structure*	Cost and Usage History*
(12) Electric power-supply infrastructure	Agency/owner, state, county, city, name, street address, facility number, function, reference longitude and latitude, power-line section	Construction number, type and date, cable material type and data, number of towers/ poles and material and size, transformer stations and equipment, meters, soil; see (1) for road and parking data; see (11) for building facility	Total/annual construction and M,R&R costs, unit cost data, operation cost, yearly kilowatt and consumers served, yearly revenue
(13) Telecomm-unications and wireless infrastructure: IT and Internet networking facilities	Owner/ operator, state, county, city, name, street address, facility number, function, reference longitude & latitude, power-line section	Construction number, type and date, fiber-optic and other cable-material type and data, number of cell towers/ TV towers/poles and material and size, ground stations and equipment, meters, soil; see (1) for road and parking data; see (11) for building facility	Total/annual construction and M,R&R costs, unit cost data, operation cost, yearly consumers and businesses served, yearly revenue

*Complete data for original construction (designated as construction number 1) and for subsequent M,R&R (designated as construction number 2 and higher). Construction number designations are used to relate all other data belonging to the particular construction project, as explained further in Sec. 5.2.5.

TABLE 5-3   Examples of Infrastructure Inventory Data Elements (*Concluded*)

The two primary functions of roads are *mobility* and *accessibility*. A functional class reflects the geometrical and structural design standards. Commonly used functional classes are arterial (such as freeway, expressway, highway), collector, and local. Local streets are generally designed for accessibility, whereas high-level arterial roads are predominantly designed for high-speed mobility. The classification by jurisdictional entity (e.g., private, municipal, county, state, federal) is also useful to know the type of M,R&R and funding resource. It is also possible to downgrade the information-quality level (IQL) of some of the data groups [Paterson 90] using the recommendation outlined in Chapter 4. For example, only pavement-surface type and thickness, collected by high-speed visual-windshield surveys and radar surveys, respectively, for highways represent IQL III for the structure and material data groups. This low-cost effort will suffice

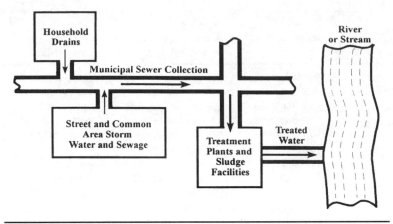

FIGURE 5-4 Configuration of urban wastewater management system.

for network-level M,R&R planning over a relatively short timeframe. Later, these data at IQL I can be collected in greater detail (material type, thicknesses, and strength of each layer using as-constructed project documents and/or coring) for overlay design at project level.

A wastewater management system, on the other hand, consists of a number of distinct components for collection and disposal of sanitary sewage and storm wastewater, where specific inventory data items can be assigned to a unit. Figure 5-4 illustrates a schematic of a typical urban wastewater management system [after Grigg 88]. The components of a comprehensive wastewater management system for sanitary sewage and storm wastewater are:

- Household drains for sanitary and stormwater sewage
- Street and common-area stormwater sewage collection (gutters, ponds, channels)
- Municipal sewer-collector system for sanitary and storm sewage (sewer pipes or open-channel ditches, manholes, pumps)
- Main sewers (pipes and ditches)
- Treatment plants
- Sludge facilities
- Disposal of treated water

5.2.5 Construction and M,R&R History

Construction and M,R&R history should be included in the database starting with the original construction. This can be very efficiently achieved by establishing the following data pertaining to *construction number* (construction number 1 for original construction, construction number 2 and above for subsequent M,R&R work, etc.): *construction type* (original construction, M,R&R type); *construction material* (primary materials used in the original construction and M,R&R

work); *construction date* (date of completion of the original construction or subsequent M,R&R work); and *construction cost* (total cost of work).

Construction number is used as a key identifier to represent any major work done that can significantly affect subsequent performance. All geometric data, structural data, materials data, and cost data should be linked to the specific construction number. Construction quality and variability affect the pavement condition; therefore, data variability should be recorded in the database if at all possible. The accumulation of all historical data, collected over time, provides the basis for assessing, developing, and calibrating performance models needed for life-cycle analysis.

5.2.6   Geometry, Structure, and Material Data

The extent of geometry, structural, and material data can vary from a few mandatory items to many hundred items, depending on the infrastructure and the intended use of the data. These data items must refer to each construction number for a given section. The mandatory geometry data include overall dimensions, length, width, and thickness of primary components. Structural and material data should include, at a minimum, material type and dimensions of the primary load-bearing component, and a list of any other components and their material type. Project-level analysis may require more detailed data for each structural component. Examples of these data for a bridge, a road section, highway-safety assets, and a building are presented in Secs. 5.6, 5.7, and 5.8. Detailed examples of the scope of needed pavement structure and material data for pavements are available in many other references [Hudson 94, Uddin 93, 95, Zhang 94].

5.2.7   Cost Data

Cost history of new construction and M,R&R strategies is an important component of the inventory database. If a computerized maintenance-bookkeeping procedure or maintenance management system already exists, then the past cost history can be electronically called up as needed. The cost-history data can be used to develop average M,R&R unit-cost models for analyses of M,R&R strategies. In addition, database files and subfiles should be developed to store unit costs of common M,R&R actions based on the past records and bid estimates, and other agency records or commercial sources, such as the means cost-estimate catalogues [Means 93]. The unit-cost database files should be updated periodically. These are primarily used to calculate agency costs for projecting the M,R&R budget of the owner agency.

User costs are also important, including costs of travel time, vehicle operation, traffic delay due to construction, user discomfort, accidents, excess costs caused by disruptions and breaks in service, user fees, and taxes. Societal costs arise from (1) medical costs associated with water contamination and air pollution and (2) CO_2 and GHG emission-reduction costs and CO_2 sequestration costs [Uddin 06, 12]. For road infrastructure, very comprehensive vehicle-operating-cost models have been developed by the World Bank [Chesher 87, Paterson 92], the Texas Research and Development Foundation (TRDF), the U.S. Federal Highway Administration [Zaniewski 82], and recently updated cost models [NCHRP 12]. These are used in programs for network-level road-management systems, such as USER [Uddin 94]. Similarly, other examples of user costs can be cited for other types of facilities:

- *Airports:* Traveler's time cost and inconvenience caused by inadequate auto parking, inefficient planning of ticket counters and gates, poor baggage-handling facilities; time cost to airlines because of delays to aircraft departures and arrivals to gates

- *Water supply:* Increased water bills to consumers because of unaccounted leakage; extra cost and inconvenience to consumers in case of loss of pressure and/or rationing of water supply

- *Electric power supply:* Consumer's lost productivity and loss to businesses caused by outages and interruptions in power supply from overloads or disasters

- *Buildings:* Occupant's lost productivity and inconvenience caused by inadequate lighting, inefficient air conditioning, poor air quality; revenue losses to owners of rental properties

- *Landline phones, mobile phones, and Internet:* Loss of service caused by poor coverage, overloads, or disasters

5.2.8   Environmental Data

Environmental conditions generally affect the life and performance of infrastructure facilities. Environmentally induced stress can contribute to premature fracture of concrete and metal structures. These failures occur at stress levels considerably below design stresses and may cause serious property damage and threat to human life. According to Fitzgerald, the majority of water-main breaks occur where the pipe wall has weakened, e.g., corrosion of cast-iron pipes [Fitzgerald 68]. Other examples of weakening and failures induced primarily by environmental factors are interruptions in water supply and loss of water due to leakage, frost-heave of buildings and roads, and soil contamination caused by leakage in underground fuel tanks. Corrosion-induced material degradation and reduced fatigue life are serious concerns for sewer mains and water mains [O'Day 84], cooling facilities at nuclear plants, steel and other industrial plants.

The conditions for environmentally assisted stress-corrosion cracking (SCC) and corrosion-fatigue cracking (CFC) are more probable in the case of heavy usage and in aging structures. Other forms of environmentally related cracking are hydrogen stress cracking and sulfide stress cracking. These phenomena can drastically reduce the

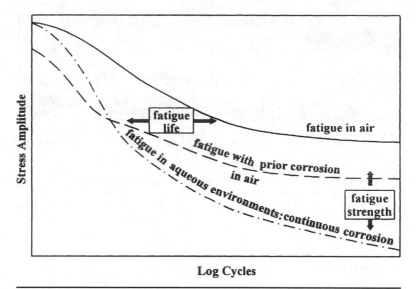

FIGURE 5-5   Schematic of the effects of corrosion on fatigue life and fatigue strength as shown in laboratory tests [after Sprowls 96].

life of a structure, and reliable service-life predictions for susceptible materials are difficult to make. A common feature of each of these processes is subcritical crack growth, during which cracks grow from existing flaws or initiation locations and increase to a size at which catastrophic failure occurs [Sprowls 96]. The fatigue strength of a given material may be significantly degraded in the presence of a harsh environment, as shown in Figure 5-5 [after Sprowls 96].

Seasonal variations of temperature and water can alter soil strength and surface or subsurface drainage characteristics and can have a direct impact on structure life and performance. It can also affect the selection and cost of M,R&R strategies. While there are many techniques for ensuring adequate or good drainage, the data characterizing drainage are usually either in terms of a porosity or permeability value, or they are subjectively recorded as good, fair, or poor [Haas 94].

Freeze-thaw conditions can significantly lower load-carrying capacity of pavements and break or disrupt water and sewer lines. For example, the appearance of potholes on roads and highways, a common scene in the United States during the spring, is affected by freeze-thaw environmental action, as shown in Figure 5-6 [Minsk 84]. Occurrence of potholes on major roads in urban America is a hot news topic each spring because of thawing and weakening of pavement layers. An estimated 77,000 potholes in Baltimore and 125,000 in New York were repaired during the first four months of 1996 [CNN 96].

If the environmental conditions vary significantly across an agency's area of jurisdiction, a record of the local environmental

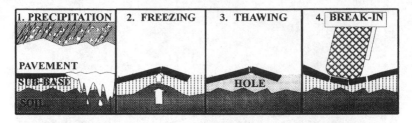

1. PRECIPITATION	2. FREEZING	3. THAWING	4. BREAK-IN
PAVEMENT			
SUB-BASE		HOLE	
SOIL			

FIGURE 5-6 A schematic of pothole formation resulting from freeze-thaw actions [after Minsk 84].

conditions are needed to predict performance and aid in the selection of appropriate M,R&R strategies. Simple indexes of environmental conditions are often used, such as the maximum- and minimum-ambient temperatures, wind speed, Thornthwaite moisture index, freeze-thaw cycles, freezing index, seasonal rainfall, or an empirical "regional factor" developed by the agency.

5.2.9 Usage History

Performance of any facility is a function of its use, load history, or traffic. Information on usage and demand on the facility are required to predict performance and to assign priorities during the selection of M,R&R projects. For water and sewer facilities, the usual measure of usage is total quantity handled per day or year. Annual consumption of kilowatt-hours is a possible measure for an electric-power supply facility. For highway agencies, the average annual daily traffic (AADT), with a breakdown into the percent of passenger vehicles and percent of trucks, is a common measure of the total traffic on the pavement section or bridge. Pavement-performance modeling and bridge deterioration, on the other hand, require an estimate of the heavy-vehicle traffic that generates the majority of the distress. For highway pavements, the total number of 80 kN (18 kip) equivalent single-axle loads (ESAL) can be used to estimate the vehicular load that the pavement or bridge has carried or is expected to carry.

Generally, usage data should be collected and recorded in the historical database using on-site instrumentation, such as flow meters, counters, scales at permanent weigh stations, weigh-in-motion equipment [TRB 86], and the application of appropriate demand-prediction models. Expected growth rates should also be included for such usage data. For airfield pavements, records should be maintained on the total number of movements of each aircraft type or class. Discharge flow of water or sewage should be recorded, if possible, on hydraulic facilities. In the case of bridges and buildings, volume usage and loading forces, such as dead and live loads, wind, and seismic forces, need to be considered.

5.3   Technologies for Inventory and Historic Data Collection

The technologies of data acquisition for infrastructure inventory items may be grouped into several classes:

- Transcription from "as-built" or "as-constructed" project records. The historical project record is the least expensive source of inventory information.

- Pedestrian observer visual survey, usually on a sampling basis. This is feasible for small-size facilities. It is also the best method to verify the information collected from other sources.

- Windshield surveys (moving-observer visual survey). This method is relatively faster but yields only estimates; however, it enables greater coverage for a fixed budget.

- Photo or video logging has the advantage of permanent records. This method can be used for roads, airports, railroad tracks, subways, sewers, large waterlines, and other facilities. Subsequent data reduction by either manual or automated image-processing techniques is, however, required, but it is costly.

- Automated measurements of geometrical and structural characteristics can be done efficiently on roads, railroad tracks, and pipelines and sewers.

- Weather records, which can provide simple, inexpensive environmental data.

- Nondestructive testing, such as ground-penetrating radar, magnetic resonance, acoustic emission, and wave-propagation methods, are sometimes useful.

- Terrestrial and kinematic laser-scanning for surface-defect mapping and three-dimensional (3D) visualization of assets at local level.

- Airborne laser scanning and digital imagery, aerial radar imaging, and spaceborne remote-sensing satellite imagery for mesoscale mapping and regional applications.

Table 5-4 summarizes several methods and options including several innovative remote-sensing technologies implemented since the late 1990s for landuse, transportation, and environmental applications. Table 5-5 identifies methods available for field-inventory data collection on roads, parking facilities, railroads, and airports. The inventory data of physical facilities are characteristics that generally do not change from year to year, and those items that do change are updated either on the completion of a project or on an

(1)	Transcription from "as-built" or "as-constructed" engineering records, such as plans, quality-control documents, and accounting and project records. The information on costs and M,R&R history is always obtained from the past records.
(2)*	Pedestrian observer visual survey, usually on a sampling basis, with manual recording, audio or photographic recording, or electronic encoding (e.g., on a handheld computer).
(3)*	Windshield survey riding on a vehicle (moving observer visual survey) with manual recording, keyboard entry to computer-encoded memory, and audio recording (voice recognition) to computer memory.
(4)*	Video or 35-mm photo logging, with permanent recording on photographic film (usually 35 mm), cinematic film (usually 16 mm, sometimes 8 mm), video film, or optical disc. Subsequent data reduction is performed by either manual or automated image-processing techniques. Digital imaging technology has been used since the mid-2000s in both van-mounted vehicles and aircraft surveys.
(5)	Automated on-board measurement of geometrical and structural characteristics.
(6)	Weather records for environmental data gleaned from published data from local or national weather stations or online Internet sources, or site-specific instrumentation and data loggers.
(7)*	Nondestructive testing, such as ground-penetrating radar, magnetic resonance, infrared imaging, acoustic emission, deflection testing, and wave-propagation methods.
(8)*	Aerial Light Detection And Ranging (LiDAR), terrestrial LiDAR, and kinematic LiDAR (scanning at road speed). Subsequent data processing is performed by specialized computer software.
(9)*	Aerial Interferometric Synthetic Aperture Radar (IfSAR) imaging. Subsequent data processing to extract digital elevation data is performed by specialized computer software.
(10)*	Spaceborne remote sensing (30-m and 15-m Landsat) and modern high-resolution 1-m or submeter commercial satellite imagery.

*May also be used for periodic condition surveys.

Table 5-4   Classifications of Inventory Data Collection Methods to Establish the Baseline History [after Paterson 90 for items 1 to 7; after Uddin 11a for items 8 to 10]

annual basis. The major data-collection effort is therefore the initial one, undertaken at the time of establishing the information system for the network. A more detailed inventory may be required at the project level for project design. Thus, the method selected for data acquisition depends on the purpose. Historical-usage data is collected by in-service monitoring on a regular periodic basis. Well-organized environmental data with periodic updates are readily

Acquisition Method	Device Type	Device Examples	Examples of User Country
Pedestrian	Handheld computers	PSION 2	Europe
	Handheld computers	Husky Hunter	U.K.
Wind shield	Keyboard	Desy 2000	France
	Keyboard	ARAN[1]	Canada
	Keyboard	RST[1]	Sweden
	Graphic tablet	RIS[2]	Norway
	Voice recognition	AREV[1]	Australia
Photo logging	35-mm continuous photography	GERPHO[1]	France
	35-mm continuous photography	ROADRECON[1]	Japan, the United States
	Video logging	ARAN[1]	Canada
	Video logging	AREV[1]	Australia
	Video logging	PAVETECH[1]	The United States

[1]Part of a multifunction dedicated vehicle.
[2]"Road Inventory System."

TABLE 5-5   Field-Encoding Methods for Inventory Data of Road and Airports [after Paterson 90]

available from the U.S. National Oceanic and Atmospheric Administration [NOAA 94].

Infrastructure-inventory applications of modern laser scanning and affordable high-resolution satellite imagery include [Uddin 08, 11a, b]: asset-footprint mapping from imagery planimetrics, terrain and built infrastructure elevations, contour maps for engineering design and floodplain mapping, airport-obstruction mapping, inventory mapping of infrastructure assets (such as road centerlines, bridges, interchanges, airport and port assets, etc.), and quality-assurance checks on Multiple Linear Referencing System (MLRS) pursued by many state highway agencies for connecting local roads with the state-highway GIS map. New van-mounted and airborne (helicopter and low-flying aircraft) methods, popularized in the 2000s, use digital-imagery sensors and LiDAR (Figure 5-7) for infrastructure-asset mapping and inventory [Uddin 11c]. Several U.S. agencies have approved LiDAR-mapping applications including state highway agencies, FEMA, and the Federal Aviation

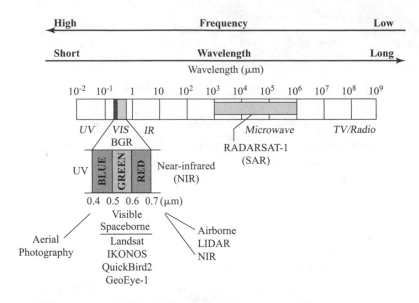

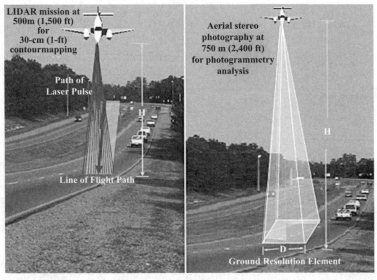

FIGURE 5-7 Electromagnetic spectrum (top) and airborne remote-sensing technologies (bottom) [Uddin 11c].

Administration (FAA) because LiDAR missions can be flown any time of day or night and are computationally significantly more efficient than traditional photogrammetry [Uddin 11b]. Table 5-6 compares selected modern remote-sensing technologies currently being used in North America and many countries worldwide [after Uddin 11c].

Satellite/Airborne	Owner/Operator, Year Launched	Spatial Resolution, Pan-Sharpened	Spectral Resolution	Temporal Resolution	Footprint (km × km)
Landsat 7	NASA, 1999	15 m	7 bands	16 days	185 × 185
ASTER (Space Shuttle)	NASA, 1999	VNIR: 15 m IR: 30–90 m	14 bands	One time	Variable
SPOT 5	Spot Image (Europe), 2002	2.5 m (20 for Mid IR)	4 bands	1–4 days	60 × 60
IKONOS	GeoEye, 1999	1 m	4 bands	3.5–5 days	11 × 11
QuickBird 2	DigitalGlobe, 2001	0.62 m	4 bands	1.5–4 days	16.5 × 16.5
WorldView 1	DigitalGlobe, 2007	0.5 m	1 band (panchromatic)	1.7–5.9 days	60 × 110 mono 30 × 110 stereo
WorldView 2	DigitalGlobe, 2009	0.46–0.52 m	8 bands	1.1–3.7 days	96 × 111 mono 48 × 110 stereo
GeoEye 1	GeoEye, 2008	0.5 m	4 bands	3 days	15.2 × 15.2
Airborne IfSAR	Inermap 1990s (estimate)	1–2 m	X band	On demand	5–10 Swath at 5,000–10,000 m height
Airborne LiDAR	Various service providers, late 1990s	up to 0.15 m	NIR band	On demand	Dense point cloud*
Aerial Photo	Various service providers	up to 0.15 m	Visible band	On demand	9 × 9 at 3,000 m

*LiDAR data at 500 m above terrain level: about 10–100 points per sq. m.

TABLE 5-6   Specifications of High-Resolution Spaceborne and Airborne Remote-Sensing Technologies [after Uddin 11a, Internet sources]

5.4   Inventory Data Collection and Processing

5.4.1   Traditional Data-Collection System

Once the section-inventory and historical-data elements are defined, the collection process begins. Depending on the sampling plan and the purpose, this task can be quick or time consuming. Electronic data collection and transfer should be used. Agencies without a centralized database may use hard-copy records and files of field-data collection. Much of the inventory information required is in historical records, but it must be automated.

Generally, the construction history begins with the as-built plans, which provide data on the layout, dimensions, and boundaries of the facility; materials used; and year of construction. In cases where the structure or facility has evolved over time or construction records are lost, such as old bridges, water and sewer facilities, and roads, it is necessary to rely on the recollection of experienced people to estimate the construction history. In some cases when the construction history is not available, physical measurements and a survey must be used to obtain information on dimensions and structure, including nondestructive testing and coring or trenching to examine the underground structure. For about one-third of the bridges in the U.S. national bridge inventory, there is no information on foundation depth, but recently nondestructive wave-propagation testing has been used to estimate this information [Aouad 96]. It is not necessary to have a separate field-investigation program to establish the construction history. The data can be collected as part of a structural evaluation over time. For example, one agency worked with the utility company using a form that work crews filled in whenever they cut a trench across a pavement to capture the thickness and material type for each layer so that this feedback information can update the inventory database [Haas 94].

Data should be collected and processed systematically using formats designed for ease of use and precise recording. Data processing in the office or the field should use a formatted data sheet or template to provide direct computer entry. A notebook or tablet computer carried into the field can provide a viable method of recording data. Procedures for quality control and quality assurance are needed to ensure data accuracy. Variability and accuracy should be estimated. For example, radar estimates of layer thickness are not nearly as accurate as actual measurements. On the other hand, a precise measurement at only one point is not an adequate sample to represent the time-mean value for a large facility.

5.4.2   High-Resolution Satellite Imagery–Based Geospatial Inventory

Geospatial-based inventories of infrastructure assets offer a fast and relatively low-cost nonintrusive alternative using high-resolution commercial-satellite imagery. Due to affordable access and world-wide coverage 1-m and submeter multispectral satellite imagery–based inventory data collection is particularly beneficial for remote areas, developing countries, and regions where lifeline-transport infrastructure inventory and GIS maps are not available. The geospatially derived spatial data can be provided for network-level linear assets (roads, railways) and other landuse assets, as well as traffic-demand volume estimates at a lower cost [Osborne 09, Uddin 10].

5.5   Institutional Issues

Dedicated staff are important for an IAMS database-development plan. To fully understand the capabilities/limitations of current technology, designated IAMS engineers should have prior experience in related areas and/or should attend a short course on infrastructure management. Visits to and interaction with other agencies of comparable size can offer useful insight into system development. However, this can be a blind-leading-the-blind situation if the agency visited is not competent in IAMS. Ongoing training of assigned IAMS staff should be an integral part of any IAMS development plan.

Lack of funding and low data priorities are obstacles to efficient inventory data collection and processing in many public-works agencies. In a recent survey of state-highway agencies, 81 respondents considered safety details to be needed but lack of resources an obstacle. One comment for example [DOT 11], "North Carolina maintains 80,000 miles of roadways. The amount of effort to collect and maintain such data would not be worth the investment .... We are looking into the possibility of adding the 'Safety Module' to our existing Agile Assets System."

5.6   Examples of Inventory-Data Systems for Transportation Assets

5.6.1   Example Inventory-Data System for Bridges

A bridge-inventory system differs from a linear facility, such as a road or a utility pipeline. A bridge can be considered to be one or more unique sections. It consists of several important groups of structural and functional or nonstructural components, as shown in Table 5-7 [Hudson 87]. In the structural bridge inventory, the three main data groups are deck, superstructure, and substructure. Figure 5-8 shows several major components of a typical bridge over a water crossing. The National Bridge Inspection Standard (NBIS) outlines a record of inventory and condition of all highway bridges and culverts with spans of 6 m (20 ft) and tunnels using a national-coding guide [FHWA 88]. The most important items in the inventory are: (1) predominant material type (such as concrete, steel, timber, etc.); (2) predominant type of design/construction, as selected from Table 5-8;

Inventory Data Groups	Number of Data Elements
Identification information	13
Environment	3
Defense-importance ratings	7
Essentiality/classification/jurisdiction	20
Navigation and waterway	3
Posting information	4
Safety inventory	8
Secondary features	20
Structural inventory	14

Note: There are six M,R&R historical variables in the NBIS database.

TABLE 5-7 Summary of Bridge-Inventory Variables [after Hudson 87]

(3) bridge-structure type; and (4) functional class. The functional classes by type of service that the bridge provides are [Xanthakos 94] highway, railroad, pedestrian, highway-railroad, waterway, highway-waterway, railroad-waterway, highway-waterway-railroad, relief for waterway, and others. Fifteen common bridge-structure types have been identified in the NCHRP study on bridge-strengthening needs in the United States [Dunker 87]. These are listed in Table 5-9.

As an example of special needs, agencies in a seismic region may need to evaluate bridges for seismic retrofitting. This would require seismic rating of each bridge. The first step in the seismic-rating process is an inventory to establish the following information [Buckle 87]: (1) structural characteristics, to determine the vulnerability rating; (2) seismicity of the bridge site; and (3) importance of the structure as a vital transportation link. These, as well as postearthquake evaluation, are further discussed in Chapter 7.

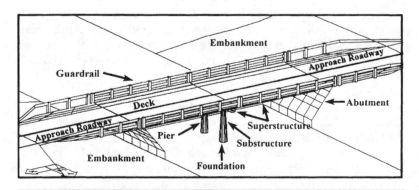

FIGURE 5-8 Major components of a typical bridge.

Code	Description	Code	Description
00	Other	12	Arch—through
01	Slab	13	Suspension
02	Stringer/multibeam or girder	14	Stayed girder
03	Girder and floor beam system	15	Movable—lift
04	T-beam	16	Movable—bascule
05	Box beams or girders—multiple	17	Movable—swing
06	Box beams or girders—single or spread	18	Tunnel
07	Frame	19	Culvert
08	Orthotropic	20	Mixed types
09	Truss—deck	21	Channel beam
10	Truss—through	22	Channel beam
11	Arch—deck		

TABLE 5-8   NBIS Code Number and Bridge Description [after FHWA 88]

NBIS Item 3	Main Structure Type	Number of Bridges	Percentage of Bridges
02	Steel stringer	130,892	27.2
702	Timber stringer	58,012	12.0
101	Concrete slab	42,450	8.8
402	Continuous steel stringer	36,488	7.6
310	Steel trough truss	31,206	6.5
104	Concrete tee	26,798	5.6
502	Prestressed concrete stringer	26,654	5.5
201	Continuous concrete slab	21,958	4.6
102	Concrete stringer	16,884	3.5
505	Prestressed concrete multiple box	16,727	3.5
303	Steel girder—floor beam	9,224	1.9
204	Continuous concrete tee	7,467	1.6
111	Concrete deck arch	6,245	1.3
501	Prestressed concrete slab	5,561	1.2
504	Prestressed concrete tee	4,687	1.0
	Total	441,253	91.8

TABLE 5-9   Distribution of 15 Common Bridge Types [after Dunker 87]

5.6.2   Example Inventory Data for a Road Section

Extensive literature on road inventory data has been published by the Federal Highway Administration [FHWA 89, FHWA 90], state highway agencies, and others. The FAA of the U.S. DOT [FAA 82] has implemented PMS on airport pavements. The PMS evolution and related technologies are well documented by Haas et al. [Haas 94]. A case study of inventory-database design for a road-management system for Dubai Emirate in the Persian Gulf [Uddin 91, 93] is presented for illustration. The first step was to identify the road network using the existing plans and databases and to establish network-partitioning criteria, homogeneous sections, and a location-referencing methodology for Dubai.

5.6.2.1   Identification and Historical Data

The following identification and historical data are included for each section: road name, sector and community numbers (planning-zone references), road number, functional class, past construction or M,R&R project data and completion date, direction, reference chainages (start and end), centerline length, carriageway type (single or dual/divided), pavement-surface type, number of through lanes in each direction, AADT, directional AADT, percentage of trucks, traffic count and axle-load data, and geographical coordinates. The inventory data-collection form provides an adequate explanation of the data ranges and/or allowable codes for use in the office as well as in the field. Inventory databases are used immediately to generate useful summary statistics and graphs, as shown in Figure 5-9.

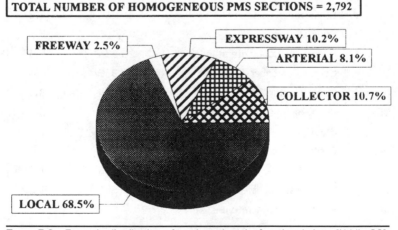

TOTAL NUMBER OF HOMOGENEOUS PMS SECTIONS = 2,792

FREEWAY 2.5%
EXPRESSWAY 10.2%
ARTERIAL 8.1%
COLLECTOR 10.7%
LOCAL 68.5%

FɪɢᴜRE 5-9   Example distribution of road sections by functional class [Uddin 93].

5.6.2.2   Geometric, Construction, and Structure Data

The geometric and construction data include the key fields of section number and the following data categories: construction number and date, carriageway geometry details (for mainline pavement, median, verge, and service road sections), sidewalk/footpath data (type, length, and width), junction types and locations (for roundabout, intersection, T-section, and interchange), parking area (type, location, length, and width), turning-lane type (left, right, U-turn, acceleration, and deceleration) and dimension, and shoulder data (inside/outside, type, length, and width). An explanation of construction number is provided in Sec. 5.2.5; this data item is used to establish and record historical reference.

Construction number and date are also used to refer to pavement-layer numbering and layer material description and thickness [Uddin 95]. The structural data form includes the key fields of section number and construction number. The data also includes types and dimensions of secondary structures, appurtenance, drainage, roadside safety structures, and details of pavement-layer material type and thickness for carriageways and shoulders/sidewalks.

5.7   Example of Transportation-Safety Assets

GASB 34 regulations shifted a dated safety-asset inventory beyond pavements and bridges. In a safety-asset management survey of state highway agencies in the United States, 29 respondents identified the most robust safety-data items included traffic signals, signs, lighting and guardrails, etc. [DOT 11].

5.8   Example of Inventory Data for Buildings

A building consists of many structural and nonstructural components. The primary classification by material type includes wood-frame, masonry, concrete, and steel-frame structures [ATC 93]. The functional classification by predominant use is a long list, covering residential, commercial, industrial, public, educational, health-related, correctional, and monumental buildings. Further classification can be by design/construction type. A building life-cycle cost program, which was developed for the American Society of Testing and Materials (ASTM), includes a large number of maintenance- and repair-cost items in the categories of carpentry, electrical, plumbing, painting, air conditioning, heating, masonry, roofing, fire safety, and steam fitting [ASTM 90]. An example of an inventory data required for a general public building is shown in Table 5-10. Separate data forms for campus data and building-specific data were designed for possible use at a college campus such as the University of Mississippi at Oxford.

Campus and Location Data	Building Specific Data
Campus name and location	Facility/building name and code; GIS code
Date of evaluation and assessor's name	Access road; parking area; foundation
Facility/building name and code (Following data for each building)	Exterior and roof data and ratings (Following data for each room/space)
GIS code	Room number and name
Location on campus	Floor level
Number of floors	Room/space use†
Facility use*	Height, dimensions, and floor area
Construction material†	Number of doors; condition rating
Construction date	Number of windows; condition rating
Last evaluation date	Last evaluation date

---

*Administration, education, health, social, plant, laboratory, other.
†Reinforced concrete, brick, timber, steel, other.
†Office, elevator, library, education, rest room, storage, corridor, exterior, car park, other.

TABLE 5-10 Example of Inventory Data Used for Buildings at a College Campus

5.9 References

[Aouad 96] M. F. Aouad, L. D. Olson, and F. Jalinoos, "Determination of Unknown Depth of Bridge Abutments Using the Spectral Analysis of Surface Waves (SASW) and Parallel Seismic (PS) Test Methods," *Proceedings, 2nd International Conference on Nondestructive Testing of Concrete in the Infrastructure*, Nashville, Tenn., 1996, pp. 147–153.

[ASTM 90] "Building Maintenance, Repair, and Replacement Database (BMDB) for Life-Cycle Cost Analysis," *A User's Guide to the Computer Program*, American Society for Testing and Materials (ASTM), Philadelphia, Pa., 1990.

[ATC 93] Applied Technology Council, "Postearthquake Safety Evaluation of Buildings Training Manual," *ATC-20-T*, funded by the Federal Emergency Management Agency, Washington, DC, 1993.

[Buckle 87] I. G. Buckle, R. L. Mayes, and M. R. Button, "Seismic Design and Retrofit Manual for Highway Bridges," *Report FHWA-IP-87-6*, Federal Highway Administration, McLean, Va., May 1987.

[Chesher 87] A. Chesher, and R. Harrison, *Vehicle Operating Costs, The Highway Design and Maintenance Standards Series*, A World Bank Publication, Johns Hopkins University Press, Baltimore, Md., 1987.

[CNN 96] *Headline News*, Cable News Network (CNN), April 27, 1996.

[Cohn 95] F. Cohn, "New York Gets Wired," *Civil Engineering*, Vol. 65, No. 9, September 1995, pp. 54–57.

[DOT 00] Department of Transportation, "Primer: GASB 34," Federal Highway Administration, Office of Asset Management, Washington, DC, November 2000.

[DOT 11] Department of Transportation, "Asset Management and Safety Peer Exchange," *Report FHWA-HIF-12-005*, Federal Highway Administration, Spy pond Partners, LLC, Arlington, MA, October 2011.

[Dunker 87] K. E. Dunker, F. W. Klaiber, and W. W. Sanders, "Bridge Strengthening Needs in the United States," in *Transportation Research Record 118*, Transportation Research Board, National Research Council, Washington, DC, 1987.

[FAA 82] "Guidelines and Procedures for Maintenance of Airport Pavements," *Advisory Circular AC: 150/5380-6*, Federal Aviation Administration, Washington, DC, 1982.

[FHWA 88] "Recording and Coding Guide to the Structure Inventory and Appraisal of the Nation's Bridges," *Report FHWA-ED-89-044*, Federal Highway Administration, U.S. Department of Transportation, Washington, DC, 1988.

[FHWA 89] "Pavement Management Systems, A National Perspective," *PAVEMENT Newsletter*, Federal Highway Administration, U.S. Department of Transportation, Issue 14, Washington, DC, Spring 1989.

[FHWA 90] "An Advanced Course in Pavement Management," Course Text, Federal Highway Administration, U.S. Department of Transportation, Washington, DC, 1990.

[Fitzgerald 68] J. H. Fitzgerald, "Corrosion as a Primary Cause of Cast Iron Main Breaks," *Journal of American Water Works Association*, Vol. 68, No. 8, 1968.

[Golabi 92] K. Golabi, P. Thompson, and W. A. Hyman, *Pontis Technical Manual*, prepared for the Federal Highway Administration, Washington, DC, January 1992.

[Grigg 88] N. S. Grigg, *Infrastructure Engineering and Management*, John Wiley & Sons, New York, 1988.

[Haas 94] R. Haas, W. R. Hudson, and J. P. Zaniewski, *Modern Pavement Management*, Krieger Publishing Company, Malabar, Fla., 1994.

[Hudson 87] S. W. Hudson, R. F. Carmichael III, L. O. Moser, W. R. Hudson, and W. J. Wilkes, "Bridge Management Systems," *NCHRP Report 300*, Transportation Research Board, National Research Council, Washington, DC, 1987.

[Hudson 94] W. R. Hudson, and S. W. Hudson, "Pavement Management Systems Lead the Way for Infrastructure Management Systems," *Proceedings, Third International Conference on Managing Pavements*, National Research Council, Vol. 2, May 1994, pp. 99–112,

[Means 93] *Means Building Construction Cost Data*, Means Company, Incorporated, R. S., Kingston, Mass., 1993.

[Minsk 84] L. D. Minsk, and R. A. Eaton, "Strategies for Winter Maintenance of Pavements and Roadways," *Infrastructure—Maintenance and Repair of Public Works, Annals of the New York Academy of Sciences*, Vol. 431, December 1984, pp. 155–167.

[NCHRP 12] National Cooperative Highway Research Program, "Estimating the Effects of Pavement Condition on Vehicle Operating Costs," *Research Report 720*, Transportation Research Board, Washington, DC, 2012.

[NOAA 94] "Climatological Data," National Oceanic and Atmospheric Administration, National Climatic Center, Asheville, NC, 1994.

[NYC 07] New York City (NYC) Mayor's Office, "Inventory of New York City Greenhouse Gas Emissions," New York City Mayor's Office of Long-Term Planning and Sustainability, April 2007. http://www.nyc.gov/planyc2030, Accessed September 15, 2010.

[O'Day 84] D. K. O'Day, "Aging Water Supply Systems: Repair or Replace," *Infrastructure—Maintenance and Repair of Public Works, Annals of the New York Academy of Sciences*, Vol. 431, December 1984, pp. 241–258.

[Osborne 09] Katherine Osborne, "GIS-Based Urban Transportation Infrastructure Management Using Spaceborne Remote Sensing Data," *M.S. Thesis*, Department of Civil Engineering, The University of Mississippi, December 2009.

[Paterson 90] W. D. O. Paterson and T. Scullion, "Information Systems for Road Management: Draft Guidelines on System Design and Data Issues," *Technical Paper INU77*, Infrastructure and Urban Development Department, The World Bank, Washington, DC, 1990.

[Paterson 92] W. D. Paterson and B. Attoh-Okine, "Simplified Models of Paved Road Deterioration Based on HDM-III," in *Transportation Research Record 1344*, Transportation Research Board, National Research Council, Washington, DC, 1992.

[Robison 96] R. Robison, "Pavement Management Pays Off," *Civil Engineering*, Vol. 66, No. 4, April 1996, pp. 44–47.

[Sprowls 96] D. O. Sprowls, "Environmental Cracking—Does It Affect You?," *ASTM Standardization News*, Vol. 24, No. 4, April 1996, pp. 24–29.

[TRB 86] J. A. Epps, and C. L. Monismith, "Equipment for Obtaining Pavement Condition and Traffic Loading Data," *NCHRP Synthesis 126*, Transportation Research Board, National Research Council, Washington, DC, 1986.

[Uddin 91] W. Uddin, "Dubai Road Pavement Management System (DRPMS)—DRPMS Manual of Road Network Partitioning & Inventory Data Collection," *U.N. Expert's Report No. PMS-12, Dubai Municipality, U.N. Project UAE/85/012*, United Nations Centre for Human Settlements (Habitat), Dubai, United Emirates, July 1991.

[Uddin 92] W. Uddin, "Highways and Urban Road Maintenance Management: Development and Operation Issues," *1992 Compendium, 6th International Pavement Management/Maintenance Exposition and Conference*, Atlanta, Ga., 1992.

[Uddin 93] W. Uddin, and M. Al Tayer, "Implementation of Pavement Management Technology for Dubai Emirate Road Network," *Proceedings, 20th World Congress of International Road Federation*, Vol. IV, Madrid, Spain, May 1993, pp. 305–314.

[Uddin 94] W. Uddin, "Application of User Cost and Benefit Analysis for Pavement Management and Transportation Planning," *Proceedings, 4R Conference and Road Show*, Philadelphia, Pa., December 1993, pp. 24–27.

[Uddin 95] W. Uddin, "Pavement Material Property Databases for Pavement Management Applications," *Computerization and Networking of Materials Databases, ASTM STP 1257*, Vol. 4, American Society for Testing and Materials, Philadelphia, Pa., 1995, pp. 96–109.

[Uddin 06] W. Uddin, "Air Quality Management Using Modern Remote Sensing and Spatial Technologies and Associated Societal Costs," *International Journal of Environmental Research and Public Health*, ISSN 1661-7827, MDPI, Vol. 3, No. 3, September 2006, pp. 235–243.

[Uddin 08] W. Uddin, "Airborne Laser Terrain Mapping for Expediting Highway Projects: Evaluation of Accuracy and Cost," *Journal of Construction Engineering and Management*, American Society of Civil Engineers, Vol. 134, No. 6, pp. 411–420, June 2008.

[Uddin 10] W. Uddin. "Spaceborne Remote Sensing Data for Inventory and Visualization of Transportation Infrastructure and Traffic Attributes," *CD Proceedings, First International Conference on Sustainable Transportation and Traffic Management and 2010 IEDC*, Karachi, Pakistan, ISBN 978-969-8620-10-3, July 1–3, 2010, pp. 3–12. http://sites.nationalacademies.org/PGA/dsc/pakistan/PGA_052872 accessed September 25, 2011.

[Uddin 11a] W. Uddin, "Transportation Management: LiDAR, Satellite Imagery Expedite Infrastructure Planning," *Earth Imaging Journal*, January/February 2011, pp. 24–27.

[Uddin 11b] Uddin, W., Bill Gutelius, and Christopher Parrish. "Airborne Laser Survey Specifications and Quality Management Protocols for Airport Obstruction Surveys." *Paper No. 11-1323, Transportation Research Record 2214, Journal of Transportation Research Board*, Washington, DC, 2011, pp. 117–125.

[Uddin 11c] W. Uddin, "Remote Sensing Laser and Imagery Data for Inventory and Condition Assessment of Road and Airport Infrastructure and GIS Visualization," *International Journal of Roads and Airports (IJRA)*, Vol. 1, No. 1, pp. 53–67.

[Uddin 12] W. Uddin, "Mobile and Area Sources of Greenhouse Gases and Abatement Strategies," Chapter 23, *Handbook of Climate Change Mitigation*, (Editors:

Wei-Yin Chen, John M. Seiner, Toshio Suzuki and Maximilian Lackner), Springer, New York, 2012, pp. 775–840.

[Xanthakos 94] P. P. Xanthakos, *Theory and Design of Bridges,* John Wiley & Sons, New York, 1994.

[Zaniewski 82] J. P. Zaniewski, B. C. Butler, G. Cunningham, G. E. Elkins, M. S. Paggi, and R. Machemehl, "Vehicle Operating Costs, Field Consumption, and Pavement Type and Condition Factors," *Texas Research and Development Foundation Final Report,* Federal Highway Administration, U.S. Department of Transportation, Washington, DC, 1982.

[Zhang 94] Z. Zhang, T. Dossey, J. Weissmann, and W. R. Hudson, "GIS Integrated Pavement and Infrastructure Management in Urban Areas," in *Transportation Research Record 1429,* Transportation Research Board, National Research Council, Washington, DC, 1994, pp. 84–89.

CHAPTER 6

In-Service Monitoring and Evaluation Data

6.1 In-Service Evaluation Data Needs

In-service monitoring and evaluation of condition is an essential part of infrastructure management. Good evaluation information is required to adequately model maintenance and rehabilitation requirements and to measure the effectiveness of various maintenance and rehabilitation methods. The application of available development and of new technologies is necessary for efficient infrastructure asset management.

6.1.1 Monitoring and Evaluation

The evaluation phase of infrastructure management involves monitoring the use and physical condition of the infrastructure assets being managed. Monitoring is the collection of field-inspection data. Evaluation involves the analysis, interpretation, and use of the information collected. The purpose of in-service evaluation in IAMS is to assess conditions periodically to provide data to:

- Update network-improvement programs
- Assess the structural integrity and possible failure of facilities
- Schedule rehabilitation and maintenance as indicated by these evaluations and updated predictions
- Check and update predicted performance
- Improve prediction models
- Provide a basis for evaluating construction and maintenance techniques

Twenty-first-century airborne and spaceborne remote-sensing and geospatial technologies can provide rapid network-level inventory and environmental-assessment processes, as elaborated in Chapter 5 and in the following sections [DOT 02, Uddin 99, 11a].

6.1.2   In-Service Condition Deterioration Process

Traditionally, infrastructure-design practices have considered initial condition, load, and material properties as the primary input variables for structural design without taking into account the effects of environment and material degradation over time. Such an approach is inadequate, and actual life is generally less than the predicted life. An early benefit of performance evaluation of highways was the identification of important environmental factors [Uddin 95a]. Figure 6-1 illustrates related factors in the case of highway and airport pavements. It is important to identify key factors, in addition to use and aging, that affect condition deterioration to develop a meaningful program for monitoring and evaluation.

Table 6-1 lists such factors in the following major categories: load, environment, material degradation, construction quality, interaction effects, and other mechanisms. Table 6-1 also lists primary- and secondary-condition deterioration mechanisms associated with possible surface defects, deformation, cracking, and failures and examples of catastrophic failures and their causes.

The primary factor in most structural deterioration and failure is load or load-accelerated distress initiated by environmental factors and material degradation. Therefore, it is important to record the loads acting on a facility, including the magnitude and repetitions. Examples of load considerations for a bridge evaluation are: (1) dead load (weight of the bridge structure without traffic); (2) primary live load, such as HS 20 Truck shown in Figure 6-2 [after AASHTO 92]; and (3) secondary live load (e.g., wind, thermal forces, braking load, earth pressure, buoyancy, centrifugal force in curves, stream flow, ice pressure, earthquake loads, and impact load).

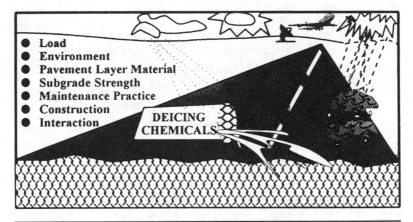

FIGURE 6-1   Key factors influencing the performance of highway and airport pavements.

Deterioration Manifestation	Deterioration Mechanisms					
	Load/Usage	Environment	Material Degradation	Construction Quality	Interaction	Other Mechanisms
1. Surface defects	Secondary	Primary	Primary	Secondary	Material/environment and extended by load	Man-made defects, maintenance patches
2. Deformation	Primary	Secondary	Primary	Secondary	Material/environment/load	
3. Cracking, disintegration, and breaks	Primary	Primary	Secondary	Secondary	Load/environmental degradation	Corrosion damage, accidents, no maintenance
4. (a) Failure (aging/inadequate structural capacity or retirement)	The facility is structurally deficient because the limiting threshold values of (1) surface defects (2) deformation and 3) cracking and disintegration are exceeded. Or the facility is retired because it is functionally obsolescent.				Nature	Capacity and safety considerations, no maintenance
4. (b) Catastrophic failure	Primary causes: (1) Natural disasters including earthquakes, floods, freeze/snow, ice, tornadoes/cyclones, wind storms, (2) accidental hazards, such as bridge damage due to crash/accidents and spills of hazardous materials from trucks/tankers, and (3) man-made disasters (examples include terrorist bombings of September 11, 2001 in the United States)				Poor construction quality/design deficiency/fire	Fire, arson, terrorist act, or accidents/crash

Table 6-1 Possible Condition Deterioration Mechanisms for Consideration n Infrastructure Evaluation

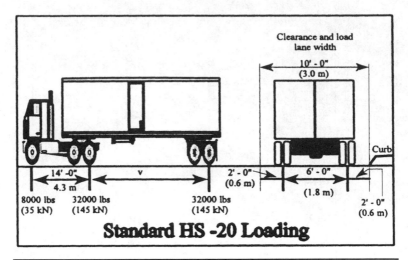

Figure 6-2   The truck-loading design for highway bridges [after AASHTO 92].

Water and sewer pipe, generally laid under roads or buried underground within the road right-of-way, are subjected to the dead load of fill and pavement and the live load of passing traffic, earth pressure, and hydrostatic pressure from all other sides, and the pressure and weight of water or waste/sewer water transported through the pipe. Just like a bridge or a building, the condition of these pipelines depends on the material of construction, the loads, and deterioration, such as corrosion. Large-diameter sewers and water tunnels made of bricks and concrete may experience structural failures as a result of erosion of bricks and mortar, and loss of support of the surrounding material. Internal visual and video-based inspections provide the best evaluation of overall integrity and location of problem areas.

As discussed in Chapter 5 and noted in Table 6-1, environmentally induced cracking can be a primary cause of failure in concrete and metal structures. This may then lead to premature failure, such as corrosion and fracture of steel components, loss of water by leakage through breaks in waterlines, failure of a high-pressure gas pipeline, plane crashes, shut down of nuclear power plants, and contamination of soil and water sources. Freeze-thaw effects can significantly lower the load-carrying capacity of pavements.

Seasonal changes in moisture and temperature can induce tensile shrinkage stresses in the structure and reduce the subsoil load-carrying capacity. However, the interaction of a load with one or more environmental mechanisms is more critical for condition deterioration than environmental mechanisms alone. Heavy use and load repetitions will always accelerate the damage caused by environmental factors and material degradation. Thermal cracking in slabs and walls constructed with concrete or brick can be caused by tensile stress due to a high-temperature differential from one surface to the other [Uddin 83, Ho 95], and low-temperature shrinkage cracking of asphalt pavement [Haas 73, Harvey 94] can be caused by low temperatures. Potholes appear more often on asphalt roads during the spring thaw season, particularly in the presence of heavy traffic. This represents a good example of interaction of all three mechanisms: load repetitions, weakening of sublayers and roadbed soil, and thaw (environmental) conditions.

Primary and secondary mechanisms of condition deterioration dictate the plan for a cost-effective and efficient condition data-collection program assessment and in-service evaluation. Table 6-1 provides guidelines for these mechanisms, which are well established for road and airport pavements constructed with asphalt, concrete, or unbound surface [FAA 82, SHRP 93, Eaton 87].

6.1.3   Approaches for In-Service Evaluation

In-service evaluation of a facility involves three steps:

- *Selecting sample sections* with inventory data and historical information on use and load
- *Monitoring condition and environmental impacts* via measurements or observations recorded in a useful format
- *Processing and evaluation* via manipulation and interpretation of the data collected to provide an overall condition rating and to make a judgment based on the results

Evaluation involves the assessment of facility conditions with respect to the *user*, the *function*, the *structural condition environmental impacts*, and *sustainability issues*.

6.1.3.1   User Evaluation

Common factors in user evaluation are convenience, comfort, health, safety, aesthetic aspects, and overall satisfaction with the quality of service. These are generally subjective in nature and often depend on visual observation or opinion. Examples include: (1) present serviceability rating (PSR) used for highways based on riding quality [HRB 62]; (2) sufficiency rating for bridges [Hudson 87]; (3) safety rating of air travel; and (4) citizens' complaints of poor water pressure and/or accumulation of garbage on streets. The subjective rating scale is generally numeric, ranging from zero (the worst or failed condition) to a specified maximum of 5, 10, or 100 for excellent condition.

6.1.3.2   Functional Evaluation

Functional evaluation is the effectiveness of a facility in fulfilling its intended functions and is similar to user evaluation. Major concerns are safety, size adequacy, sufficiency based on capacity and demand requirements, serviceability or quality of service, and physical appearance. Use of objective measurements is preferred because they lead to more productivity and better repeatability. Examples include: (1) a present serviceability index (PSI) based on pavement roughness measurements used for highways; (2) a Pavement Condition Index (PCI) based on distress measurements used for highways and airports [FAA 82, Shahin 90]; (3) a condition index for bridge decks based on detailed distress mapping and analysis; (4) a track-quality index for railroad tracks based on objective measurement of track geometry, cross slope, and alignment [Fazio 80]; (5) water quality and health hazard; (6) a performance index for waterlines and gaslines based on leakage analysis from supply and consumption records or nondestructive leakage testing of utility lines; (7) level-of-service and congestion indicators for transportation facilities; and (8) hazards and safety problems caused by reduced skid resistance, hydroplaning, and poor drainage on roads. Where possible, the condition or performance indices are based on objective measurements of selected performance indicators, or an analysis of maintenance records.

6.1.3.3 Structural Evaluation

Structural evaluation is used to assess structural integrity by conducting tests and structural analysis of test data. Structural evaluation includes load-carrying capacity, structural integrity, seismic or fire hazard, and safety of users. Example are: (1) the structural capacity index, used for highways based on pavement-deflection testing and load-carrying capacity analysis [Haas 94, Uddin 95b]; (2) the structural rating and remaining life for bridge structures based on nondestructive-deflection testing and structural-fatigue analysis; (3) the track modulus index for railroad tracks based on nondestructive-deflection testing, or remaining life based on fatigue testing; and (4) frequency of breakdowns and failures of utility lines. Structural-condition indices, based on objective measurements, indicate structural integrity and load-carrying capacity. Nondestructive testing and evaluation methods are most desirable.

Harty and Peterson [Urban 84a] point out that reliable information on the condition of infrastructure enables local officials to accomplish the following:

- To rate and rank facilities according to their current physical condition and performance.
- To determine deterioration rates and the best time to undertake maintenance for each facility. To build a more effective constituency for capital upkeep by providing the public with reliable information on the current facility condition and projections of the consequences of underspending on capital-facility preservation.

6.1.3.4   Environmental Sustainability Evaluation

Evaluation of the environmental impacts of a facility due to construction and long-term usage are equally important considering diminishing natural sources and preservation of water and air quality, as well as sustainability related to GHG emissions [DOT 02, Uddin 12].

6.1.4   Data Details and Data Collection Frequency

The first basic requirement for condition data collection is a proper location reference system, as outlined in Chapter 5. It is desirable to have a common location-referencing scheme across an agency so that data of all types can be linked among inventory, construction, maintenance, and so forth. It is imperative that evaluation measurements be properly indexed by section, subsection, and date for efficient data management. The selection of data types and details can be linked to the information-quality level (IQL) concept discussed in Chapter 5. The precision and extent of data items depend on the information-quality level being sought.

Evaluation databases vary from a few mandatory elements to a comprehensive list of data elements, depending on the functional level of the IAMS at network-level or project-level design. As shown in Figure 2-4, the data details and system complexity increases as one moves from the network-level to project-level IAMS functions. At the network level, high productivity and low cost are essential; however, the data must provide insight into M,R&R needs of the network. Evaluation at the project level is more detailed, involving causes of deterioration and subsequently predicting appropriate M,R&R strategies. For research objectives, very detailed and more frequent measurements and detailed analyses are required.

The next step is to select the condition attributes that will be evaluated. The final selection of the factors should be made by a group of experts familiar with local infrastructure conditions. Carefully written definitions of the data items should be prepared and reinforced with photographs. If possible, the definitions should include instructions for rating the severity and measuring the extent of each condition attribute.

There are no absolute standards for the evaluation frequency. Guidelines can be based on available resources and data-processing speed, as recommended in Table 5-2 for different decision-support system (DSS) components. Table 6-2 shows the recommended collection frequency for project-level usage and evaluation data. At the network level, major roads and bridges should be evaluated annually, but no less than once every three years. Priority should be given to aged facilities or those exhibiting problems. One example is the sewer system in St. Louis, Missouri, which is reportedly among the oldest in the United States, with some parts

	Database Update		
DSS Component	**Mandatory**	**Desirable**	**Database File**
Usage (current usage record)	Annually	Monthly and seasonal in addition to yearly total	Project
Condition evaluation	Sample once every 1–3 years	Annually (on 100% network)	Project

TABLE 6-2   In-Service Evaluation Frequency and Database Update

dating to 1850 [Collins 95]. Since most sewers are built under road pavement or in right-of-ways, sewer failure often results in cave-in of the overlying road with a huge repair cost. To comply with clean-water regulations, a project was designed to implement an overflow-regulation system and eliminate sewage overflows during high-river stages of the Mississippi River. As a part of a major structural reha-bilitation project, criteria were established to identify those sewer sec-tions most at risk, to define pumping station modification needs and avoid external hydrostatic pressure [Collins 95].

Because condition assessments are costly, trouble-prone facilities should receive the most frequent assessment. Local agencies should identify the facilities where problems are most likely to occur, or where failures will have the most costly consequences. Other facilities can be inspected less often and/or on a sampling basis [Urban 84a, b, c]. The GASB 34 guidelines for public infrastructure asset management, adapted by the U.S. DOT, recommend asset-condition evaluation every three years [DOT 00].

6.1.5   User Interface and System Operation Issues

The rate of physical deterioration is often a critical element of in-service evaluation, and it is important to keep records that are con-sistent from year to year. Database entries on usage should be updated regularly as new data become available, and in-service monitoring and evaluation database files should be maintained by survey date so that all historical evaluation data files are readily accessible for M,R&R needs assessment, improvement of prediction models, and other project-level applications.

6.2   In-Service Evaluation of Physical Assets

Typical examples of recommended evaluation data elements for various infrastructure groups are shown in Table 6-3. As constructed quality affects facility life, assessment immediately after construction should be included in the database.

In-Service Condition Monitoring and Evaluation Data Elements			
Facility	User Evaluation	Functional Evaluation	Structural Evaluation
(1) Highways, roads, streets, parking areas	Present serviceability rating (PSR), ride quality, vehicle operating costs, user satisfaction ratings based on congestion and pollution, traffic data	Present serviceability index (PSI), international roughness index (IRI), Pavement Condition Index (PCI) based on distress data, pavement condition rating (PCR), and security rating	Deflection testing, remaining life, structural capacity index
(2) Airports: pavements, buildings, aviation facilities	Ride quality, user satisfaction ratings for other facilities, aircraft operations data	PCI (pavements), ride-quality index (RQI); see (11) for buildings, air-traffic control facilities, airport lighting, and security rating, etc.	Deflection testing, remaining life, structural capacity index; see (11) for buildings and other facilities
(3) Bridges: interchange, overpass, underpass flyover, rail/river crossing	Ride quality, user-satisfaction ratings, traffic data	Bridge inventory and appraisal ratings, sufficiency rating based on inventory and appraisal ratings, deck rating, and security rating	Vibration and seismic testing, nondestructive evaluation, load rating, remaining life
(4) Railroad: terminals, track, bridge	Service quality and efficiency; user satisfaction ratings, train operations data	Track quality index (TQI); see (3) for bridges; see (11) for buildings and other facilities, and security rating	Deflection testing, track modulus, remaining life; see (3) for bridges; see (11) for buildings and other facilities
(5) Waterborne transport: ports, inland terminals, intermodal facilities	Container handling and intermodal efficiency rating, track/pavement quality rating, user satisfaction ratings	Ratings for safety, barge/bridge accidents, container/intermodal operations, capacity, and security	NDE (deflection testing, vibration and seismic testing), load rating, remaining life for each asset type
(6) Mass transit: subway and metros, tracks, stations, vehicle stocks, maintenance facilities	Ride quality, user satisfaction ratings	Ratings for safety, ride quality, crossing accidents, operations, efficiency, capacity, and security	NDE (deflection testing, vibration and seismic testing), load rating, remaining life for each asset type

TABLE 6-3 Examples of Infrastructure In-Service Monitoring and Evaluation

In-Service Condition Monitoring and Evaluation Data Elements			
Facility	User Evaluation	Functional Evaluation	Structural Evaluation
(7) Supply chain: container terminals, manufacturing sites, trucks and intermodal fleet, distribution centers, retail stores	Container handling rating, intermodal efficiency rating, track/ pavement quality rating, user satisfaction ratings	Ratings for safety, accidents, container/ intermodal operations, efficiency, capacity, and security	NDE (deflection testing, vibration and seismic testing), load rating, remaining life for each asset type
(8) Water: pumping stations, mains and supply pipe network	Health hazard, water treatment, user satisfaction, water pressure, and consumption	Number of breaks per year, leakage per year, number of repairs, efficiency and capacity ratings, and security rating	Failures/breaks per year, damage to roads and other structures, remaining life
(9) Waste water and sewer: sewer lines, treatment plants	Health hazard, water treatment, user satisfaction, discharge data	Number of breaks per year, leakage per year, number of repairs, efficiency and capacity ratings, and security rating	Failures/breaks per year, damage to roads and other structures, remaining life
(10) Solid-waste facilities: landfills, methane gas collection, and energy production	Health hazard, fire hazard, odor and treatment, user satisfaction, waste load	Spilled trash, missed collections, soil and water contaminations, efficiency and capacity ratings, and security rating	Failures, damaged and inadequate collection and disposal facilities
(11) Buildings: public, commercial, industrial, multipurpose complexes, landmarks	Complaints, discomfort, user satisfaction ratings, live load and occupancy data	Interruptions in utility services, heating and cooling effectiveness, air quality, condition ratings, and security rating	Structural and ground failures, posting, remaining life, safety hazards
(12) Electric power-supply infrastructure: traditional and renewable	Fire hazard, interruptions in service, user satisfaction, power generated and distributed	Frequency of service breaks, power surges, condition ratings of poles and cables, grid capacity rating, and security rating	Failures, remaining life; see (1) for roads; see (11) for buildings and other facilities

TABLE 6-3 Examples of Infrastructure In-Service Monitoring and Evaluation (*Continued*)

In-Service Condition Monitoring and Evaluation Data Elements			
Facility	User Evaluation	Functional Evaluation	Structural Evaluation
(13) Telecomm- unications & wireless infrastructure: IT and Internet networking facilities	Ratings for interruptions in service, user satisfaction	Frequency of service breaks, power surges; ratings for fire hazards, operations, and protection from poor or non-coverage, capacity, and security	Failures/service breaks, power surges, condition ratings of fixed assets

TABLE 6-3   Examples of Infrastructure In-Service Monitoring and Evaluation (*Concluded*)

6.3   Technologies for In-Service Monitoring and Evaluation

6.3.1   Overview

Monitoring data, whether traditional, nondestructive, or innovative may be grouped into the following categories.

6.3.1.1   General Automated Records

Some data are available in electronic and online databases, as well as published documents for the location of interest. Good examples are the weather and climate data recorded on airports and other local weather channels available from NOAA [NOAA 94] and the Internet for most of the United States. An airport will have aircraft-traffic data available through records of air-traffic control. Traffic history, loading, axle configuration, and volume data for roads and bridges are often collected and stored in central transportation-agency databases.

6.3.1.2   Visual Inspection

Visual observation and inspection is the most widely used method of condition monitoring. Manual visual inspections are labor-intensive, expensive, and subject to inspectors' judgments. However, in some cases they are necessary, such as postearthquake evaluations [ATC 93] or after other catastrophes (hurricane, floods). Costs can be reduced through sampling and/or summary methods, such as windshield surveys.

6.3.1.3   Terrestrial Photographic and Optical Methods

Terrestrial photographic and optical methods include videotape recording, 35-mm photologging, borescope inspections, internal pipe video, and other optical methods with permanent records [Hudson 87a, b]. The records may be obtained from selected positions, slow-walk speed, or high-speed dedicated vehicles. Photo records must be read or interpreted to produce data. Automated pattern-recognition techniques that can identify and quantify data are improving with better computing hardware and software. Video-based methods are also being used successfully for traffic studies. These passive sensor technologies record reflections from the surface of objects. With advances in computers and IT infrastructure the trend is to adopt digital-imaging technologies and digital videography [Al-Turk 99, Uddin 11a, b].

6.3.1.4 Automated Geometrical and Safety Data Collection
Automated measurements of geometrical characteristics can be cost-effective for network-level evaluation, for example, longitudinal and cross-fall deficiencies on paved surfaces and slabs, and track geometry [Hudson 87b, Fazio 80]. Such data collection can be especially useful and efficient on roads, railroad tracks, and pipelines. Use of modern laser surveys and digital imagery technologies in the last two decades have enhanced automated condition monitoring of infrastructure assets, as well as collection of safety attributes of pavements, rail tracks, and pipelines [Uddin 09, 11b].

6.3.1.5 Nondestructive Structural Evaluation Methods
Structural integrity of a facility can be checked using nondestructive-evaluation (NDE) methods, which do not subject the facility to actual loading or destructive testing. These NDE methods include seismic evaluation, such as wave propagation; vibration methods, such as modal analysis; acoustic and ultrasonic methods; dynamic deflection testing; X-ray diffraction techniques; electromagnetic methods, and electrical resistivity methods [Hudson 87a, c, d, Matzkanin 84, Metzger 93, Uddin 94, 96a, 96b, NDT 96]. Corrosion damage in steel and concrete can be estimated by electrical resistivity methods [Chaker 96, Bridge 96]. Other noncontact and nondestructive testing (NDT) technologies for structural evaluation include ground-penetrating radar, infrared thermography, laser scanning, high-speed video and related optical methods, the Moire technique, acoustic sensors, microwave radar sensors, and high-resolution GPS receivers [Uddin 94, NDT 96, Mayer 10, Uddin 06, 11b].

6.3.1.6 "Smart" Sensors
"Smart" sensor technologies rely on the embedment of fiber optics, ultrasound sensors, or piezoelectric sensors for lifelong monitoring and evaluation of structures [Ansari 96, Metzger 93]. Fiber-optic technology offers a potential application to monitor deformation and stress responses to loads.

6.3.1.7 Airborne and Spaceborne Remote Sensing
Modern airborne and spaceborne remote-sensing technologies are affordable and computationally efficient for nonintrusive evaluation of physical assets [Al-Turk 99, Uddin 06, 11a, b, c]. These include active

airborne LiDAR laser scanning and passive multispectral high-resolution imagery, as discussed in Chapter 5. Both provide georeferenced digital data for easy computerized processing and software manipulation to extract built-infrastructure features, topographic data, and airport obstacles in navigation space [ACRP 10, Gutelius 12, Uddin 06, 09, 11c].

6.3.2 Destructive Testing and Evaluation

Although the in-service condition can be evaluated by surface or NDT observations, it occasionally becomes necessary to remove portions of a facility to ascertain more specifically where problems are occurring and why. This is called *destructive testing* because a portion of the original structure is removed and must be repaired or replaced. In general, such procedures are used where there is evidence of unexpected deterioration; however, destructive methods have been used on full-scale tests of roads, dams, pipelines, underground tanks, and pile foundations as well. For example, they have been used on the AASHO Road Test [HRB 62] and form an important part of the SHRP-LTPP study [SHRP 89a, b]. Surface defects can be used as general guides to the underlying conditions; however, it is often necessary to open a facility internally to determine the true position and cause of failure for a completely reliable analysis.

Minimum destructive-testing techniques include coring and coupon sampling of structural members, such as bridge decks. On highways and airports or industrial floors, destructive testing is conducted by coring or cutting into the facility and removing samples of the various layers, examining the samples in the field, and then testing them in the laboratory. In some cases, full destructive tests are carried out until the structure yields or fails; for example, a pile-load test. Such tests are performed on a sample basis because of the time and cost involved in the destructive testing. The results then are inferred to the remaining units.

6.3.3 Traffic, Usage, and/or Occupancy

An important part of in-service data is measurement of traffic or other usage factors, such as volume of flow or number of take-offs and landings. Generally, each type of infrastructure has its own type of usage. For airfields, this includes the number of take-offs and landings and the number of individual enplaned counts. These data are recorded by commercial airlines and by the FAA for airplane traffic.

For highways, the data are collected on a sampling basis using traffic counters at specific locations for several days per year. These data are then expanded to estimate the annual average-daily traffic for each highway location. Load information is estimated by taking a sample weighing of axle loads by various classes of trucks and by doing automatic classifications of vehicles in various classes, such as passenger cars, light trucks, three-axle trucks, and so forth, through five-axle tandem trucks. A wide variety of automatic data-collection devices are used for these data (electromagnetic sensors, radar, and video). For remote areas and developing regions with less resource to use these field devices, an innovative satellite imagery-based geospatial methodology has been developed and implemented with fairly accurate results [Osborne 09, Uddin 10, 12].

Flow of sewage and water are measured by flow meters in selected pipes and then expanded to the entire flow-pipe system. Flow capacity and quantity are also recorded at water plants and sewage-treatment plants. Sampling is used effectively.

For parks and other such facilities, entrance access is usually controlled and recorded. These data can be used for IAMS purposes.

6.3.4   Evaluation Methods and Applications

Table 6-4 lists several evaluation methods, with examples of application, to evaluate various types of infrastructure.

Technology	Applications	Infrastructure
Visual inspection	Manual record of observed distresses and defects	All types of infrastructure assets
Digital continuous photography (35-mm replaced by digital)	Permanent record of physical condition, inspection, distress, and defect data acquisition (35-mm technology mostly replaced by modern digital technology)	Road and airport pavements, all other assets
Video recording, real-time video surveillance	Permanent record of physical condition, inspection, distress, and defect data; road vehicular traffic data; ITS and security surveillance	Road and airports, railroad tracks, sewer and waterlines, bridges, dams, nuclear power plants, malls, sports complexes, and all other assets
Seismic, deflection, and vibration test methods	In situ material characterization, structural integrity evaluation, seismic risk assessment	Road and airport pavements, bridges, all other structures
GPR	Structural integrity evaluation, delamination, layer delineation, voids and moisture damage, dielectric constant properties, buried pipe/tunnel location	Road and airport pavements, bridges, all other concrete and masonry structures; underground cavities and tunnels
Infrared thermography	Temperature measurement, detection of moisture leakage, delamination, defective areas, stress mapping	Road and airport pavements, bridges, all underground assets, leakage and breaks of water and sewer lines

TABLE 6-4   Application of Traditional and Innovative Technologies for In-Service Evaluation and Condition Assessment of Infrastructure Facilities

Technology	Applications	Infrastructure
Acoustic and ultrasonic testing	Cracking and defect detection, longitudinal roughness and rut depth measurement on paved surfaces, stress, strain, displacement measurement	Road and airport pavements, bridges, underground utilities, all other structures; infrasensing for remote monitoring of seismic activities and nuclear explosions
Fiber optics	Stress and strain measurement	Bridges, tunnels, "smart" structures
Piezoelectric sensor	Stress, strain, traffic measurement	Road and bridges
Electrical resistivity	Detection of surface and subsurface material changes, corrosion damage	Soils characterization, concrete corrosion
Electromagnetic conductivity	Detection of cracking, voids, and subsurface defects	Bridges, tunnels, other structures
Topographic surveying, GPS, gyroscope	Location identification, alignment, longitudinal and cross falls, routing, real-time tracking	Road and airport pavements, bridge decks, railroad tracks, utilities, water and sewer
Laser (single beam or multiple beams mounted on vehicles; moving at low speed during testing)	Cracking and defect detection, profile measurement, longitudinal roughness and rut depth measurement, faulting of transverse joints, surface texture and displacement measurement	Road and airport pavements, bridges, railroad tracks, and other structures
Terrestrial laser scanning (LiDAR mounted on vehicle tops or tripods; stationary during testing)	Pavement-surface mapping and profile measurements for area defects, building information management modeling, reconstruction of accident scenes, and digital 3D modeling of large and historical structures	Road and airport pavements, bridges, railroad tracks and road crossings, port and marine terminals, buildings, and other structures
Kinematic laser scanning (LiDAR mounted on vehicle tops or tripods; moving at highway speed during testing)	Rapid data collection for pavement surface mapping and profile measurements for area defects, building information management modeling, reconstruction of accident scenes, and digital 3D modeling of large and historical structures	Road and airport pavements, bridges, railroad tracks and road crossings, port and marine terminals, buildings, and other structures

TABLE 6-4 Application of Traditional and Innovative Technologies for In-Service Evaluation and Condition Assessment of Infrastructure Facilities (*Continued*)

Technology	Applications	Infrastructure
Airborne laser scanning (LiDAR mounted on aircraft; flying at low altitude during testing)	Rapid data collection for large areas of topographic surveys, surface mapping for area defects, building footprint modeling, airport obstruction mapping, power transmission grid mapping, digital 3D modeling of large infrastructure, and floodplain mapping for risk assessment	Road corridors and airports, railroad tracks and road crossings, pipeline infrastructure, port and marine terminals, power grid transmission lines, urban infrastructure, and environmental sites
Airborne radar scanning (IfSAR mounted on aircraft; flying at high altitude during testing)	Rapid data collection for large areas of topographic surveys, surface mapping, and floodplain mapping for risk assessment	Flood-risk mapping, forest-area estimation, and wetland sites
Satellite imagery (panchromatic and/or multispectral; orbiting at 400–600 km in space)	High-resolution pan-sharpened multispectral imagery for coverage of large regions, infrastructure footprint modeling and inventory, land-use studies, and disaster damage assessment from pre- and post-disaster imagery scenes	Base GIS imagery for infrastructure and environmental projects, road corridors, airports, railroads, pipeline infrastructure, port and marine terminals, city's built infrastructure, forest area estimation, and wetland sites

TABLE 6-4   Application of Traditional and Innovative Technologies for In-Service Evaluation and Condition Assessment of Infrastructure Facilities (*Concluded*)

6.4   Inspection, Photographic, and Optical Evaluation

Assessment of physical condition and defects is an integral part of effective in-service evaluation. Video and 35-mm continuous photography methods are being used for pavement inventory and distress-data acquisition in pavement management applications. Photologging techniques have been used by several agencies for geometric-inventory purposes and safety evaluation. Videotape recording is also very useful for traffic data collection and inspecting bridges, tunnels, water mains, sewers, and other structures that are hard to access by a human body. The use of photographic methods and video provide a permanent record of the image at the time of evaluation, and comparison of these records over time can be useful in interpreting condition deterioration rates and mechanisms. In some cases, accuracy and information on some cases of observed distress are lost when reducing the condition-deterioration data from photo records, because the interpretation is based on two-dimensional images and current pixel size.

Continuous 35-mm strip photographic techniques for distress surveys of highway and airport pavements and right-of-way inspections have been used in the United States since 1986 and in France and Japan for several years [Hudson 87b]. Video technology for recording images of the pavement surface and collecting distress data (including interpretation) is now available commercially [Hudson 87a]. Photographic methods are becoming more popular because of the technological advances in computer-based digital-image processing. For example, videotaped images of 33,600 km (21,000 mi) of main railroad track and a GIS database in the Engineering Facilities Management System (EFMS) were used by the Union Pacific railroad company to assess the performance of rail track, to program on-time maintenance, to improve safety of operations, and to assist in real-estate management [Gerard 91]. Video imaging is also being used for roads, airport pavements, and bridges where manual on-site inspection is difficult and risky in highly trafficked sections. Video also provides remote assessment of physical condition, structural integrity, and emergency-maintenance needs for buildings and sewer and other pipeline systems.

In the post-2000 era, most service providers use digital imaging and digital video technologies, which are easy to store and analyze.

6.5   Nondestructive and Noncontact Structural Evaluation

Nondestructive and noncontact testing and evaluation techniques are the primary methods used for structural integrity and load-carrying capacity evaluations. These techniques are listed in Table 6-4. In this section, some of the successful and promising methods are discussed in more detail.

6.5.1   Vibration, Seismic, and Ultrasound Methods

Several types of vibration and seismic equipment developed for industrial vibration problems and earthquake engineering are now used for structural evaluation of pavements, foundations, and structures. Stress-wave propagation using spectral-analysis-of-surface waves (SASW) [Heisey 82, Nazarian 83, Aouad 96], CEBTP's transient dynamic-response technique [Uddin 94, Hertlein 96], and acceleration monitoring for structure-modal analysis [O'Leary 96] have been applied with success. The advantages of such techniques are the elimination of holes in the structure, tests results under realistic field conditions without altering the mechanical state of stress, and reduced traffic disruption.

6.5.1.1   Principles of Wave-Propagation Tests

The wave-propagation test methods primarily measure wave velocity at or near the surface of a structure subjected to a small transient or vibratory force. Generally, accelerometers and/or velocity transducers (geophones) are used to measure the surface response. The tests

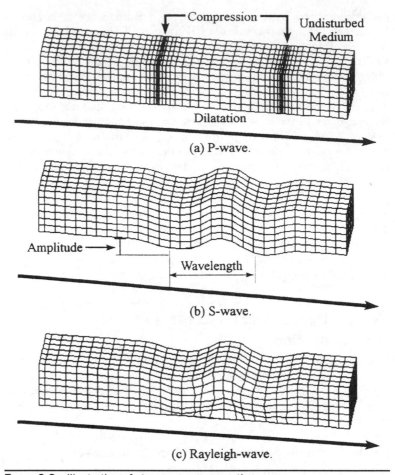

Compression — Undisturbed Medium

Dilatation

(a) P-wave.

Amplitude →

Wavelength

(b) S-wave.

(c) Rayleigh-wave.

FIGURE 6-3 Illustration of stress-wave propagation.

are used to measure the velocity of compression waves (P-wave), shear waves (S-wave), or Rayleigh waves (surface wave or R-wave), depending upon the mode of excitation forces [Lysmer 70, Abbiss 81]. Figure 6-3 illustrates these waveforms [Hudson 97]. The wave velocities are related to material moduli. These low strain-amplitude nondestructive tests can provide an estimate of linear elastic-material properties. The subsoil properties can be used to evaluate liquefaction potential and soil support.

Steady-state vibration techniques have been traditionally used for stress-wave testing using time-domain analysis. Shear-wave tests are particularly useful for soils and foundations because the shear-wave velocity (V_S) is directly related to the shear modulus (G) by:

$$G = \rho(V_S)^2 \tag{6.1}$$

where ρ = mass density, kg/m^3.

Young's modulus (E) for a homogeneous linear elastic material can be calculated from G by μ:

$$E = 2G(1 + \mu) \tag{6.2}$$

In the case of the P-wave, the modulus depends on the Poisson's ratio:

$$G = \rho(V_p)^2[(1 - 2\mu)/2(1 - \mu)] \tag{6.3}$$

The dynamic Poisson's ratio can be found from the ratio of P-wave and shear-wave velocities using Eqs. (6.1) and (6.3).

Rayleigh waves travel on the surface with an elliptical particle motion at a velocity (V_R), close to that of a shear wave. The amplitude of Rayleigh waves attenuates with depth, and most of the energy is found within one wavelength from the surface. The wavelength, L_R, is related to the distance, d, between the two geophones, and the phase difference (degrees) between the two signals received during a test in which the surface is excited by a vibrator oscillating with a vertical steady-state harmonic motion. A typical test setup is shown in Figure 6-4. The Rayleigh-wave velocity is assumed to correspond to that of the material at a depth of one-half wavelength.

6.5.1.2   The Spectral-Analysis-of-Surface-Wave Method

The spectral-analysis-of-surface-wave (SASW) method uses frequency-domain analysis of Rayleigh-wave propagation to determine pavement moduli and layer thicknesses [Nazarian 83]. Recently, it has been used to estimate the depth of bridge foundations [Aouad 96, TRB 95]. The test setup is shown in Figure 6-5. The excitation source is a hammer blow that applies a transient vertical impact, which generates a group of Rayleigh waves of various frequencies. The propagation

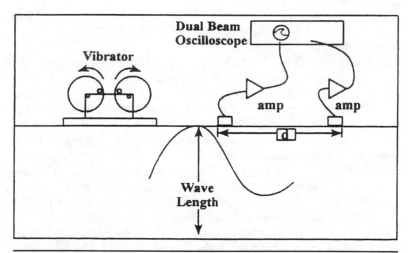

Figure 6-4   Rayleigh-wave test [after Abbiss 81].

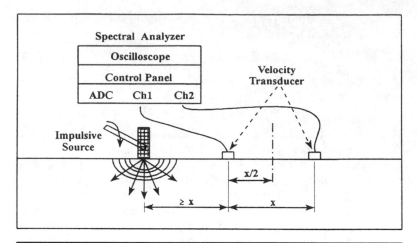

FIGURE 6-5   Schematic of the SASW test setup [Nazarian 83].

is monitored by two geophones on the surface. The output of the receivers are simultaneously recorded on a Fourier spectral analyzer that decomposes the transient waveforms into a group of simple harmonic waveforms that are analyzed individually to determine R-wave velocity and wavelength associated with each frequency.

6.5.1.3   Transient Dynamic Response Method
The transient dynamic response method has been used to locate voids under concrete pavements [Uddin 94] and to study the integrity of shaft foundations [Hertlein 96]. A hammer equipped with a load cell is used to generate small stress waves, and the response is measured by a geophone. The time-domain data is converted into the frequency domain, which is interpreted to provide information on dynamic stiffness and concrete quality.

6.5.1.4   Impact Echo (IE) and Ultrasonic Pulse Velocity (UPV) Methods
The impact echo (IE) method is a popular and common ultrasonic technique in which a hammer is used to excite the surface axially, and the response of the reflected waves is measured by an accelerometer. The low-strain compression waves travel through the thickness or depth of the structure and are partially or completely reflected by changes in the conditions, such as varying cross sections, concrete defects, voids, delaminations, and so forth. This test is used for concrete structures, including beams, columns, foundations, pavements, and bridge decks, to evaluate structural integrity, defects (such as voids, honeycombing, cracks), and delamination [Olson 96, Hertlein 96]. Advanced scanning techniques have been developed in Germany [Schickert 96].

Figure 6-6 shows a schematic of the IE method and a mobile IE scanner developed for bridge decks [after Olson 96]. The ultrasonic

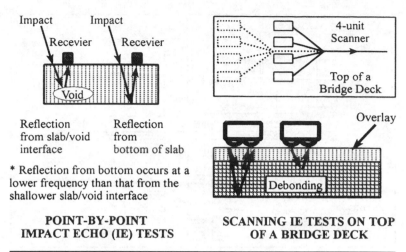

FIGURE 6-6   A schematic of impact echo test setup [after Olson 96].

pulse velocity (UPV) method is based on compression-wave analysis if two surfaces are accessible, as shown in Figure 6-7 [after Olson 96]. Currently, a superscanner is being developed by U.S. Army Corps of Engineers [Alexander 96] to detect delamination of bridge decks.

6.5.1.5   Parallel Seismic Test

The parallel seismic test is an effective method to evaluate piles, drilled shafts, and deep foundations [Hertlein 96]. The test is very suitable where the head of pile or shaft is inaccessible. It has been successfully used to measure the depth of bridge foundations [Aouad 96, TRB 95]. Figure 6-8 illustrates the test setup [Hertlein 96]. A bore hole is made in the ground near to, and parallel with, the foundation. The bore hole is lined with a PVC pipe and filled with water to provide an acoustic response that will be monitored by hydrophones.

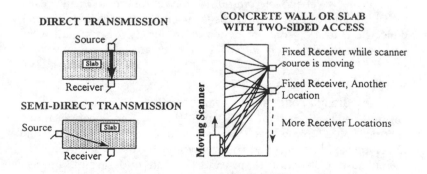

FIGURE 6-7   Ultrasonic pulse velocity (UPV) test configuration [after Olson 96].

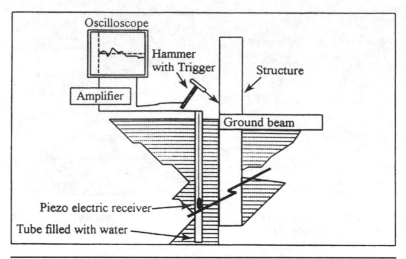

FIGURE 6-8   The test configuration for parallel seismic test [after Hertlein 96].

6.5.2   Dynamic-Deflection Equipment

Commercially available dynamic-deflection equipment measures and records the dynamic response of a road or airport pavement subjected to steady-state vibratory or transient-loading forces. Examples are Dynaflect, Road Rater, and falling-weight deflectometer (FWD), as discussed in detail by Haas et al. [Haas 94]. Improvements have been made to automate the testing sequence and provide on-board data processing. Table 6-5 shows guidelines for the sampling rate. Several models of noncontact dynamic deflection devices are being tested for network level.

6.5.3   Time-Domain Reflectometry

The time-domain reflectometry (TDR) method was originally developed by the power and telecommunications industries to locate breaks in cables, and the method has been extended to measure material dielectric constant, concrete material properties, discontinuities in concrete and rock [Su 96], and to monitor unstable slopes [Kane 96].

6.5.4   Noncontact Methods

6.5.4.1   Ground-Penetrating Radar Technology

Ground-penetrating radar (GPR) used for civil-engineering applications is an adaption of the vehicle-mounted mine-detection radar developed for the U.S. Army. This technology has been used to identify the subsurface structure of soil, and to locate undergroundutilities, pipelines, and tunnels. With improved antenna and other hardware developments, the technology has been applied to pavement. Radar testing devices are available commercially for identifying and locating

Application	Sampling Rate	Information Quality Level (IQL)	Evaluation Method
Network planning	1–5% network length (stratified random sample)	IV or III	Dynamic deflection, construction record, DCP profile, visual survey
Network programming	0.3–1.0 km intervals (min. 5 points per section) JRP: At slab center and joints	III or II	Dynamic deflection, deflectograph deflection
Project design	20–200 m (min. 5 points per section) each wheelpath. RP: 20–40 m in outer wheelpath	Minor roads: III or II Major roads: II or I	Dynamic deflection, deflectograph deflection, DCP profile + SN calculation [+field sampling + material testing as required]
Research and special investigation	3–20 m intervals RP: every joint or crack	II or I	NDT deflection + field sampling + material testing

Notes: Dynamic deflection devices = Dynaflect and FWD (falling weight deflectometer); NDT = nondestructive testing; DCP = dynamic cone penetrometer; RP = rigid pavements; JRP = jointed rigid pavements; SN = structural number

TABLE 6-5 Recommended Sampling Rates and Methods for Pavement Nondestructive Evaluation [Paterson 90]

voids under concrete pavement, locating steel reinforcement, evaluating effectiveness of grouting work under concrete pavements, and other related applications [Uddin 94, 96b]. During the post-Katrina recovery period after August 29, 2005, the approach slabs of several bridges were scoured below the pavement, developing cavities that endangered traffic safety. These sites were identified by GPR surveys with some success [Uddin 06]. The interpretation of the response outputs is the most critical element in the correct and effective use of radar. The test monitors the dielectric constant profile, which is then used to estimate layer thickness and material properties. The advent of color-output monitors has improved the value of radar tests, but the true effectiveness of these methods remains to be proven.

A radar unit emits a short pulse of electromagnetic energy and examines the reflected energy from that pulse. In the case of a pulse fired into a pavement structure, the electromagnetic wave travels until it meets a discontinuity caused by a change in the dielectric constant of the material in the path of the wave. A portion of the wave is reflected by this discontinuity and a portion continues to travel through subsequent layers.

The amount of energy reflected at the discontinuity is a function of the wave impedance of the two materials. At the interface between materials with similar dielectric properties, such as two lifts of an asphalt concrete pavement, most of the energy passes through the interface and very little is reflected back to the transmitter. Conversely, where the difference in dielectrics is significant, such as in an asphalt layer over concrete or over a granular base, much of the energy is reflected to the transmitter and very little is passed to the next layer. This reflection phenomenon is the theoretical basis for the various radar signatures produced by different subsurface anomalies [Pulse 94]. Basic GPR principles of test and data interpretation [Uddin 06] are as follows:

- EM waves propagate in a medium in short pulse duration of ≤ 1 ns (1×10^9 sec) and velocity, v.

- In a vacuum or in air, EM waves travel at the velocity of light, c, at 3×10^8 m/sec.

- EM velocity $v = c/\varepsilon^{0.5}$, where ε is the relative dielectric constant, which is a material property.

  - The dielectric constant is a measure of the ability of the material to pass electromagnetic radiation through it.

  - Layer thickness, $d = v(t/2)$, where t = time between reflections.

- Factors affecting the dielectric constant are material composition, water content, saturated clays, saline water, and freezing temperature.

GPR data can be collected at up to normal highway speed at a rate of 50 data points per second. The time from the firing of the pulse to its return to the antenna is very accurately timed. This pulse travel time is a function of the thickness of each layer and the *dielectric constant*. Thickness of each layer can be estimated by the expression:

$$d_n = 6t_n/[\,\varepsilon_n]^{0.5} \tag{6.4}$$

where d_n = thickness of layer n

   t_n = the two-way travel time (in nanoseconds) of the pulse in layer n

   ε_n = the dielectric constant of layer n

The methodology is described in *ASTM D4748-87, Standard Test Method for Determining the Thickness of Bound Layers Using Short Pulse Radar* [ASTM 91]. The data is acquired using a short-pulse radar with a center frequency of 1 GHz. Longitudinal position is determined by a transmission-mounted encoder. The data is digitally stored in a microcomputer for processing, and is displayed in color on a monitor for an expert analyst to review [Pulse 94].

The Federal Communication Commission (FCC) Part 15 regulations of July 15th, 2002 required all radar devices in the U.S. operating in the ultrawide EM band to register, which affected the road GPR equipment working in 1- to 2-GHz range. Final resolutions to use GPR for pavement evaluation include [Uddin 06]:

- Existing GPR devices were allowed to be "grandfathered" if registered with FCC by October 15, 2002.

- On February 13, 2003, FCC adopted an amendment to the Part 15 rules to specifically allow for the operation of new GPR systems.

An industry and user agency survey [Uddin 06] revealed that FCC certified 1-GHz antenna is more noisy and susceptible to RF interference than the older antennas.

Layer modulus and thickness variations show significant effects on pavement deflection responses and rehabilitation strategies. Therefore, GPR testing is now used by some highway agencies to estimate in situ pavement thickness at each FWD test location. Accurate pavement thickness significantly improves the accuracy of layer-modulus values and rehabilitation thickness calculations. A benefit/cost ratio of over 80 is estimated [Uddin 06] if GPR-thickness surveys are conducted side-by-side with the deflection testing.

6.5.4.2   Acoustic and Ultrasonic Testing

Nondestructive sonic testing has been applied to detect delamination, honeycombing, and other discontinuities in concrete slabs. In the pulse-echo method, a compression wave is sent through the material, and the operator monitors the reflection pattern of the wave caused by discontinuities in the material. Acoustic methods have been particularly useful to evaluate cracks, corrosion, and other defects in concrete [NDT 96].

In recent years, noncontact ultrasonic sensors have been used for pavement evaluation for longitudinal profile and rut-depth measurements [Haas 94]. Ultrasonic sensors are, however, sensitive to environmental factors like humidity and moisture and may produce bad data under some conditions [Hudson 87a].

6.5.4.3   Laser Scanning

Single- and multiple-beam laser technologies have been used for several commercial applications in pavement evaluation. It has been commercially applied near highway speeds for counting transverse cracks and measuring joint faulting, longitudinal roughness, and surface texture [Hudson 87a, b]. Laser-based pavement evaluation equipment should be used more often in the future because of its high level of automation, leading to greater productivity and reduced operating constraints (like traffic interference, environmental factors, and speed dependency).

Modern laser scanning using 3D laser point cloud technology has been implemented on terrestrial and airborne platforms since the late

1990s and evolved into high-speed kinematic laser mapping in the last decade [Al-Turk 99, Gutelius 12, Uddin 08, 11a, b].

6.5.4.4 Thermal Infrared Photography

Thermal infrared photography can be used to locate defects and voids in concrete structures, bridge decks, and buried pipes [Uddin 96b, Delahaza 96, Del Grande 96]. Infrared thermography is a noncontact, noninvasive means of obtaining visible images from the invisible heat energy emitted by an object. Television-like images are produced with gray-scale variations, or different colors, representing various temperatures. Thermography has limitations, as do all temperature-sensing and measurement techniques. The observed radiometric temperature is affected by such things as the object's absolute temperature, the ambient temperature, the object's emissivity, the radiant energy of nearby surroundings, the atmospheric filtering of energy, and the distance from the object to the scanner, all of which require that the thermographic information be adjusted [EnTech 94].

An infrared imaging system consists of an infrared scanner, which records primarily infrared radiation, and a real-time computer display monitor (similar to a portable television with a small screen coupled to a microprocessor). The scanner unit converts the radiated heat that passes through the optics into an electronic signal. The signal produces a real-time thermal image on the computer screen composed of a gray scale with continuous tones ranging from black through gray to white. Areas of higher-relative temperature appear lighter and areas of lower-relative temperature appear darker. A color monitor and an additional microprocessor can also be used to display the thermal image. In this case, each color represents a particular temperature range.

The infrared-scanning systems used in a concrete pavement study [Uddin 96b] allow for rapid data collection and analysis. With this equipment it is possible to observe, quantify, and record the thermal image of the surface of an object whose temperature is between −20°C and 2,000°C. The sensitivity of the imaging equipment is such that it has the capacity to detect surface-temperature differences between two given objects to an accuracy of ±0.05°C at an ambient temperature of 30°C [EnTech 94].

6.6 Combined Evaluation Data

In many types of infrastructure, more than one method of evaluation are important. In the case of pavements, for example, roughness, cracking, and deflection are all used. In such cases, the various response data may be used separately or aggregated into one composite measure of overall response. A combined index is one useful method to accomplish this task.

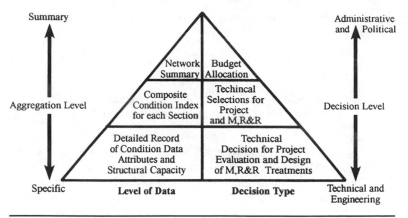

FIGURE 6-9   Data aggregation and level of decision [after Haas 94].

6.6.1   Reasons for Use of a Composite-Condition Index

A composite condition index is sometimes used to combine two or more evaluation attributes and to communicate summary evaluation information to administrators, elected officials, and the public. Figure 6-9 illustrates the levels of data that are appropriate to various levels of decisions for infrastructure management.

At the first, or disaggregated, level, involving specific activities and technical engineering decisions, detailed evaluation data attributes for each section are required. The second level involves aggregation and decisions for network-level management. At this level, composite measures and combined indices are useful for establishing priorities for the selection of projects and M,R&R strategies. Composite measures are absolutely required at the third level, which involves administrative and political decisions. Personnel at this level are faced with large amounts of information, and therefore need aggregated data to portray the overall quality of the network and to project future quality as related to budgets. Generally, the condition index is reported on a scale of 100 (excellent condition or new construction) to 0 (worst condition). The condition index concept is also useful for developing performance equations that model the trend of condition deterioration and effect of M,R&R intervention, as discussed in Chapter 8.

6.6.2   Methods of Developing a Condition Index

Each infrastructure facility requires a condition index (CI) that relates to the data being collected, the local conditions, and the objectives of the agency. There is no single best engineering or analytical formula for establishing a CI. Techniques for developing simple linear and nonlinear equations have been borrowed from the social-sciences field, using subjective data. A composite CI combines two or more condition attributes and represents an aggregation of the different measures of condition; for example:

$$CI = W_1C_1 + W_2C_2 + W_3C_3 \tag{6.5}$$

where CI = composite condition index
W_1, W_2, W_3 = weighting factors for condition measures 1, 2, and 3
C_1, C_2, C_3 = values for condition measures 1, 2, and 3

Other methods use the deduct-value concept. In this method, a facility in excellent condition, such as new construction, is given a score of 100. Any defects or distress present results in some deduction, depending upon the defect severity and extent, which are subtracted from 100 to produce the CI. Adjustments are made for multiple defects so the total deducted values can never exceed 100. This approach was first used by the Washington Department of Transportation for highway pavements and was later adapted by the U.S. Army Corps of Engineers to establish the PCI for airport pavements [Haas 94, FAA 82, Shahin 90]. This approach also was used to develop a sufficiency index for bridges [Hudson 87].

The Delphi technique can also be used to establish a composite CI by capturing expert opinion. It involves selecting a panel of experts who are asked to rank a series of hypothetical situations that describe a facility with different combinations of physical deterioration manifestations and structural capacity. The initial rankings are analyzed, and the results are presented to the panel for their assessment. An eventual convergence to a common ranking is reached, and the information can then be statistically modeled to produce a composite CI [Haas 94].

6.6.3   Precautions in the Use of Composite CI

The key to developing a combined or composite index of infrastructure condition and service quality is to recognize the subjective nature of the problem and the associated techniques for quantifying subjective information. The methodology may be transferable but not the specific models developed. Thus the index will require calibration or redevelopment for each agency or geographic region.

Due to the aggregate nature of a composite CI, caution must be exercised in interpreting the index in selecting an M,R&R treatment. For example, the CI may produce a specific estimate at the network level for the selection of a treatment strategy for the M,R&R budget. But at the project level, using precisely detailed evaluation data, a particular defect may ultimately require the use of a different M,R&R treatment. This does not imply that composite indices are inapplicable at the project level, but it does mean that their primary use is to convey summary information at the network level [Haas 94].

6.7   Institutional Issues

Equipment, staff, garage, and office space are needed to develop and implement in-service monitoring, evaluation plans, and IAMS evaluation database. The agency-resource requirements and the system complexity for data collection and database development will be influenced by the size of the infrastructure network and the short- and long-term goals for the data and M,R&R analysis. The amount of detail required for these various applications increases progressively from the overall summary statistics, through planning and programming, to design and research. Collecting and processing of data will also be increasingly more complex and costly. In the face of limited budgets, compromises and wise decisions must be made.

6.8   References

[AASHTO 92] "Standard Specifications for Highway Bridges," 15th ed., American Association of State Highway and Transportation Officials (AASHTO), Washington, DC, 1992.

[Abbiss 81] C. P. Abbiss, "Shear Wave Measurement of the Elasticity of the Ground," *Geotechnique*, Vol. 31, No. 1, 1981, pp. 91–104.

[ACRP 10] Airport Cooperative Research Program (ACRP), "Light Detection and Ranging (LIDAR) Deployment for Airport Obstructions Survey," *Research Results Digest 10*, Transportation Research Board, The National Academies, Washington, DC, July 2010. http://144.171.11.40/cmsfeed/TRBNetProjectDisplay.asp?ProjectID=135, accessed November 7, 2011.

[Al-Turk 99] E. Al-Turk, and W. Uddin, "Infrastructure Inventory and Condition Assessment Using Airborne Laser Terrain Mapping and Digital Photography," *Transportation Research Record 1690*, Transportation Research Board, National Research Council, Washington, DC, 1999, pp. 121–125.

[Alexandar 96] A. M. Alexander, R. W. Haskins, and D. E. Wilson, "Development of the Superscanner for the Detection and Mapping of Delaminations in Concrete Bridge Decks," *Proceedings, 2nd International Conference on Nondestructive Testing of Concrete in the Infrastructure*, Nashville, Tenn., 1996, pp. 39–47.

[Ansari 96] F. Ansari, Y. Libo, I. Lee, and H. Ding, "A Fiber Optic Embedded Crack Opening Displacement Sensor for Cementitious Composites," *Proceedings, 2nd International Conference on Nondestructive Testing of Concrete in the Infrastructure*, Nashville, Tenn., 1996, pp. 268–277.

[Aouad 96] M. F. Aouad, L. D. Olson, and F. Jalinoos, "Determination of Unknown Depth of Bridge Abutments Using the Spectral Analysis of Surface Waves (SASW) and Parallel Seismic (PS) Test Methods," *Proceedings, 2nd International Conference on Nondestructive Testing of Concrete in the Infrastructure*, Nashville, Tenn., 1996, pp. 147–153.

[ASTM 91] ASTM D4748-87, "Standard Test Method for Determining the Thickness of Bound Pavement Layers Using Short-Pulse Radar," *1991 Annual Book of ASTM Standards*, Vol. 04.03, pp. 557–562.

[ATC 93] Applied Technology Council, "Postearthquake Safety Evaluation of Buildings Training Manual," *ATC-20-T*, funded by the Federal Emergency Management Agency, Washington, DC, 1993.

[Bridge 96] "Calculating Corrosion," *BRIDGE Design and Engineering, London*, No. 3, May 1996, p. 78.

[Chaker 96] V. Chaker, "Measuring Soil Resistivity," *ASTM Standardization News*, Vol. 24, No. 4, April 1996, pp. 30–33.

[Collins 95] M. A. Collins, and C. T. Stude, "Masonry Sewer Rehab," *Civil Engineering*, Vol. 65, No. 9, September 1995, pp. 65–69.

[Del Grande 96] N. K. Del Grande, P. F. Durbin, C. M. Logan, D. E. Perkins, P. C. Schaich, "Determination of Dual-Band Infrared Thermal Imaging at Grass Valley Creek Bridge," *Nondestructive Evaluation of Bridges and Highways, Proceedings SPIE 2946,* The International Society for Optical Engineering, Scottsdale, Ariz., pp. 166–177.

[Delahaza 96] A. O. Delahaza, "Nondestructive Testing of the Concrete Roof Shell at the Kingdome in Seattle, Washington," *Proceedings, 2nd International Conference on Nondestructive Testing of Concrete in the Infrastructure,* Nashville, Tenn., pp. 256–267.

[DOT 00] Department of Transportation, "Primer: GASB 34," Federal Highway Administration, Office of Asset Management, Washington, DC, November 2000.

[DOT 02] Department of Transportation, "Achievements of the DOT-NASA Joint Program on Remote Sensing and Spatial Information Technologies: Application to Multimodal Transportation, 2000-2002," U.S. DOT Research and Special Program Administration, Washington, DC, April 2002.

[Eaton 87] R. A. Eaton, S. Gerard, and D. W. Cate, "Rating Unsurfaced Roads," *Special Report 87-15,* U.S. Army Corps of Engineers, Cold Regions Research and Engineering Laboratory, 1987.

[EnTech 94] EnTech Inc., *Infrared Thermographic Investigation for the Location of Highway Pavement Subsurface Anomalies,* report prepared for The University of Mississippi/Mississippi Department of Transportation, May 1994.

[FAA 82] "Guidelines and Procedures for Maintenance of Airport Pavements," *Advisory Circular AC: 150/5380-6,* Federal Aviation Administration, Washington, DC, 1982.

[Fazio 80] A. E. Fazio, and R. Prybella, "Development of an Analytical Approach to Track Maintenance Planning," in *Transportation Research Record 744,* Transportation Research Board, National Research Council, Washington, DC, 1980, pp. 46–52.

[Gerard 91] S. V. Gerard, "UP Links Video and Graphic Images for Greater Clarity," *Railway Track and Structures,* April 1991.

[Gutelius 12] Bill Gutelius, "Airborne LiDAR for Obstruction Mapping: Enabling Flight Safety," *LiDAR Magazine,* Spatial Media, Vol. 2, No. 1, 2012, pp. 50–55. http://www.lidarnews.com/, accessed December 20, 2012.

[Haas 73] R. C. G. Haas, "A Method for Designing Asphalt Pavements to Minimize Low-Temperature Shrinkage Cracking," *Research Report 73-1,* Asphalt Institute, Riverdale, Maryland, 1973.

[Haas 94] R. Haas, W. R. Hudson, and J. P. Zaniewski, *Modern Pavement Management,* Krieger Publishing Company, Malabar, Fla., 1994.

[Harvey 94] J. Harvey, T. Lee, J. Sousa, J. Park, and C. L. Monismith, "Evaluation of Fatigue and Permanent Deformation Properties of Several Asphalt Aggregate Field Mixes Using Strategic Highway Research Program A-300A Equipment," in *Transportation Research Record 1454,* Transportation Research Board, National Research Council, Washington, DC, 1994, pp. 123–133.

[Heisey 82] S. Heisey, K. H. Stokoe II, W. R. Hudson, and A. H. Meyer, "Determination of In Situ Shear Wave Velocities from Spectral Analysis of Surface Waves," *CTR Research Report 256-2,* Center for Transportation Research, The University of Texas at Austin, 1983.

[Hertlein 96] B. Hertlein and C. N. Baker, "Practical Experience with Nondestructive Testing of Deep Foundation," *ADSC,* The International Association of Foundation Drilling, March/April 1996, pp. 19–26.

[Ho 95] D. Ho, "Temperature Distribution in Walls and Roofs," *Journal of Architectural Engineering,* Vol. 1, No. 3, September 1995, pp. 121–132.

[HRB 62] "The AASHO Road Test, Report 5—Pavement Research," *Special Report 61E,* Highway Research Board, National Research Council, Washington, DC, 1962.

[Hudson 87] S. W. Hudson, R. F. Carmichael III, L. O. Moser, W. R. Hudson, and W. J. Wilkes, "Bridge Management Systems," *NCHRP Report 300,* Transportation Research Board, National Research Council, Washington, DC, 1987.

[Hudson 87a] W. R. Hudson and W. Uddin, "Future Evaluation Technologies: Prospective and Opportunities," *Proceedings, 2nd North American Pavement Management Conference,* Toronto, Ont., 1987.

[Hudson 87b] W. R. Hudson, G. E. Elkins, W. Uddin, and K. Reilley, "Improved Methods and Equipment to Conduct Pavement Distress Surveys," *Report No. FHWA-TS-87-213,* ARE Inc. Report for the Federal Highway Administration, April 1987.

[Hudson 87c] W. R. Hudson, G. E. Elkins, W. Uddin, and K. Reilley, "Evaluation of Deflection Measuring Equipment," *Report No. FHWA-TS-87-208,* ARE Inc. Report for the Federal Highway Administration, March 1987.

[Hudson 87d] W. R. Hudson, W. Uddin, and G. E. Elkins, "Smoothness Acceptance Testing and Specifications for Flexible Pavements," *Proceedings, 2nd International Conference on Pavement Management,* Toronto, Ont., 1987.

[Hudson 94] W. R. Hudson and S. W. Hudson, "Pavement Management Systems Lead the Way for Infrastructure Management Systems," *Proceedings, 3rd International Conference on Managing Pavements,* National Research Council, Vol. 2, May 1994, pp. 99–112.

[Hudson 97] W. R. Hudson, R. Haas, and W. Uddin, *Infrastructure Management,* McGraw-Hill, New York, 1997.

[Kane 96] W. F. Kane and T. J. Beck, "Rapid Slope Monitoring," *Civil Engineering,* Vol. 66, No. 6, June 1996, pp. 56–58.

[Lysmer 70] J. Lysmer, "Lumped Mass Method for Rayleigh Waves," *Bulletin, Seismological Society of America,* 1970, Vol. 60, No. 1, pp. 89–104.

[Matzkanin 84] G. A. Matzkanin, L. S. Fountain, and O. Tranbarger, "Nondestructive Evaluation of Infrastructure Conditions," *Infrastructure—Maintenance and Repair of Public Works, Annals of the New York Academy of Sciences,* Vol. 431, December 1984, pp. 268–303.

[Mayer 10] L. Mayer, B. Yanev, L. D. Olson, and A. Smyth, "Monitoring of the Manhattan Bridge for Vertical and Torsional Performance with GPS and Interferometric Radar Systems," *CD Proceedings, Annual Meeting of the Transportation Research Board,* Washington, DC, January 2010, pp. 1–13.

[Metzger 93] D. S. Metzger, C. Barnes, and E. K. Miller, "Smart Roads," *Proceedings of the Smart Pavement Conference,* convened by ASTM Committee E-17 on Vehicle Pavement Systems, proceedings presented by Alliance for Transportation Research, Dallas, Tex., December 1993, pp. 4–18.

[Nazarian 83] S. Nazarian, K. H. Stokoe II, and W. R. Hudson, "Use of Spectral Analysis of Surface Waves Method for Determination of Moduli and Thickness of Pavement Systems," in *Transportation Research Record 930,* Transportation Research Board, National Research Council, Washington, DC, 1983, pp. 38–45.

[NDT 96] R. A. Miller, S. E. Swartz, and S. P. Shah, editors. *Proceedings, 2nd International Conference on Nondestructive Testing of Concrete in the Infrastructure,* Nashville, Tenn., Society for Experimental Mechanics, Inc., 1996.

[NOAA 94] "Climatological Data," National Oceanic and Atmospheric Administration, National Climatic Center, Asheville, NC, 1994.

[O'Leary 96] P. N. O'Leary, R. R. Sartor, Y. Fu, and J. T. DeWolf, "Nondestructive Testing of Columns in Reinforced Concrete Bridge," *Proceedings, 2nd International Conference on Nondestructive Testing of Concrete in the Infrastructure,* Nashville, Tenn., 1996, pp. 222–229.

[Olson 96] *Condition Assessment of Buildings and Bridges,* equipment brochure, Olson Engineering Inc., Golden, Colorado, 1996.

[Osborne 09] Katherine Osborne, "GIS-Based Urban Transportation Infrastructure Management Using Spaceborne Remote Sensing Data," *M.S. Thesis,* Department of Civil Engineering, The University of Mississippi, December 2009.

[Paterson 90] W. D. O. Paterson, and T. Scullion, "Information Systems for Road Management: Draft Guidelines on System Design and Data Issues," *Technical Paper INU77,* Infrastructure and Urban Development Department, The World Bank, Washington, DC, 1990.

[Pulse 94] Pulse Radar Inc., *Ground Penetrating Radar Surveys, MDOT Study—US78 PCC Pavement,* report prepared for The University of Mississippi/Mississippi Department of Transportation, May 1994.

[Schickert 96] M. Schickert, "The Use of Ultrasonic A-Scan and B-Scan and SAFT Techniques for Testing Concrete Elements," *Proceedings, 2nd International Conference on Nondestructive Testing of Concrete in the Infrastructure*, Nashville, Tenn., 1996, pp. 135–142.

[Shahin 90] M. Y. Shahin, and J. A. Walter, "Pavement Maintenance Management for Roads and Streets Using the PAVER System," *USACERL Technical Report M-90/05*, Champaign, Ill., 1990.

[SHRP 89a] "SHRP-LTPP Guide to Field Sampling and Handling," *SHRP 5021*, Strategic Highway Research Program, National Research Council, Washington, DC, 1992.

[SHRP 89b] "SHRP-LTPP Laboratory Guide for Testing Pavement Samples," *SHRP 5025*, Strategic Highway Research Program, National Research Council, Washington, DC, 1992.

[SHRP 93] "Distress Identification Manual for the Long-Term Pavement Performance Project," *Report SHRP-P-338*, Strategic Highway Research Program, National Research Council, Washington, DC, 1993.

[Su 96] M. B. Su, and Y. J. Chen, "Multiple Reflection of Metallic Time Domain Reflectometry," *Proceedings, 2nd International Conference on Nondestructive Testing of Concrete in the Infrastructure*, Nashville, Tenn., 1996, pp. 60–65.

[TRB 95] L. D. Olson, F. Jalinoos, and M. F. Aouad, "Determination of Unknown Subsurface Bridge Foundations," *NCHRP Report 21-5*, Transportation Research Board, Washington, DC, 1995.

[Uddin 83] W. Uddin, S. Nazarian, W. R. Hudson, A. H. Meyer, and K. H. Stokoe II, "Investigations into Dynaflect Deflections in Relation to Location/Temperature Parameters and in Situ Material Characterization of Rigid Pavements," *CTR Report 256-5*, Center for Transportation Research, The University of Texas at Austin, 1983.

[Uddin 94] W. Uddin, and W. R. Hudson, "Evaluation of NDT Equipment for Measuring Voids under Concrete Pavements," *2nd International Symposium on Nondestructive Testing and Backcalculation of Moduli, ASTM STP 1198*, Vol. 2, June 1994, pp. 488–502.

[Uddin 95a] W. Uddin, "Pavement Material Property Databases for Pavement Management Applications," *Computerization and Networking of Materials Databases, ASTM STP 1257*, Vol. 4, Philadelphia, Pa., 1995, pp. 96–109.

[Uddin 95b] W. Uddin, and V. Torres-Verdin, "Structural Capacity and User Cost Analyses for Road Investment Planning," *Proceedings, 2nd International Conference on Roads and Road Transport (ICORT-95)*, Delhi, India, December 1995.

[Uddin 96a] W. Uddin, R. M. Hackett, P. Noppakunwijai, and Z. Pan, "Three-Dimensional Finite-Element Simulation of FWD Loading on Pavement Systems," *Proceedings, 24th International Air Transportation Conference*, Louisville, Ky., 1996, pp. 284–294.

[Uddin 96b] W. Uddin, R. M. Hackett, P. Noppakunwijai, and T. Chung, "Nondestructive Evaluation and In Situ Material Characterization of Jointed Concrete Pavement Systems," *Proceedings, 2nd International Conference on Nondestructive Testing of Concrete in the Infrastructure*, Nashville, Tenn., 1996, pp. 242–249.

[Uddin 06] W. Uddin, "Ground Penetrating Radar Study, Phase I—Technology Review and Evaluation," *Report No. FHWA/MS-DOT-RD-06-182*, Final Report UM-CAIT/2006-01 State Study SS 182, Center for Advanced Infrastructure Technology, University of Mississippi, for the Mississippi Department of Transportation, December 2006.

[Uddin 08] W. Uddin, "Airborne Laser Terrain Mapping for Expediting Highway Projects: Evaluation of Accuracy and Cost," *Journal of Construction Engineering and Management*, American Society of Civil Engineers, Vol. 134, No. 6, pp. 411–420, June 2008.

[Uddin 09] W. Uddin, Carla Brown, E. Scott Dooley, and Bikila Wodajo, "Geospatial Analysis of Remote Sensing Data to Assess Built Environment Impacts on Heat-Island Effects, Air Quality and Global Warming," *Paper No. 09-3146, CD Proceedings, 85th Annual Meeting of The Transportation Research Board*, Washington, DC, January 11–15, 2009.

[Uddin 10] W. Uddin, "Spaceborne Remote Sensing Data for Inventory and Visualization of Transportation Infrastructure and Traffic Attributes," *CD Proceedings, First International Conference on Sustainable Transportation and Traffic Management and 2010 IEDC*, Karachi, Pakistan, ISBN 978-969-8620-10-3, July 1–3, 2010, pp. 3–12. http://sites.nationalacademies.org/PGA/dsc/pakistan/PGA_052872, accessed September 25, 2011.

[Uddin 11a] W. Uddin, "Transportation Management: LiDAR, Satellite Imagery Expedite Infrastructure Planning," *Earth Imaging Journal*, January/February 2011, pp. 24–27.

[Uddin 11b] W. Uddin, "Remote Sensing Laser and Imagery Data for Inventory and Condition Assessment of Road and Airport Infrastructure and GIS Visualization," *International Journal of Roads and Airports (IJRA)*, Vol. 1, No. 1, pp. 53–67.

[Uddin 11c] W. Uddin, Bill Gutelius, and Christopher Parrish. "Airborne Laser Survey Specifications and Quality Management Protocols for Airport Obstruction Surveys." Paper No. 11-1323, *Transportation Research Record 2214, Journal of Transportation Research Board*, Washington, DC, 2011, pp. 117–125.

[Uddin 12] W. Uddin, "Mobile and Area Sources of Greenhouse Gases and Abatement Strategies," Chapter 23, *Handbook of Climate Change Mitigation*, (Editors: Wei-Yin Chen, John M. Seiner, Toshio Suzuki, and Maximilian Lackner), Springer, New York, 2012, pp. 775–840.

[Urban 84a] The Urban Institute, *Guides to Managing Urban Capital*, Vol. I: A Summary, H. P. Harty and G. E. Peterson (eds.), Washington, DC, 1984.

[Urban 84b] S. R. Goodwin, and A. E. Peterson, *Guides to Managing Urban Capital Series*, Vol. 2: *Guide to Assessing Capital Stock Condition*, The Urban Institute, Washington, DC, 1984.

[Urban 84c] H. P. Harty, and B. G. Steinthal, *Guides to Managing Urban Capital Series*, Vol. 4: *Guide to Selecting Maintenance Strategies for Capital Facilities*, The Urban Institute, Washington, DC, 1984.

CHAPTER 7

Uses of Monitoring Data and Examples of In-Service Evaluation

In-service evaluation of infrastructure facilities by nondestructive testing is an important way to estimate:

1. The physical and structural condition of aging structures
2. The effects of environmental and corrosion stresses
3. Evaluation of structural integrity after natural disasters, such as floods and earthquakes, and accidents, such as vehicular collisions or fire
4. Seismic safety and retrofitting needs

The available sensors and nondestructive-evaluation (NDE) technologies were briefly reviewed in Table 6-4. Some examples of in-service evaluation for selected infrastructure facilities, listed below, are presented in the following sections:

- Road and airport pavements
- Railroad tracks
- Bridges
- Water pipelines
- Gas pipelines
- Buildings

7.1 In-Service Evaluation of Road and Airport Pavements

Pavement management systems for roads and highways have been established since 1970 and are supported by the Federal Highway Administration (FHWA) [FHWA 89] and many state highway agencies.

Pavement Management System (PMS) has also been implemented on airport pavements by the Federal Aviation Administration (FAA) [FAA 82]. The activities of these agencies, and pavement-evaluation technologies, are well documented by Haas and Hudson [Haas 78, 94]. Therefore, only a limited discussion of road infrastructure is presented here. These principles and databases are also used to manage roadway signs, traffic-control devices, and traffic incidents on roadways.

7.1.1 Pavement Monitoring and Evaluation

The evaluation of pavements involves monitoring of one or more of the following: distress, structural capacity, riding comfort, safety, and appearance. Table 7-1 lists several evaluation measures covering pavement condition and overall performance. Not all of these are essential for pavement management. At the network level, roughness is the primary measure used by some agencies, but at the project level, several major objective measurements are used by some agencies.

Monitoring	Evaluation
Longitudinal roughness	Serviceability
Surface distress and defects (cracking, deformation, patches, disintegration, surface defects)	Deterioration, overall composite index, and maintenance needs
Deflection testing	Material properties and structural capacity
Skid resistance or surface friction	Safety against skidding
Ride quality	User evaluation of overall pavement quality
Appearance	Aesthetics
Traffic	Performance and remaining life
Costs (construction, maintenance, user)	Unit-cost summaries for economic evaluation
Location reference, geometric and structure data, longitudinal and cross-fall deficiency, coring for layer thickness	Verification of inventory database, inputs for structural evaluation, safety against potential hydroplaning
Environment (climate, pavement temperature, drainage, water below surface, freeze/thaw)	Material degradation, distress and defects progression, structural integrity, performance
Environmental impacts on ecosystems and air quality	Wetland loss, air pollution, CO_2 emission, contamination of water and soil, deforestation, etc.

TABLE 7-1 Concepts for In-Service Monitoring and Evaluation of Road Pavements

Pavement condition changes with time; therefore, periodic measurements are required to develop a distress history.

7.1.2   Pavement Nondestructive Evaluation

The major type of pavement nondestructive evaluation measured and used to evaluate structural capacity and residual life is deflection testing. The Benkelman deflection beam was developed at the WASHO Road Test in the mid-1950s and is still widely used around the world. For over two decades, most pavement-rehabilitation design procedures were based on Benkelman beam deflections. The ever-growing demand for faster, easier to use, and more automated nondestructive testing (NDT) equipment resulted in the development of vibratory loading devices like the Dynaflect and Road Rater, and the impact-loading falling-weight deflectometer (FWD). Improvements have been made to automate the testing sequence and on-board data processing. Table 6-5 shows guidelines for the deflection data sampling rate for roads.

Deflection test data are also used to estimate in situ material properties, and to evaluate the structural integrity of concrete pavements [Hudson 87a]. The in situ material properties and structural capacity are calculated from the dynamic deflection data using static-load analytical procedures and empirical performance relationships [Uddin 86], as shown in Figure 7-1. A significant step in this area is the development of three-dimensional finite-element dynamic-analysis procedures [for example, Uddin 96] for interpretation of the dynamic-deflection data.

7.1.3   The Serviceability and Performance Concept

Serviceability is the ability of a specific section of pavement to serve traffic in its existing condition. Present Serviceability Index (PSI) can be estimated from roughness data on a scale of 5 to 0. Performance is a measure of the accumulated service provided by a facility, i.e., the adequacy with which a pavement fulfills its purpose.

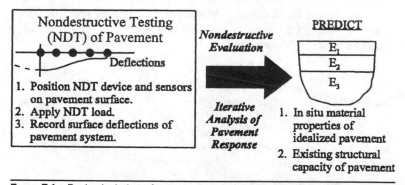

FIGURE 7-1   Backcalculation of pavement material properties from deflection data.

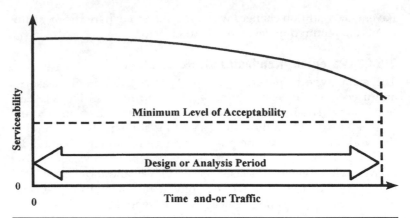

FIGURE 7-2 Deterioration of ride quality or serviceability over time [after Haas 94].

The evaluation of pavement performance involves studying the functional behavior of a section or length of pavement. For a functional or performance analysis, information is needed on the history of the riding quality of the pavement section for the time period chosen and the traffic during that time. This can be determined by periodic measurements of the pavement riding quality, coupled with records of traffic history and time. This history of deterioration of the serviceability provided to the user defines pavement performance (see Figure 7-2).

Until a measure of pavement serviceability was developed in conjunction with the AASHO Road Test [HRB 62], inadequate attention was paid to the evaluation of pavement performance. A pavement was considered to be either satisfactory or unsatisfactory (i.e., in need of repair or replacement). Pavement-design technology did not directly consider performance. Design engineers have varied widely in their concepts of desirable performance. For example, at one extreme an engineer asked to design a pavement for a certain expected traffic level for 20 years might consider the job properly done if little or no cracking occurred during the 20 years. On the other hand, a second designer might be satisfied if the pavement had reached a totally unacceptable level of serviceability at the end of the 20 years.

7.1.4 Relating Roughness to Serviceability

The primary use of objective roughness measurements is to estimate the pavement serviceability, which is a subjective (i.e., user) rating of pavement-ride quality. The first and most widely used method for this purpose was the PSI, at the AASHO Road Test [HRB 62]. The original functional form of the PSI equation is:

$$PSI = C + (A_1 R_1 + \cdots) + (B_1 D_1 + B_2 D_2 + \cdots) \pm e \qquad (7.1)$$

where C = coefficient (5.03 for flexible pavements and 5.41 for rigid pavements)

A_1 = coefficient (−1.91 and −1.80 for flexible and rigid, respectively)

R_1 = function of profile roughness [log $(1 + SV)$], where SV = mean slope variance obtained from the CHLOE profilometer

B_1 = coefficient (−1.38 for flexible and 0 for rigid)

D_1 = function of surface rutting (RD), where RD = mean rut depth as measured by simple rut-depth indicator

B_2 = coefficient (−0.01 for flexible; −0.09 for rigid)

D_2 = function of surface deterioration ($C + P$), where $C + P$ = amount of cracking and patching, determined by procedures developed at the AASHO Road Test

e = error term

Given this general form of equation, it is necessary to determine the coefficients for a particular set of input variables. This was done at the AASHO Road Test for several sets of variables [Carey 60]. It is important to understand that the result is a best-fit equation based on all observed data used in the equation. Other variables were candidates for inclusion in the equation, but they added no significance in predicting PSI.

Additionally, a regression equation is not a causative relationship, and covariance between terms can account for very small coefficients on a variable that alone is only slightly less well correlated with the dependent variable. For example, if the observed roughness in a pavement is caused by cracking, the two factors are correlated. Consequently, once the roughness term is included in the equation, little variation remains to be explained by adding the cracking terms, and thus the coefficient is small. This does not indicate a lack of concern for cracking, but if users of the equation alter it arbitrarily because they intuitively "feel" that cracking is more important, then erroneous and unpredictable results will occur.

7.1.5 Evolution of the Serviceability-Roughness Concept

Any change in measurement methods or units will result in a modified equation. This can be done either by performing an entirely new regression if all data are available, or by comparing the old measurement to the new and making an appropriate substitution in the equation. For example, at the AASHO Road Test [HRB 62] the BPR roughometer output, R, in inches per mile, was correlated to slope variance, SV, and substituted into Eq. (7.1). The resulting equation for rigid pavements was:

$$PSI = 5.41 - 1.80 \log(0.40R - 30) - 0.09\sqrt{(C+P)} \qquad (7.2)$$

It must be emphasized that any PSI model, such as that in Eq. (7.1), is not an end in itself. Carey and Irick made this quite clear in pointing out that it is intended only to predict PSR to a satisfactory approximation [Carey 60].

7.1.6 Distress Monitoring and Evaluation

Pavement distress is observable deterioration or damage in the pavement. Distress data attributes include cracking, permanent deformation, and disintegration caused by load, age, environment, material deficiency and degradation, inadequate pavement design or poor construction quality, and interaction among these factors. Because maintenance may have been performed on some of the distress, the evidence of this maintenance in the form of patches and sealed areas is a distress and also should be monitored. Distress data (type, severity, and extent) are useful measures for selecting a proper M,R&R alternative. As discussed in a later section, the distress data collected on a pavement section can be combined into a single numeric, such as the Pavement Condition Index (PCI) on a scale of 100 to 0 [Shahin 90, Uddin 05]. Figure 7-3 shows an example of the PCI and PSI scales related to overall pavement condition and maintenance needs [OECD 87].

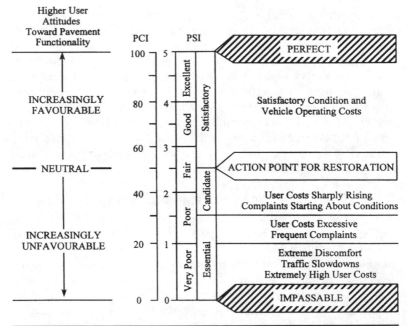

FIGURE 7-3   Significance of PCI and PSI for pavement evaluation [after OECD 87].

7.1.7   Safety Evaluation

Safety may be measured by skid resistance or friction measurements or in an empirical fashion through determination of those locations with high accident rates. However, safety hazards may not be caused by pavement-related factors, but could, for example, indicate an alignment problem, driver's distraction, and/or wet weather. Such factors may be included in the roadway management system, at the discretion of the agency involved. The typical current practice is to use skid resistance as indicated by pavement-surface friction as the primary measure of pavement-related safety.

Computer simulations have emerged as viable methods in combination with field crash tests to assess crashworthiness of road-safety appurtenance and to enhance the safety of roadway systems [Uddin 97].

7.1.8   Selection of Pavement-Evaluation Technologies

The selection of a specific type of technology for each data category depends on the data quality needed, data-collection speed/productivity, and available budgets. The equipment and methods can be grouped into three classes [Hudson 87a]: (1) visual distress and rutting (detailed slow-walk survey or high-speed photographic and video survey); (2) stand-alone automated equipment (e.g., dedicated roughness equipment like Maysmeter, Bump Integrator, or laser profilometer), deflection devices like Dynaflect or FWD, or skid-resistance measuring equipment; and (3) multipurpose automated vehicles, high-speed equipment used to collect distress, roughness, rutting and other condition/video data during a single pass. Some multipurpose equipment is listed in Table 5-5.

Multifunction automated-evaluation equipment can offer the desired productivity, cost-effectiveness, and electronic data-processing capabilities for both network-level and project-level pavement management applications, and they are still being improved upon [Haas 94, Hudson 97].

7.1.9   Airport-Pavement Evaluation

Surface distress is often used as the condition measure of airfield pavements. Deflection tests are used for NDE. Factors contributing to the use of distress in preference to roughness for airfield-pavement evaluation include [Haas 94]:

1. The relatively limited surface area of an airfield permits the collection of more detailed distress data and therefore an improved ability to rate pavement quality.

2. The military and FAA support the PAVER method to evaluate PCI.

3. There is lack of development of a serviceability concept comparable to the highway application.

4. Damage potential is possible if loose pavement pieces are ingested into an aircraft engine, such as foreign-object damage (FOD).

These factors, however, do not mean roughness should be excluded from indicators of airfield pavement-performance measures. In fact, rough runway pavements contribute to the fatigue of aircraft structural parts. Profiler devices are required to meet the critical-roughness criteria for airfields to simulate the aircraft landing and take-off patterns. The recently developed aircraft ride quality index (RQI), based on aircraft landing or take-off simulations on pavement longitudinal profiles, provides an improved procedure for airfield-pavement roughness evaluation. The RQI is calculated from the vertical acceleration experienced at the pilot seat and the aircraft center of gravity. If the acceleration exceeds 0.4 g, the ride quality is in the area of human discomfort [Gerardi 96].

7.1.10   Remote-Sensing LiDAR Evaluation of Airport Infrastructure

The FAA-supported National Academies project reviewed a number of airport projects in the United States, where airborne LiDAR surveys were conducted in combination with aerial imagery for obstruction mapping and developing data collection and processing specifications [ACRP 10]. Nonintrusive LiDAR surveys provide millions of georeferenced 3D points and raster imagery over the survey area [Uddin 11a]. Traditionally, LiDAR point clouds are collected at a ground-point density of up to 20 per sq. m but advanced airborne and terrestrial kinematic LiDAR sensor systems can collect a high ground-point density of over 100 to 200 or more per sq. m [Uddin 11b]. Geospatial analysis of dense LiDAR point-cloud data and intensity images on an airport GIS map can facilitate accurate assessment of pavement surface condition, distresses, and maintenance quantities. Furthermore, the use of LiDAR remote sensing air port survey with the imagery capture is now allowed by the FAA [Uddin 13], which can facilitate easy processing and delivery of required obstruction maps and other engineering products (including airfield pavement condition and maintenance maps) through the FAA-GIS web portal.

7.2   In-Service Evaluation of Railroad Tracks

An effective track-management system requires periodic monitoring and evaluation of track structure. In the 1990s, Union Pacific embarked on high-tech video and GIS use for in-service inspection [Gerard 91]. Inventory elements include details of gauge width, type of ballast, type of joints, date of construction, and so forth.

7.2.1   Track Quality Index

The evolution of track condition assessment is summarized by Fazio and Prybella [Fazio 80]. In general terms, track quality can be defined as "the ability of the track structure to meet its functional requirements." According to this definition, the type of operations supported by the track will influence the selection of the parameters used to measure quality. Although many railroads have in-house estimators of track quality—such as the "condition index" used by Conrail—existing estimators either lack universality or require the collection of excessive amounts of field data to calculate the parameter for a given track segment.

A measure of track quality, called track quality index (TQI), was developed by the U.S. Federal Railroad Administration (FRA). The TQI statistic can be derived primarily from data collected by automatic track geometry cars. Parameters that can be measured by these cars include gauge, cross level, warp (rate of change of cross level), and alignment. The FRA-owned cars are capable of recording measurements every 0.7 m (2 ft) at speeds up to 242 km/h (150 mph) and carry on-board computers so that an electronic record of the collected data, and a strip chart supplied to maintenance-of-way personnel, are available at the conclusion of the survey. Each of these parameters of track geometry can be used to generate various summary statistics. The TQI parameters also can include other track data, such as bad-tie counts collected by manual inspections. This might be avoided by developing the capability to estimate track modulus (the ability of the track to resist deflection under load) by using automated track geometry cars.

Operating and loading conditions will affect the required level of TQI. For instance, different TQIs would be applicable to tracks that support 80 km/h (50 mph) freight traffic and tracks that support 160 km/h (100 mph) Metroliner service. The FRA-Conrail project focused on operating conditions that are common to many North American freight railroads: mixed freight trains of 50 to 100 cars operating on conventional track (wood tie and stone ballast) at speeds of 64 to 80 km/h (40 to 50 mph). In Europe, the Netherlands Office of Research and Experiments (ORE) also has attempted to develop an easily measurable quantifier of track condition based on automatic track geometry cars [Fazio 80].

7.2.2   Railroad Track Condition Evaluation

The Construction Engineering Research Laboratory (CERL) in the United States has developed a track maintenance management program, RAILER, for inventory, inspection, evaluation, and maintenance/ repair cost analysis. Defects are rated according to five pre-established condition levels: no defects, no restrictions, 10 mph, 5 mph, and no operation. It also allows automated track geometry data input. The maintenance and repair needs are assessed by comparing the visually collected data and rail-flaw inspection information with the railroad track standards [Uzarski 88].

Helicopter based LiDAR and train mounted mobile GPR sensor systems are now also being used to evaluate alignment and structural integrity of track infrastructure.

7.3   In-Service Evaluation of Underground Utilities

The need for locating determining depth below ground and assessing structural integrity of underground utilities is important during any construction or maintenance project. Available nondestructive evaluation methods include the use of seismic, acoustic, electromagnetic, and inertial navigation sensors. Ground-coupled GPR is the mostly widely used equipment and its interpretation has improved over the last two decades.

For monitoring condition of large subways and tunnels, survey speed and data quality are important. Terrestrial LiDAR systems, stationary and kinematic mobile, offer tremendous capabilities by providing 3D maps at 50 to 200 point density per sq. m and georeferenced LiDAR intensity image and/or digital imagery of the entire surface. Additionally, thermal infrared mapping can provide high resolution temperature profiles revealing weak or leaking sections. Extensive published literature is available for NDE evaluation in these underground infrastructure areas [Gikas 12].

7.4   Evaluation of Bridges

The first national bridge management study [Hudson 87] included an in-depth review of the existing NBIS procedures of bridge-structure inventory and condition appraisal (SI&A) standards, identified several improvements in these procedures to enhance in-service evaluation for meaningful M,R&R programming, and recommended a model bridge management system (BMS).

7.4.1   Federal SI&A Data and Limitations

The bridge condition variables, included in the federal SI&A standards, generally exist in the form of a single severity rating for each of the major components of a bridge. These ratings are used during visual bridge inspections. Table 7-2 lists these major components and the detailed condition variables recommended by the BMS study.

The following limitations in SI&A data were identified in NCHRP Report 300 [Hudson 87]:

- Lack of distress extent and type data.
- No flagging mechanism for urgent actions.
- Ratings do not reflect maintenance needs (type and quantity); for example, a deteriorated deck condition may be a costly major maintenance item, but it has little influence on the structural condition severity ratings.

Federal SI&A Rating Group	BMS Recommended Condition Variables	
	Number	**Examples**
Roadway condition rating	8	Deck, wearing surface, joints, drainage, curb/sidewalk/parapet, median barrier, railing, delineation
Superstructure condition rating	9	Main members and connections, floor members and connections, secondary members and connections, expansion and fixed bearings, steel protective coating
Substructure rating	7	Abutments, intermediate supports, collision protection system, steel protective coating, retaining walls, culverts, concrete protective system
Channel and channel protection rating	5	Banks, bed, rip rap, dikes and jetties, sub-structure foundation erosion
Approaches rating	6	Embankments, pavements, relief joints, drainage, guard fence, delineation markers
Estimated remaining life	1	Estimated remaining life
Date and other information inspection	5	Dates of last inspection and unusual inspection features, frequency and date of last unusual inspection, inspector

TABLE 7-2 Summary of Bridge Condition Variables Recommended for an Ideal BMS

- Low precision and reliability of the ratings and their associated definitions, especially in gray areas of intermediate condition ratings (it is difficult to differentiate between needs for minor maintenance, major maintenance, and minor rehabilitation).
- Lack of objective instrumentation that is reliable and measures valid parameters.
- Lack of maintenance items on SI&A forms (e.g., painting is not considered in the inspection).
- Field inspection is geared to rehabilitation and replacement (structural, adequacy, safety considerations), and not to minor or preventive maintenance.
- Inadequate maintenance data, cost-effectiveness data, and inadequate overall condition index for network-level prioritization.

The Federal Sufficiency Rating (*SR*) is based on the SI&A ratings and used to determine the eligibility of bridges for funding under the Federal Highway Bridge Replacement and Rehabilitation Program (HBRRP). Other ratings are also used—load capacity in terms of Inventory Rating (IR) and Operating Rating (OR). The SI&A ratings are made on a scale of 9 (excellent condition) to 0 (worst condition). SR ranges from 100 to 0, and is calculated as follows [Hudson 87b]:

$$SR = S1 + S2 + S3 - S4 \tag{7.3}$$

The variables $S1$, $S2$, $S3$, and $S4$ are defined below:

1. $S1$ = Structural adequacy and safety (55 maximum, 0 minimum)

$$S1 = 55 - (A + I) \tag{7.4}$$

where A is used to deduct values for deterioration, based on superstructure rating and substructure rating, and I is used for reduction for load capacity

$$I = (36 - AIT)^{1.5} \times 0.2778 \tag{7.5}$$

where AIT is adjusted inventory rating.

2. $S2$ = Serviceability and functional obsolescence (30 maximum, 0 minimum)

$$S2 = 30 - [J + (G + H) + I] \tag{7.6}$$

where J shall not exceed 13; $(G + H)$ shall not exceed 15; and I shall not exceed 2. J is a linear function of roadway condition, structural condition, roadway geometry, and underclearance; G and H are functions of roadway width insufficiency; and I is a function of vertical clearance insufficiency.

3. $S3$ = Essentially (15 maximum, 0 minimum)

$$S3 = 15 - (A + B) \tag{7.7}$$

where A and B are reductions for public and military uses, respectively.

4. $S4$ = Special reductions (to be used only when $S_1 + S_2 + S_3$ are equal to or more than 50)

$$S4 = A + B + C \tag{7.8}$$

where A is detour-length reduction; B is structure-type reduction; and C is highway-safety-feature reduction. A and B are reductions for public and military uses, respectively. Maximum value of A, B, or C is 5.

7.4.2 Survey of Bridge Condition Evaluation Methods

The AASHTO manual for condition evaluation of bridges is the result of cooperative research conducted under NCHRP project 12-23 in cooperation with the Federal Highway Administration [AASHTO 94]. The manual is intended to establish standard inspection procedures and load-rating practices. Tables 7-3, 7-4, and 7-5 are based on the

Method Based on	Capability of Defect Detection						
	Cracking	Scaling	Corrosion	Wear and Abrasion	Chemical Attack	Voids in Grout
Strength	Unsuitable	Unsuitable	Poor	Unsuitable	Poor	Unsuitable
Sonic	Fair	Unsuitable	Good[†]	Unsuitable	Unsuitable	Unsuitable
Ultrasonic	Good	Unsuitable	Fair	Unsuitable	Poor	Unsuitable
Magnetic	Unsuitable	Unsuitable	Fair	Unsuitable	Unsuitable	Unsuitable
Electrical	Unsuitable	Unsuitable	Good	Unsuitable	Unsuitable	Unsuitable
Nuclear	Unsuitable	Unsuitable	Fair	Unsuitable	Unsuitable	Unsuitable
Thermography	Unsuitable	Good[*]	Good[†]	Unsuitable	Unsuitable	Unsuitable
Radar	Unsuitable	Good[*]	Good[†]	Unsuitable	Unsuitable	Unsuitable
Radiography	Fair	Unsuitable	Fair	Unsuitable	Unsuitable	Fair

[*]Beneath bituminous surfacing.
[†]Detects delaminations.

TABLE 7-3 Capability of Evaluation Techniques for Concrete Structures [AASHTO 94]

		Capability of Defect Detection								
		Cracks								
Method Based on	Minor Surface	Deeper Surface	Internal	Fatigue	Internal Voids	Welds A	Thick B	Stress C	Blist D	Corr E
Radiography	N*	Fair†	Fair†	Poor	Good	Good	Fair	Fair	Poor	Good
Magnetic particle—wet	Good	Good	N	Good	N	N	N	Good	N	N
Magnetic particle—dry	Fair	Good	N	Good	N	N	N	Fair	N	Poor
Eddy current	Fair	Good	N	N	N	Poor	Poor	N	N	N
Dye penetration	Fair	Good	N	Good	N	N	N	Good	N	Fair
Ultrasonic‡	Poor	Good	Good	Good	Good	Fair	Good	Fair	Fair	Poor

Note: A = porosity and slag in welds; B = thickness; C = stress corrosion; D = blistering; E = corrosion pits.

*N = Not suitable.
†If beam is parallel to cracks.
‡Capability varies with equipment and operating mode.

TABLE 7-4 Capability of Evaluation Techniques for Steel Structures [AASHTO 94]

	Capability of Defect Detection				
Method Based on	**Surface Decay and Rot**	**Internal Decay and Voids**	**Weathering**	**Chemical Attack**	**Abrasion and Wear**
Penetration	Good	Good	Fair	Fair	Unsuitable
Electrical	Fair	Fair	Unsuitable	Unsuitable	Unsuitable
Ultrasonic	Unsuitable	Good	Good	Unsuitable	Unsuitable

TABLE 7-5   Capability of Evaluation Techniques for Timber Structures [AASHTO 94]

AASHTO manual and provide a comparison of the capabilities of the test methods for the evaluation of concrete, steel, and timber structures.

At the 20th World Road Congress of the Permanent International Association of Road Congresses (PIARC) in Montreal in 1995, an international survey of the techniques for assessing in-service condition of bridges was reported [PIARC 95]. The following information is based on a summary of that survey information from Europe, the United States (New Jersey), and Japan.

Inspection intervals vary according to the importance of the bridge, the age, and type of structure (steel or concrete), the degree of damage, and traffic conditions. The number of data items for bridge inspections and condition assessment range from 20 (France) to 250 (United Kingdom). Visual inspections and equipment-aided inspections are carried out in most countries. Inspection walkways or movable platforms are installed as permanent inspection facilities in all countries. Boom vehicles and over-the-fence operation vehicles are also used. The survey included a review of primary damage mechanisms and inspection techniques in the following categories of bridge components: steel members, concrete members, piers, abutments, foundations, and ancillary equipment. Determination of causes of damage is an important part of visual inspection and testing. Types of condition deterioration and inspection methods for these components are summarized in Table 7-6.

Scouring around bridge piers is a major concern of bridge [NCHRP 07]. Remote sensing monitoring of scour depth and damage at piers and abutments include:

- GPS/Total Station Survey
- Underwater photography
- Acoustic mapping with hydrophones (underwater sonar system)
- Green band LiDAR, such as SHOALS (Scanning Hydrographic Operational Airborne LiDAR Survey)

Bridge Component	Deterioration and Damage	Inspection Method
Steel members	(1) Paint-film aging and premature damage caused by rust and corrosion	(1) Visual and photo inspection; tape pull-off and adhesion tests
	(2) Loss of joint bolts or rivets and weld damage; cracks caused by fatigue and corrosion; buckling; brittle fracture	(2) Detection of abnormal noise or vibration; radiograph and ultrasonic testing; vibration tests
Concrete members	(1) Reinforcement corrosion and cracking caused by salt damage	(1) Visual inspection for cracks and all other observable damage
	(2) Surface cracking and corrosion caused by alkali-silica reaction (ASR)	(2) Visual inspection; concrete strength test; gel examination, chemical ASR tests; x-ray analysis (not common) of mortar bar and aggregate
	(3) Reinforcement corrosion and cracking caused by concrete carbonation	(3) Phenolphthalein method to detect carbonation (most common method)
	(4) Prestress steel corrosion caused by deteriorated duct grouping	(4) Drilling and visual inspection; nondestructive testing by x-ray, ultrasonic, acoustic emission
	(5) Concrete slab fatigue cracking, spalling, delamination, and damage	(5) Visual inspection; concrete strength testing; corrosion condition evaluation
	(6) Cracks caused by excessive loading (near the center of span, at quarter point of span, the periphery of supports)	(6) Visual inspection; ultrasonic testing or coring to check crack depth
Piers, abutments, foundations	(1) Deterioration of steel and concrete members	(1) Visual inspection and other methods listed above
	(2) Damage caused by accidents (impact of ships, floating logs, vehicles) or natural disasters (earthquake, floods)	(2) Site and incident specific inspection; seismic evaluation and liquefaction potential
(Critical for overall safety of the bridge)	(3) Deformation and damage caused by ground changes (erosion and scouring, settlement. Tilting, substructure problems)	(3) Frequent monitoring of: erosion and scouring by using poles and drivers and ultrasonic testing; settlement by survey, depression meter, ultrasonic testing; tilting by survey, tilt meter; substructure inspection by test excavation and machine boring

TABLE 7-6 Summary of Bridge Condition Evaluation Methods Based on a Survey of 17 Countries [PIARC 95]

Bridge Component	Deterioration and Damage	Inspection Method
Ancillary equipment	(1) Damage to bearings (restriction to rotation and movement, lack of mobility, rust and corrosion, deformation of bearing seat or anchor bolts or sole plates)	(1) Mostly checked by visual inspections; mobility gages
	(2) Damage to expansion joints (bumps between traveling surfaces, abnormalities in joint openings, loosening of bolts, fracture of fingers, water leaks, abnormal noise and vibrations)	(2) Mostly checked by visual inspection; mobility checks; anchor-bolt torque test; hammer test

Note: Concrete strength by core testing, Schmidt hammer, ultrasonic testing (see Chapter 6 for new and innovative NDE evaluation methods).

TABLE 7-6 Summary of Bridge Condition Evaluation Methods Based on a Survey of 17 Countries [PIARC 95] (*Concluded*)

During the 2005 Hurricane Katrina disaster on the U.S. coast of the Gulf of Mexico several bridges in Louisiana and Mississippi were totally destroyed because of the strong winds and water surges. Important lessons gained from disasters were to: (1) protect levees particularly along approach slabs to avoid creating cavities, which cause failure of approach slabs and (2) tie slab decks down to girders and piers for resistance against strong winds and uplift pressure from floodwater. These lessons are now important considerations in design and construction of bridges in coastal areas.

7.5 Evaluation of Water Pipelines

The following section focuses on leakage, corrosion, and structural deterioration of water mains. O'Day groups water-distribution-system deterioration problems into the following areas [O'Day 84]:

- Water-quality problems related to tuberculation and internal pipe corrosion
- Low-pressure and high head-loss problems caused by main tuberculation
- Distribution system leakage caused by joint leaks, main breaks, and service leaks
- Main breaks caused by main deterioration from internal and/or external corrosion
- Valves and hydrants inoperable caused by neglect and/or unavailable replacement parts

The forces acting on the mains are: (1) internal pressure; (2) external pressure from surrounding earth, truck load, and frost in cold

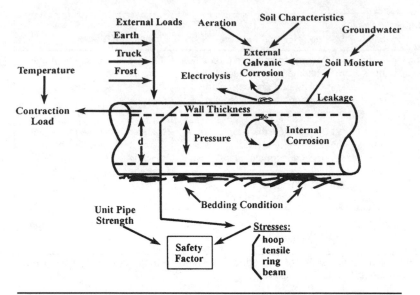

FIGURE 7-4   Conceptual model of water-main structural condition [after O'Day 84].

climates; (3) bending stress if the pipe is not uniformly supported; and (4) thermal stresses if the main is restricted from expansion and contraction. Internal, external, and electrolytic corrosion of pipe material leads to the loss of wall thickness. Soil resistivity and water content play a major role in corrosion rates. Leakage from water mains tends to increase corrosion deterioration. Figure 7-4 shows the factors that affect the forces on the mains and those that affect the loss of pipe wall [O'Day 84].

Leakage and break frequency are important evaluation parameters for water mains. Leakage monitoring is performed by actual flow test, sonic listening. Nondestructive and noncontact testing by infrared thermography is another method to locate areas of potential breaks and leakage.

The following five levels of analysis of leakage and break data are routinely collected by utility companies [O'Day 84]:

1. Analysis of break trends over time to determine the problem areas.

2. Analysis of break patterns by main type (considering material type, diameter, age, joint type, importance of the area, and other urgency factors) to define the main types and areas that experience the greatest rate of failure.

3. Analysis of break patterns by break type to find out the break type that is linked to main diameter, age, material, and location. This should be recorded at the time of the repair, to know the cause of structural deterioration and failure. Circular breaks

are generally common on small-diameter (equal to or less than 25.4 mm or 10 in) pipes, and longitudinal breaks are frequently recorded on large-diameter pipes.

Break Type	Cause of Structural Failure
Circumferential	Thermal contraction, bending
Longitudinal	Excessive crushing load
Hole	Internal pressure
Bell crack	Thermal contraction, joint material expansion

4. Visual inspection of break samples is used to determine the source, extent, and depth of corrosion.

5. Microscopic examination of break samples can identify the corrosion type by the microstructure analysis of the metal.

7.6 Evaluation of Gas Pipelines

Natural gas is transported under high pressure through steel pipelines with protective coating. The Natural Gas Pipeline Safety Act of 1968, the Hazardous Liquid Pipeline Act of 1979, and the Pipeline Safety Reauthorization Act of 1988 require the U.S. Department of Transportation (DOT) to develop and enforce safety regulations [GAO 92]. There are more than 2.6 million km (1.6 million mi) of natural gas pipelines in the United States, including 427,000 km (267,000 mi) of interstate natural gas pipelines. According to a report by the General Accounting Office (GAO), the safety of these pipelines is becoming a concern as they age; for example, 9 percent of the network was constructed before 1940, 10 percent before 1950, and most were constructed during the 1950s and 1960s. The following discussion is based on the GAO report.

From 1985 to 1991, 1726 natural-gas pipeline accidents involving 131 fatalities and 634 injuries were reported. The leading cause of pipeline failure is outside-force accidental damage caused by outside force such as a third-party excavation; the second cause is internal and external corrosion. The pipeline must be protected during its service life from damage and degradation, and this is generally accomplished by external corrosion controls and inspection techniques. The pipeline inspection and evaluation methods include:

- Visual inspection, line walking, line walking with a flame detector, portable ultrasonic sensor to check the structural integrity, and use of light aircraft or helicopter to check for evidence of leaking, such as dying vegetation, water bubbles, and cavities in the ground

- Hydrostatic pressure testing of the pipeline by forcing water into a segment of pipeline and monitoring its resistance against rupture

- X-ray analysis to check pipe welds
- Placing an instrumented device, such as smart pig, inside the pipe to record flaws as the instrument is transported through the pipe by the natural gas

Smart pigs have been used since the 1960s. They carry ultrasonic or magnetic flux leakage-measuring instruments to monitor internal and external corrosion and other flaws. This technology is very cost-effective and does not require a break in service for hydrostatic pressure testing. The GAO report provides details about conventional and smart-pig technologies for inspection and testing of gas pipelines [GAO 92].

7.7   Evaluation of Buildings

Buildings are the most vital infrastructure assets for proper functioning of public life at private, commercial, and public levels. Proper functioning of buildings requires inspections and maintenance of structural aspects, as well as heating and cooling, ventilation, water and drainage, emergency and security systems, horticulture, parking, and other nonstructural components. Unfortunately, as-built records of building construction and M,R&R actions generally are not adequately preserved and accessible.

7.7.1   Example of Building Condition Rating and Evaluation

The inspection and building assessment approach also depends upon the insight and management approach of owners, operators, and regulating agencies. Table 7-7 shows a public facility rating form used by the U.S. Postal Service for its decision analysis [after Urban 84].

Building components have wide ranges of life expectancy, depending upon the material, design and construction practices, soil and environment, and occurrences of natural hazards. By far, fire has been the most common type of accidental hazard to ordinary buildings.

7.7.2   Postearthquake Evaluation

Earthquakes are important natural disasters that are difficult to predict. Figure 7-5 shows areas of significant earthquakes in the United States. The plot shows contours of effective peak acceleration (expressed as a decimal fraction of gravity, g) that might be expected (with odds of one in ten) to be exceeded during a 50-year period [FHWA 81, Buckle 87]. New Madrid fault zone in the central and southern United States experienced most severe earthquakes in the 1800s, and it is easy to assess such regional impacts using geospatial evaluation. Figure 7-6 shows a spatial seismic risk map of the region with 46.7 million people in eight states, for an affected 200 km radius around the fault epicenter.

General Mail Facility—Working Conditions Summary, Lancaster, Pennsylvania										
	Condition			Area of Interest						
Components	**A**	**P**	**I**	**H**	**M**	**Sa**	**Se**	**E**	**C**	**Remarks**
Air conditioning			X	X				X		Window units only
Electrical		X		X						Octopus system difficult to maintain
Heating	X									Recent major modification
Lighting			X				X	X		Office levels of 20–30 ft candles
Plumbing		X						X		Reduced flows
Offiooc		X						X		Lighting, HVAC
Lobl ıy		X				X			X	Lighting levels low
Lockers			X	X		X	X	X		All-round deficiencies
Lunch room				X					X	All-round deficiencies
Rest rooms		X		X				X		Deficient by criteria
Support area	X									—
Work room		X					X	X		Poor lighting levels
Platform			X				X	X		Canopy repairs required
Parking area			X					X	X	No carrier parking
Maneuvering area			X							Lack of sufficient depth
Structure	X									—
Miscellaneous	X									—

Note: A = adequate; P = poor; I = inadequate; H = health; M = maintenance; Sa = safety; Se = security; E = employee accommodation; and C = customer accommodation.

TABLE 7-7 Example of U.S. Postal Service Public Facility Rating Form [after Urban 84]

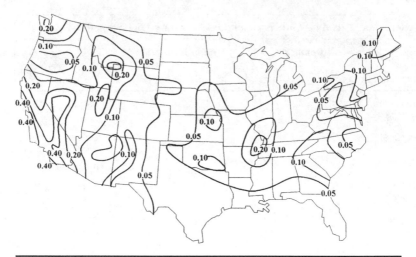

FIGURE 7-5   Areas of significant earthquakes in the United States [adapted from FHWA 81, Buckle 87].

Postearthquake safety evaluation, posting, and condemnation of buildings are now desired by federal and state emergency management agencies, and staff members of concerned agencies are being trained in earthquake-prone areas. This can be a mammoth task for a large urban area; for example, San Francisco has over 150,000 buildings, 30 building inspectors, and 15 civil planning and structural engineers working with the government agency. If 10 percent of buildings need inspection, then the buildings/inspector ratio is 300 to 1 [ATC 93]. The Applied Technology Council (ATC) training manual, prepared with funding from the U.S. Federal Emergency Management Agency (FEMA), outlines some of the key concerns, as summarized below:

- *Principal safety concerns* (collapse, falling hazards, geotechnical hazards, hazardous materials, field safety)
- *Posting system* (timely, consistent, visible, authority of jurisdiction); inspected (green)—appears safe for lawful occupancy; limited entry/restricted use (yellow)—some restriction on use, controlled by tenant; unsafe (red)—entry controlled by jurisdiction
- *Evaluation procedures* (windshield survey to assess scope of damage; rapid, to assess obvious and smaller buildings and recommend detailed evaluation where needed; detailed, to assess difficult and large buildings closely; and engineering, using consultants engaged by owner)
- *Evaluation steps* (for rapid and detailed evaluation)
  1. Examine entire outside of building.
  2. Examine ground for distress.
  3. Enter if safe and continue inspection.

FIGURE 7-6 Seismic risk map of New Madrid fault zone in the United States (spatial map credit: Carrissa Beasely/W. Uddin, The University of Mississippi, May 2012).

4. Discuss observations; evaluate by criteria.
5. Post building.
6. Inform occupants of hazards.

- *Detailed evaluation of essential facilities* (health-care facilities, police and fire stations, jails and detention centers, emergency operations centers, high-occupancy shelters)
- *Detailed evaluation criteria* (for hazardous condition)
  1. Vertical load capacity not significantly decreased
  2. Lateral load capacity not significantly decreased
  3. No falling or other hazards present
  4. No evidence of foundation damage or ground displacement
  5. Main exits are useable
  6. No other unsafe condition
- *Hazards of nonstructural elements*
  1. Parapets, chimneys, ornaments
  2. Cladding and glazing
  3. Partitions
  4. Suspended ceilings, raised floors
  5. Piping and duct work
  6. Equipment, furnishings
  7. Portable buildings
- *Geotechnical hazards* (increased ground shaking, liquefaction, subsidence, landslide)

7.7.3   ASCE Standards

The American Society of Civil Engineers (ASCE) Committee on Structural Condition Assessment and Rehabilitation of Buildings developed standard guidelines for structural condition assessment in 1990 [ASCE 90, Pillared 96] with the support of the Buildings and Fire Research Laboratory of the National Institute of Standards and Technology (NIST). The ASCE standard was the first coordinated effort to provide guidelines for assessing the structural condition of existing buildings constructed of concrete, masonry, metals, and wood. Due to potential costs, a multilevel approach is recommended, which starts with a preliminary assessment for initial structural adequacy and establishes the need and priority for detailed assessment. This is followed by a detailed assessment that recommends a course of action and alternatives. Post-September 11, 2001 investigations by NIST showed the vulnerability of concrete-steel composite structured frames used in the World Trade Center Building 7 in New York City, which collapsed under extreme fire, failure of sprinkler systems, and super heat conditions [NIST 03].

Another ASCE subcommittee is preparing guidelines for condition assessment of building envelopes. Building envelope is defined as "the exterior surface of a building providing protection from weather." The proposed standards work, which was motivated by concern for public safety, outlines the following six envelope categories [Pillared 96]:

- Roof systems are designed to weatherproof and insulate the top surface of a building.
- Balcony systems are exterior platforms supported at the exterior walls of a building.
- Plaza deck systems are structures with a waterproof membrane applied to a horizontal surface that is intended to receive vehicular or pedestrian traffic.
- Bearing wall systems support gravity and vertical loads.
- Nonbearing wall systems are not intended to support vertical loads other than their own; for example, curtain and panel walls.
- Foundation wall system is the part of the building in contact with the soil. It is usually a load-bearing wall, often with waterproof coating and drainage components.

The ASCE Subcommittee on Seismic Rehabilitation of Buildings was established to assist in the consensus review of the seismic rehabilitation document ATC-28, prepared by the Structural Engineers Association of California with funding from FEMA [Pillared 96]

7.8 Landuse Evaluation for Heat-Island Effects and Flood Risk Assessment

Man-made infrastructure surfaces such as buildings and asphalt constructed surfaces absorb more heat than natural ground surfaces caused by the low solar reflectivity of darker surfaces. This heat-island effect causes an increase in surface temperature and air temperature in urban areas. Urban sprawl and associated transportation-related emissions also tend to increase the area temperature. Built surfaces increase energy demand (to cool) and heat-island effects, adversely impact air quality, and produce greenhouse gases that contribute to global warming [Uddin 09]. Accurate landuse maps help assess these heat-island effects to develop mitigation plans such as increased green spaces and "cool" and "green" construction materials for building roofs and pavements.

Natural disasters require rapid response to the most devastated areas to save lives. Recovery and restoration planning requires accurate quantitative information about damaged infrastructure such as commercial facilities, housing stock, disrupted transportation network (roads and bridges, railways), and storm debris, and eroded materials [Uddin 12]. Accurate landuse maps are required for assessment of flood risk maps.

Geospatial and GIS maps, used to assess heat-island effects and flood risks, were created traditionally using intensive field surveys or by the U.S. National Land Cover Datasets (NLCD) at 15 or 30 m ground resolution. The NLCD maps are out of date and inadequate for detailed engineering analysis [Uddin 12]. Current worldwide coverage of high resolution satellite imagery, makes it feasible to create accurate landuse classification maps to estimate heat-island effects and flood disaster risks.

7.9   References

[AASHTO 94] *Manual for Condition Evaluation of Bridges*, American Association of State Highway and Transportation Officials (AASHTO), Washington, DC, 1994.

[ACRP 10] Airport Cooperative Research Program (ACRP), "Light Detection and Ranging (LIDAR) Deployment for Airport Obstructions Survey," *Research Results Digest 10*, Transportation Research Board, The National Academies, Washington, DC, July 2010.

[ASCE 90] *ASCE 11-90, Standard Guidelines for the Structural Condition Assessment of Existing Buildings*, American Society of Civil Engineers, New York, 1990.

[ATC 93] Applied Technology Council, "Postearthquake Safety Evaluation of Buildings Training Manual," *ATC-20-T*, funded by the Federal Emergency Management Agency, Washington, DC, 1993.

[Buckle 87] I. G. Buckle, R. L. Mayes, and M. R. Button, "Seismic Design and Retrofit Manual for Highway Bridges," *Report FHWA-IP-87-6*, Federal Highway Administration, McLean, Va., May 1987.

[Carey 60] W. N. Carey, and P. E. Irick, "The Pavement Serviceability—Performance Concept," *Highway Research Bulletin 250*, Highway Research Board, National Research Council, Washington, DC, 1960.

[FAA 82] "Guidelines and Procedures for Maintenance of Airport Pavements," *Advisory Circular AC: 150/5380-6*, Federal Aviation Administration, Washington, DC, 1982.

[Fazio 80] A. E. Fazio, and R. Prybella, "Development of an Analytical Approach to Track Maintenance Planning," in *Transportation Research Record 744*, Transportation Research Board, National Research Council, Washington, DC, 1980, pp. 46–52.

[FHWA 81] "Seismic Design Guidelines for Highway Bridges," Applied Technology Council, *Report FHWA/RD-81/081*, Federal Highway Administration, Washington, DC, 1981.

[FHWA 89] "Pavement Management Systems, A National Perspective," *PAVEMENT Newsletter*, Federal Highway Administration, U.S. Department of Transportation, Issue 14, Spring 1989.

[GAO 92] "Natural Gas Pipelines—Greater Use of Instrumented Inspection Technology Can Improve Safety," GAO Report to Congressional Committees, *Report No. GAO/RCED-92-237*, United States General Accounting Office, Washington, DC, 1992.

[Gerard 91] S. V. Gerard, "UP Links Video and Graphic Images for Greater Clarity," *Railway Track and Structures*, April 1991.

[Gerardi 96] T. Gerardi, "The Importance of Maintaining Smooth Airport Pavements," *Proceedings, 24th International Air Transportation Conference*, Louisville, Ky., pp. 295–305.

[Gikas 12] Vassilis Gikas, "Three-Dimensional Laser Scanning for Geometry Documentation and Construction Management of Highway Tunnels during Excavation," *Sensors 2012*, Vol. 12, No. 8, pp. 11249–11270.

[Haas 78] R. C. G. Haas and W. R. Hudson, *Pavement Management Systems*, McGraw-Hill, New York, 1978.

[Haas 94] R. Haas, W. R. Hudson, and J. P. Zaniewski, *Modern Pavement Management*, Krieger Publishing Company, Malabar, Fla., 1994.

[HRB 62] "The AASHO Road Test, Report 5—Pavement Research," *Special Report 61E*, Highway Research Board, National Research Council, Washington, DC, 1962.

[Hudson 87] S. W. Hudson, R. F. Carmichael III, L. O. Moser, W. R. Hudson, and W. J. Wilkes, "Bridge Management Systems," *NCHRP Report 300*, Transportation Research Board, National Research Council, Washington, DC, 1987.

[Hudson 87a] W. R. Hudson, and W. Uddin, "Future Evaluation Technologies: Prospective and Opportunities," *Proceedings, 2nd North American Pavement Management Conference*, Toronto, Ont., 1987.

[Hudson 87b] W. R. Hudson, C. Boyce, and N. Burns, "Improvements in On-System Bridge Project Prioritization," *Research Report 439-1*, Center for Transportation Research, The University of Texas at Austin, 1987.

[Hudson 97] W. R. Hudson, R. Haas, and W. Uddin, *Infrastructure Management*, McGraw-Hill, New York, 1997.

[NCHRP 07] National Cooperative Highway Research Program, "Countermeasures to Protect Bridge Piers from Scour," *REPORT 593*, Transportation Research Board, Washington, DC, 2007.

[NIST 03] National Institute of Standards and Technology, "NIST Response to the World Trade Center Disaster: World Trade Center Investigation Status," U.S. Department of Commerce, December 2003.

[O'Day 84] D. K. O'Day, "Aging Water Supply Systems: Repair or Replace," *Infrastructure Maintenance and Repair of Public Works, Annals of the New York Academy of Sciences*, Vol. 431, December 1984, pp. 241–258.

[OECD 87] *Pavement Management Systems*, Road Transport Research, Organization for Economic Cooperation and Development (OECD), Paris, France, 1987.

[PIARC 95] "Road Bridges," *Reference 20.11.D, Report of the Committee, 20th World Road Congress, Permanent International Association of Road Congresses (PIARC)*, Montreal, P.Q., September 1995.

[Pillared 96] J. Pillared, C. Baumert, and M. Green, "ASCE Standards on Structural Condition Assessment and Rehabilitation of Buildings," *Standards for Preservation and Rehabilitation, ASTM STP 1258*, American Society for Testing and Materials, Philadelphia, Pa., 1996, pp. 126–136.

[Shahin 90] M. Y. Shahin and J. A. Walter, "Pavement Maintenance Management for Roads and Streets Using the PAVER System," *USACERL Technical Report M-90/05*, Champaign, Ill., 1990.

[Uddin 86] W. Uddin, A. H. Meyer, and W. R. Hudson, "Rigid Bottom Considerations for Nondestructive Evaluation of Pavements," in *Transportation Research Record 1070*, Transportation Research Board, National Research Council, Washington, DC, 1986, pp. 21–29.

[Uddin 96] W. Uddin, R. M. Hackett, P. Noppakunwijai, and Z. Pan, "Three-Dimensional Finite-Element Simulation of FWD Loading on Pavement Systems," *Proceedings, 24th International Air Transportation Conference*, Louisville, Ky., 1996, pp. 284–294.

[Uddin 97] W. Uddin and R.M. Hackett, "Three-Dimensional Finite Element Modeling of Vehicle Crashes Against Roadside Safety Barriers," *IJCrash 1999—International Journal of Crashworthiness*, Woodhead Publishing Ltd, Vol. 4, No. 4, 1999, pp. 407–417.

[Uddin 05] W. Uddin, "Chapter 18—Pavement Management Systems," *CRC Handbook of Highway Engineering*, Editor T.F. Fwa, ISBN 9780-84931-9860, Pearson-CRC Press, Inc, September 2005.

[Uddin 09] W. Uddin, Carla Brown, E. Scott Dooley, and Bikila Wodajo, "Geospatial Analysis of Remote Sensing Data to Assess Built Environment Impacts on Heat-Island Effects, Air Quality and Global Warming," *Paper No. 09-3146, CD Proceedings, 85th Annual Meeting of Transportation Research Board*, Washington, DC, January 11–15, 2009.

[Uddin 11a] W. Uddin, "Transportation Management: LiDAR, Satellite Imagery Expedite Infrastructure Planning," *Earth Imaging Journal*, January/February 2011, pp. 24–27.

[Uddin 11b] W. Uddin, "Remote Sensing Laser and Imagery Data for Inventory and Condition Assessment of Road and Airport Infrastructure and GIS Visualization," *International Journal of Roads and Airports (IJRA)*, Vol. 1, No. 1, 2011, pp. 53–67.

[Uddin 12] W. Uddin and Katherine Osborne, "A Geospatial Methodology for Rapid Assessment of Disaster Impacts on Infrastructure," *Online Proceedings, 91st Annual Meeting of Transportation Research Board*, Washington, DC, January 22, 2012.

[Uddin 13] W. Uddin and Catherine Colby Willis, "Airborne Laser Surveys for Expediting Airport Obstruction Mapping," *Journal of Airport Management*, London, U.K., Vol. 7, No. 2, Spring 2013, pp. 179–194.

[Urban 84] H. P. Harty and B. G. Steinthal, *Guides to Managing Urban Capital Series, vol. 4: Guide to Selecting Maintenance Strategies for Capital Facilities*, The Urban Institute, Washington, DC, 1984.

[Uzarski 88] D. R. Uzarski and D. E. Plotkin, "Interim Method of Maintenance Management for U.S. Army Railroad Track Network," in *Transportation Research Record 1177*, Transportation Research Board, National Research Council, Washington, DC, 1988, pp. 84–94.

CHAPTER **8**

Performance Modeling and Failure Analysis

8.1 Performance Evaluation

8.1.1 Performance Concepts

An infrastructure facility is intended to serve its owners and users at an acceptable level of quality. The constructed level of quality usually does not stay constant over time but deteriorates throughout the service life. Based on the discussions of service life and performance in Chapter 3, a general definition of *performance* related to an infrastructure facility would be "the degree to which a facility serves its users and fulfills the purpose for which it was built or acquired, as measured by the accumulated quality and length of service that it provides to its users" [Hudson 97]. Figure 8-1 illustrates the concept of *performance*, in which performance is represented by the area under a time series of a quality measurement, such as condition index (CI), on a scale of 100 (such as for a new facility in the best possible condition) to 0 (totally unacceptable). Of course, other scales may also be used, depending on the facility, the measure, and other factors. A common scale is the 0 to 5 scale of the Present Serviceability Index (PSI) for pavements.

In-service evaluation and condition assessment of infrastructure facilities, shown in Figure 8-1, were discussed in detail in Chapters 6 and 7. The area under each curve in Figure 8-1 represents the accumulated service or performance. Using this graphical interpretation, several concepts associated with performance are identified:

- Initial condition, level of service, and quality are near the top of the scale in Figure 8-1 (i.e., 100), similar to a newly constructed facility after successful acceptance testing and commissioning. A minimum-acceptable level must also be set.

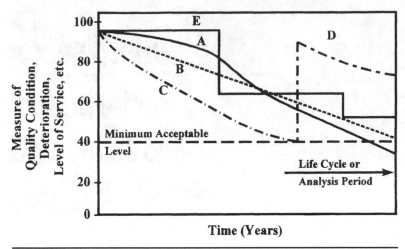

FIGURE 8-1 Conceptual illustration of several forms of *performance* curves.

In Figure 8-1 this is illustrated as 40, but of course it will vary with many factors, such as class of facility, agency policy, safety, and economics.

- There is a deterioration of condition with time. This may reflect not only deterioration in physical condition, but also level of service, quality, function, structural capacity, and/or safety.

- The facility may or may not reach a minimum-acceptable level before the end of the life cycle or analysis period. This can happen because of continuing deterioration, functional obsolescence, safety reasons, or a catastrophic event like fire, flood, earthquake, or any other disaster.

- The slope and form of the condition deterioration history determines the performance. A good-performing facility should provide a high service level and should remain in relatively good and acceptable condition for most of the service life, such as shown by Curve A in Figure 8-1 (curvilinear form), Curve B (linear form), or Curve E (discontinuous form). Curve C shows a poorer-performing facility caused by a relatively higher deterioration rate in the initial years.

- The in-service evaluation and review of performance trends can lead to an effective intervention policy for M,R&R based on a preselected level of deterioration. A major restoration and rehabilitation intervention can extend the service life of a facility, as Curve D in Figure 8-1 does to Curve C.

A facility is the result of interrelated components, designed and built in a logical sequence, that deliver a final product that serves its intended purpose. Deterioration of physical condition is a complex process, as shown by wear and aging caused by loading history and use, degradation of materials of construction as affected by the environment, and the interaction of these mechanisms. Traditional design practices have relied on a preselected service life using safety factors to take into account uncertainties in design and demand inputs, without the concept of performance, as shown by the curves in Figure 8-1. This old practice is unrealistic, and appropriate performance measures should be identified and monitored so that historical data can be used to develop and improve performance-prediction models.

Sustainability, reduced carbon footprint, and resilience against natural disasters are three more areas requiring attention for appropriate performance measures and modeling.

8.1.2   Deterioration Process

A number of factors involved in the physical deterioration of a facility are listed in Table 6-1 in five major categories: load, environment, material degradation, construction quality, and interaction affects and other mechanisms. The table also lists primary and secondary condition deterioration mechanisms associated with possible surface defects, deformation, cracking or breaks, and failures. Deterioration is a function of properties of the materials used in the facility, demand on the facility, and the operating environment. The properties of a material can be grouped into physical, chemical, mechanical, thermal, and hydraulic. The behavior of the material can be further characterized under static-loading and dynamic-loading conditions. Chemical actions and additives associated with a given environment can produce drastic changes in these properties. For example, addition of lime and fly ash to cement in ordinary Portland cement concrete can markedly increase the compressive strength. Polymer modifiers in asphalt can significantly decrease creep strain and therefore improve rutting resistance of asphalt pavements. Special alloys can reduce steel corrosion. On the other hand, aggregates showing alkali-silica reaction will have durability problems. Saline environments and the use of salt on bridge decks lead to corrosion in steel and reinforced concrete structures. Wood structures have problems related to water and termites. It is possible to develop reliable material models and predict the performance of a material under specific operating conditions and environment. An empirical model can be developed based on the accumulation of experience with that material, such as wood, steel, and concrete in different environments.

To develop a performance model, it is imperative to identify the deterioration mechanisms and deterioration timeline. For many situations a curvilinear, S-shaped curve, as shown by Curve S in Figure 8-2, can be used to define performance. Various parts of the

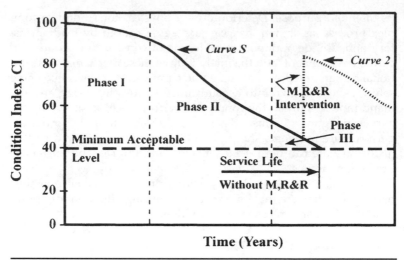

Figure 8-2   Effect of M,R&R action on performance.

S-shape characterize the deterioration process. Based on the systems concepts for pavement management [Haas 94] and for building assessment [Harris 96], the deterioration process and performance curve can be characterized by three distinct phases of the facility life in the absence of any major M,R&R treatment. The phases I, II, and III are shown in Figure 8-2 for a typical performance curve, Curve S. It has a small slope in phase I, followed by accelerated deterioration in phase II, and then a decelerated deterioration rate in phase III (which is often the result of increased maintenance). Throughout the life the slope is never zero or negative, and the upper and lower bounds are known.

Phase I of the performance curve commences immediately following construction, or when all the components are new, or after a major M,R&R as in the case of Curve 2. The slope of the curve and time period in this phase will influence the overall performance. A smaller slope and longer period indicates good performance. The rate of deterioration and duration of phase I depends upon the inherent properties of the construction materials used, the quality of construction, demand on the facility, and the interaction with the operating environment. The operators of the facility would like to see this period equal to the expected design life, ideally with minimum maintenance. A higher rate of deterioration will indicate poor performance and the possibility of early "failure" before reaching the design life. A sharp increase in the rate of deterioration indicates the start of phase II and possible deterioration of one or more structural components. Without appropriate maintenance and repair, the deterioration may accelerate in a relatively short period to the minimum-acceptable level. This initiates phase III—decelerating deterioration with incipient functional and structural failure. The condition is generally stable because it cannot get much worse. Some examples of this last phase are a badly cracked and potholed road, a silted dam, a nearly unusable building, and an unsafe tunnel. An earlier intervention, at the beginning of phase III as shown in Figure 8-2, usually would be desirable. If deterioration is allowed to continue to the end of phase III, reconstruction or replacement may be the only effective option.

Intervention analysis is needed for catastrophic failures caused by natural disasters (such as bridge failures during August 2005's Hurricane Katrina on the U.S. Gulf Coast and the nuclear plant destruction after Japan's record earthquake and tsunami in March 2011) and man-made disasters (such as the destruction of the World Trade Center in New York City after the terrorist attacks on September 11, 2001).

8.1.3 Need for Performance Evaluation

Performance evaluation of a facility in an infrastructure management system is essential to:

- Assess the service quality and functionality of the facility related to the materials and construction methods used
- Predict the rate of deterioration and possible failure of the facility
- Verify performance predictions and improve prediction models
- Measure the effectiveness of M,R&R strategies and the value of extended life
- Update and improve life-cycle cost-and-benefit analysis methodologies
- Schedule M,R&R strategies and future in-service inspections and evaluations
- Plan M,R&R budgets for short- and long-term work programs and update network improvement programs

The development and study of performance models have received major attention from public works agencies, which are responsible for preserving the transportation infrastructure and providing expected service to the traveling public at local, state, regional, and national levels. In particular, a number of pavement-performance models have been developed over the last three decades [Haas 94], and a national pavement-performance database has been developed through the Strategic Highway Research Program (SHRP) conducted from 1987 to 2002 [SHRP 92, Uddin 95].

8.1.4 Service Life

Prediction of service life is an important and fundamental aspect of performance modeling. This may or may not coincide with the design life. For an infrastructure facility constructed with different materials for different structural components and subjected to varying demand and environment conditions, service-life prediction can be complex. Therefore, a facility must have to be partitioned into several parts or segments based on traffic, material, construction, and operating conditions. A logical partitioning of the facility may lead to predictable estimates of expected service life in terms of functional life. Traffic marking and signposts on a road, painting of steel girders and other components of steel in bridges, and painting of the exterior of residential buildings are examples of functional life span that can be established using an empirical database. It may not be necessary to establish a performance equation for these examples. For many widely used materials and construction methods, the expected service-life estimate is available from prior experience. However, the overall service life of a facility is a complex matter. Service life is the period of time from the completion of the facility to the date that the facility or any of its components reach a state where the facility cannot provide acceptable service. Some typical expectations of infrastructure service life based on Table 2-1 are listed in Table 8-1. These lives are mostly 100 years or less, with the exception of landfills that are legally mandated to be safe for 250 years [Krinitzsky 93]. Table 3-2 shows

Infrastructure Facility and Components		Expected Service Life
Airports	Buildings/structures	50–70 years
	Runways/taxiways/aprons	40–50 years
Bridges	Decks	20–30 years
	Substructure/ superstructure	50–100 years
Dams, tunnels	For auto traffic, water	80–100 years
Nuclear power plants	—	20–40 years
Ports, marine terminals	—	70–100 years
Public buildings	—	30–100 years
Electricity transmission/ telephone lines	—	50–100 years

TABLE 8-1 Typical Expectations of Infrastructure Service Life

a comparison of service lives of buildings based on codes and recommendations extracted from relevant documents from Japan, Canada, and the United Kingdom.

Evaluation of overall service life of infrastructure assets should be based on the life of critical structural components. Service life can be estimated from: (1) empirical experience; (2) historical databases using survivor techniques; (3) established performance models; (4) laboratory testing; and (5) accelerated field testing. The prevailing load, material, and method of construction, as well as environmental, ground, and operating conditions should always be considered to estimate the service life.

8.2 Performance Modeling

Performance modeling is an important part of infrastructure management at both project and network levels. A performance model relates a selected performance indicator to a set of causal variables such as age, load, load repetitions, usage history, material properties, environmental factors, and M,R&R history.

8.2.1 Performance Indicators

Overall functional performance of a facility is generally assessed from the user perspective, with such measures or indicators as quality or level of service, system effectiveness, productivity and efficiency, and resource utilization and cost-effectiveness [Urban 84]. On the other hand, the technical evaluation of performance is vital to engineers because it includes measures of mechanistic behavior and physical deterioration that lead to the selection of correct M,R&R alternatives. Therefore, it is important to include both user and engineer evaluation to formulate a comprehensive performance indicator. Humplick and Paterson recommend that the performance indicators for road-transportation infrastructure should include measures of sufficiency, efficiency, and effectiveness, as well as institutional aspects of effectiveness [Humplick 94]. Performance indicators can be grouped into the following broad categories:

- Service and user rating
- Safety and sufficiency
- Physical condition
- Structural integrity/load-carrying capacity

To establish an overall performance indicator for an infrastructure facility, the highest weight should be given to structural integrity and service to users. Some examples of performance indicators are provided in Chapter 3. Table 8-2 shows specific examples of performance indicators as related to the categories just listed.

	Performance Indicators			
Facility	**Service and User Rating**	**Safety, Sufficiency, and Security**	**Physical Condition**	**Structural Integrity/ Load Capacity**
(1) Highways, roads, streets, parking areas	Present serviceability rating (PSR), ride quality, vehicle-operating costs	Ratings based on skid resistance, crashes/accidents, congestion, pollution, and/or security	Present serviceability index (PSI), international roughness index (IRI), Pavement Condition Index (PCI) based on distresses	Structural rating based on deflection testing, remaining life, load capacity
(2) Airports: pavements, buildings, aviation facilities	Ride quality, user-satisfaction ratings for other facilities	Operating ratings for aircraft, passengers and baggage, flow and capacity, and security	PCI (pavements), ride-quality index (RQI); see (8) for buildings, air-traffic-control facilities, air-port lighting, etc.	Structural rating based on deflection testing, remaining life, load capacity; see 8 for buildings and other facilities
(3) Bridges: interchange, overpass, underpass, rail/river crossing	Inventory and appraisal ratings for deck and other components	Ratings for safety, capacity, sufficiency, congestion, and security	Ratings based on distress and appraisal data, overall condition rating	Ratings of structural components based on visual inspection, vibration and seismic testing, nondestructive evaluation, load rating, and remaining life
(4) Railroad: terminals, track, bridge	Track-quality rating, user satisfaction ratings	Ratings for safety, crossing accidents, operations, capacity, and security	Track-quality index (TQI), track-structure condition index (TSCI); see (3) for bridges and (11) for buildings	Track modulus, load rating, remaining life; see (3) for bridges and (11) for buildings
(5) Waterborne transport: ports, inland terminals, intermodal facilities	Container-handling rating, intermodal efficiency rating, track/ pavement quality rating, user satisfaction ratings	Ratings for safety, barge/bridge accidents, container/intermodal operations, capacity, and security	Quality index (QI) and condition index (CI) for each asset type; see (3) for bridges and (11) for buildings	Load rating and remaining life for each asset type; see (3) for bridges and (11) for buildings

TABLE 8-2 Examples of Performance Indicators

Facility	Performance Indicators			
	Service and User Rating	Safety, Sufficiency, and Security	Physical Condition	Structural Integrity/Load Capacity
(6) Mass transit: subway and metros, tracks, stations, vehicle stocks, maintenance facilities	Track-quality rating, user satisfaction ratings	Ratings for safety, crossing accidents, operations, capacity, and security	Pavement/track-quality index (QI), pavement CI, track-structure condition index (TSCI); see (3) for bridges and (11) for buildings	Load rating and remaining life for each asset type; see (1) for roads/parking area, see (3) for bridges and (11) for buildings
(7) Supply chain: container terminals, manufacturing sites, trucks and intermodal fleet, distribution centers, retail stores	Container-handling rating, intermodal-efficiency rating, track/pavement quality rating, user satisfaction ratings	Ratings for safety, accidents, container/intermodal operations, capacity, and security	QI and CI for each asset type; see (1) for roads/parking area, (3) for bridges and (11) for buildings	Load rating and remaining life for each asset type; see (1) for roads/parking area, see (3) for bridges and (11) for buildings
(8) Water: treatment plants, storage, pumping stations, mains and supply-pipe network	Water quality, user satisfaction ratings	Ratings for health hazard, water pressure, supply, consumption, operations, capacity, and security	Number of breaks per year, leakage per year, and number of repairs per year	Failures, damage to roads and other structures, remaining life, and demand
(9) Wastewater and sewer: sewer lines, treatment plants	Treatment quality, user satisfaction ratings	Ratings for health hazard, discharge, operations, capacity, and security	Number of breaks per year, leakage per year, and number of repairs	Failures, damage to roads and other structures, remaining life, and demand
(10) Solid-waste facilities: landfills, methane gas collection and energy production	Odor and treatment, user satisfaction ratings	Ratings for health hazard, fire hazard, discharge, operations, capacity, and security	Ratings for spilled trash, missed collections, soil and water contaminations, and facility condition	Failures, damaged and inadequate collection and disposal facilities, remaining space, and demand

TABLE 8-2 Examples of Performance Indicators (*Continued*)

Performance Indicators				
Facility	Service and User Rating	Safety, Sufficiency, and Security	Physical Condition	Structural Integrity/Load Capacity
(11) Buildings: public, commercial, industrial, multipurpose complexes, landmarks	Discomfort and user satisfaction ratings	Ratings for fire hazard, utilities, occupancy, operations, capacity, and security	Interruptions in utility services, heating and cooling effectiveness, air-quality rating, and condition ratings	Structural and ground failures, posting, remaining life, safety hazards, and demand
(12) Electric power-supply infrastructure: traditional and renewable	Ratings for interruptions in service, and user satisfaction ratings	Ratings for fire hazard, operations, and power generated and distributed, grid capacity, and security	Frequency of service breaks, power surges, and condition ratings of fixed assets	Failures, remaining life, safety, and demand
(13) Telecomm-unications and wireless infrastructure: IT and Internet networking facilities	Ratings for interruptions in service, and user satisfaction ratings	Ratings for fire hazard, operations, and protection from poor or non-coverage, capacity, and security	Frequency of service breaks, power surges, and condition ratings of fixed assets	Failures, remaining life, safety, and demand

TABLE 8-2   Examples of Performance Indicators (*Concluded*)

8.2.2 Modeling Techniques

Performance models can be developed by a variety of techniques, including the following:

- Expert system incorporating a knowledge base of empirical experience
- Regression analysis and time series analysis
- Markov-transition probabilities
- Artificial neural-networks analysis
- Bayesian methods
- Econometrics methods
- Geospatial methods

Expert systems and regression-analysis techniques were discussed in Sec. 4.4. A more comprehensive discussion of knowledge-based expert-systems technology (KBES) applications, comprising an entire chapter, has been provided by Haas et al. in *Modern Pavement Management* [Haas 94]. The authors illustrate that a significant feature of KBES, when compared to conventional techniques, is its ability to use knowledge when it is needed.

A large body of information is available on regression analysis and time series modeling techniques, ranging from relatively simple and introductory texts to advanced books. The reader can access any technical library to see the numerous texts that are available. Generally, regression analysis for performance modeling uses historical data to develop a relationship between an independent variable, such as a condition index, and a set of independent or causal variables, such as those noted in Sec. 8.1.2. For network-level purposes, simplified time-series models may be sufficient, whereas on a project-level basis a more comprehensive set of data on materials properties, loads, environment factors, and so forth, may be needed to develop a sufficiently accurate model.

Markov-transition probability models are particularly useful where an historical database does not exist. They capture the experience of engineers or technologists in a structured way by using different classes or combinations of situations and condition states. For example, one class for developing a road-deterioration prediction model might be high traffic combined with a minimum thickness structure combined with a strong foundation. Of course, boundaries for the variables comprising the classes must be provided (say high traffic is greater than 10,000 AADT). Condition states, using the vertical-scale example of Figures 8-1 and 8-2 and a CI measure could be $CI_1 = 100$ to 90, $CI_2 = 90$ to 80, and so forth (10 condition states). Many references are available on Markov modeling, including applications in the transportation field [Li 96].

Bayesian-modeling techniques have existed for many years and have been applied in a variety of infrastructure areas. Essentially, they use prior knowledge and/or experience (preposterior), which can be combined with actual observations (usually in cases of limited data) to produce posterior estimates of deterioration [Larson 69]. Figure 8-3 illustrates the general procedure in the Bayesian approach. An example of extensive use of this approach is represented in the long-term pavement-performance modeling in the SHRP studies [C-SHRP 95]. The main advantage of using a Bayesian approach, compared to classical-regression analysis, is that Bayesian can be used even if a comprehensive database is lacking.

Artificial neural networks (ANN) analysis, which has gained popularity in engineering modeling in recent years, involves parallel computational models for knowledge representation and information processing [Begley 95, Ghaboussi 92]. The processing units model the action of the synapses and dendrites in the human brain. Because of their fundamental similarity to the human brain in terms of hardware,

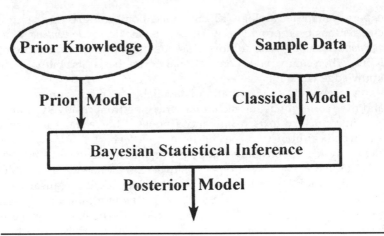

Figure 8-3 General approach for Bayesian modeling.

neural networks have some unique, human-like capabilities in information processing. Neural networks are capable of learning complex, highly nonlinear relationships and associations from a large body of data. The information and the knowledge learned by the neural networks is encoded and stored in the connection weights of the network. The retrieval of the stored information is done routinely by providing the network with an input pattern, which acts as a key. Detailed discussion on ANN modeling is provided in Sec. 4.4.

Econometrics methods can also be effectively used in certain situations to predict performance [Madanat 95]. The authors contend that simple regression models, which are commonly used to model deterioration or distress propagation, suffer from selectivity bias because of the nonrandom nature of the estimation sample used. An example application has involved pavement-cracking progression, using a method known as Heckman's procedure and a full-information maximum-likelihood method, both to correct for selectivity bias [Madanat 95].

Geospatial methodologies are relatively new techniques and are possible due to enhanced geospatial/GIS programs and easily available high-resolution multispectral and hyperspectral imagery. Although not discussed here for brevity, geospatial methodologies and quantitative disaster-damage assessment models have been developed and applied successfully to assess Hurricane Katrina damage and elsewhere [Uddin 09, 10].

8.2.3 M,R&R Intervention Performance Modeling

During the life cycle of an infrastructure facility, an M,R&R intervention may be required, as illustrated in Figure 8-2. The following alternative interventions are possible:

- Do-nothing policy or deferring M,R&R action beyond the minimum-acceptable level until the expected time of replacement.

- Corrective maintenance to keep the condition at the minimum-acceptable level.

- Repair, rehabilitation, restoration, and renovation to improve the condition and extend the life at or before the minimum-acceptable level is reached.

- Replacement and reconstruction.

It is generally assumed that routine and minor emergency maintenance does not have any significant effect on condition and only a major M,R&R action improves the condition, as shown by Curve 2 in Figure 8-2.

Given the foregoing intervention alternative actions, the question is how to model the postintervention performance. Basically, the same techniques identified in Sec. 8.2.2 are applicable. Quite often it is sufficient to simply use the same performance trend or model for Curve 2 as for Curve S in Figure 8-2, especially if the basic deterioration pattern of the facility is not expected to change. Of course, if a major structural or functional improvement is achieved by the intervention, then a new Curve 2 model would likely be needed, even though the same basic techniques still apply.

8.3 Failure Analysis

Failure analysis is an important way to examine failed facilities or components and determine the possible reasons of a failure. Some failures are catastrophic, resulting in loss of human life and/or property. Examples are bridge collapses, off-shore oil-rig disasters, foundation failures, failure of dams and dikes, corrosion-related failures, and earthquake damages. Sometimes a full-scale instrumented facility is tested to failure, to learn the failure mechanisms and their interaction effects. Failure analysis is also the most important component of forensic engineering [Bodzay 91]. Valuable lessons for improving design and construction methods have been learned from catastrophic failures caused by:

- Inadequate structural strengthening (2007 collapse of I-35 bridge on Mississippi River)

- Natural disasters (2005 Hurricane Katrina in the United States, 2011 Japan's nuclear disaster from the earthquake and tsunami)

- Man-made disasters (September 11, 2001 terrorist attack on New York's World Trade Center and collapse of adjacent building in the complex)

8.3.1 Approach and Technologies for Failure Analysis

Failure analysis requires a collaborative approach among structural and design engineers, material scientists, and metallurgists, as well as the use of the latest available NDE methods. As discussed previously, the normal deterioration process starts soon after construction, and deterioration is generally not visible for some period of time. A catastrophic failure is triggered by a sudden increase in critical stress, strain, or displacement caused by corrosion, dynamic loads and shock waves, hydraulic pressure, and/or ground failure.

The knowledge of the precise reasons of failure can be used to improve performance models. Environmentally associated cracking and stress-corrosion cracking (SCC) play important roles in lifetime predictions because their catastrophic character can affect a wide variety of metals and alloys exposed to specific environments [Roberge 95]. It is now believed that SCC, fatigue, and general fracture should not be considered separately, because there is a continuum in their relative-failure boundaries as shown in Figure 8-4. An expert-system's approach may be used for failure prediction in this case because of the lack of any obvious relationship between these phenomena.

A thorough examination of the design criteria, construction details, service conditions, and usage history provides the clues to unexpected and premature failures. Therefore, the inventory and in-service monitoring and evaluation principles discussed in Chapters 4 through 7 are important for failure analysis. Inadequate analysis of soils and ground conditions is a primary reason for slab

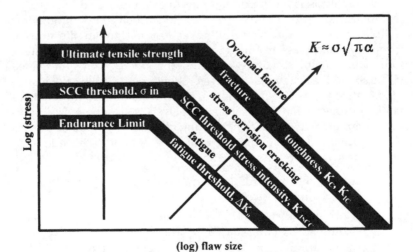

Figure 8-4   Relative-failure boundaries among fractures, SCC, and fatigue [Roberge 95].

cracking and settlement; for example, the slab-on-grade foundation problems in areas of expansive clay soils. Many costly failures on expansive soils result from a failure to consider expansive soil movements or the failure of builders to incorporate the design considerations for structural integrity of the foundation on such soils [Godwin 93]. As one moves to new materials frontiers and tries to avoid catastrophic failures, one needs to rely more on advanced NDE technologies; analysis methods, such as fracture mechanics and neural-network analysis; and advanced computational tools like three-dimensional finite-element analysis. The following sections present some case studies of failure analysis.

8.3.2 Lessons from the Schoharie Creek Bridge Failure

The Schoharie Creek Bridge built in the early 1950s on the New York State Thruway collapsed on April 5, 1987, in a few hours during floods [Thornton 88]. As the authors noted, Pier 3 failed by scour caused by stream-bed erosion, dropping spans 3 and 4 into the creek followed by failures of other piers and spans. A thorough investigation of this failure indicated poor communication between the designer's original assumptions of soil properties and riprap and the construction and maintenance teams, as well as inadequate inspections. Some of the key recommendations from the study are [Thornton 88]:

- An expanded bridge-inspection program by engineers familiar with bridge design

- A structural-integrity check on a periodic basis (at least once every 10 years)

- Inspection on at least the following schedule: (1) complete structure and foundation once a year; (2) water crossings and stream-bed changes after each storm; (3) movements of bearings and expansion joints during at least one high-temperature period, low-temperature cracking of tension members or flanges, and drainage of salt-laden runoff during a low-temperature period (every year); and (4) after each accident on the bridge

- Thorough supervised underwater inspection

- A scour-indicator system for foundations with running water (checked during each inspection)

- Review of all maintenance reports and repair/rehabilitation records before and after their implementation

- Frequent contact with the bridge owner for knowledge of deficiencies and urgency of required repair

Scouring and erosion of stream-beds around piers and columns are primary reasons for the undermining of bridge foundations and resulting bridge failures. The mechanisms and analysis of scouring and erosion of rocks and soils have not received adequate attention in the past. A new quantitative methodology for an erodibility index, proposed by Annandale et al., depends on the erodibility threshold of a material and the erosive power of water [Annandale 96].

8.3.3   Fire and Explosion Investigation

Calligiuri describes the following investigation of a fire and explosion case [Calligiuri 91]. The initial analysis of the remaining evidence of the investigation provided no immediate reason for the accident; however, attention was given to the crushed segment of a natural gas pipeline that ran beneath the explosion site. Detailed visual examinations, metallurgical tests, and pressure vessel tests were conducted on undamaged pieces. Welds were examined by ultrasonic testing and video probe. Computer simulations of the buried pipe were conducted by the DANTE finite-element code for no internal pressure and for 2,067 MPa (300 psi) internal pressure. The computed deformations with time supported the conclusion that the pipeline was depressurized prior to the fire [Calligiuri 91].

Post-September 11, 2001 investigations of the World Trade Center Building 7 in New York City showed the vulnerability of concrete-steel composite-structured frames to extreme fire hazards. This building collapsed under extreme fire, failure of sprinkler systems, and super heat conditions [NIST 03].

8.3.4   Lessons from Earthquakes

Most of the current earthquake-resistant design codes and retrofitting of older structures have been based on in-depth postearthquake inspections and structural seismic analysis of damaged facilities. However, recent earthquakes bring newer concerns and provide opportunities to assess the performance of existing procedures. In an award-winning article "The Lessons from Kobe" in the *Los Angeles Times,* Hotz and Reich summarize the results of their interviews with many earthquake experts after the 6.8 Richter scale earthquake on January 17, 1995 in Kobe [Hotz 95]. The magnitude of the 1994 Northridge earthquake in Los Angeles measured 6.7 on the Richter scale. Table 8-3 is based on the Hotz and Reich article.

Japan's nuclear disaster of Fukushima Daiichi Nuclear Power Station after the 9.0-magnitude earthquake and tsunami of March 11, 2011 clearly showed the vulnerability of these critical power-infrastructure assets. It is a wake-up call for the need of periodic monitoring of these assets and evaluation of structural integrity as an integral part of IAMSs.

Concerns	What Happened in Kobe	Japan's Preparedness	California's Preparedness
Highways	The ground shaking exceeded design limits by so much that hundreds of support pillars were seriously damaged. Four railways and three major highways were cut.	Huge brittle bridge and highway supports are designed to withstand shaking at about one-third the force of gravity. In the Kobe quake, the force exceeded that limit and pillars snapped.	Highway engineers emphasize smaller, more flexible supports that can sway with ground shaking up to 70% of the force of gravity (g) without breaking. In the Northridge earthquake, shaking exceeded 1 g in some areas.
Steel-frame buildings	Most newer high-rise buildings appeared to withstand the shocks, while precode buildings collapsed in many cases.	High-rise buildings constructed after 1981 incorporated flexible design meant to safely absorb seismic energy, but are still stiffer than in California.	Flexible-frame design is favored. After the Northridge earthquake, structural engineers found dangerous cracks in more than 120 buildings. The newest buildings with shock-absorbing foundations, called base-isolation systems, emerged without damage.
Water and gas	Miles of water, sewer, and gas pipes broke. That, compounded with Kobe's high population density, resulted in 300 fires, twice as many as occurred after the Northridge earthquake.	Japanese utility companies had developed what they believed would be an automatic cutoff system to safely shut down natural-gas lines in the event of a serious earthquake.	Unprotected oil and gas lines, along with sewer and water systems, remain a major concern.

TABLE 8-3   The Lessons from the Kobe Earthquake (after [Hotz 95])

Concerns	What Happened in Kobe	Japan's Preparedness	California's Preparedness
Routine residential construction	Thousands of homes simply collapsed, killing or injuring the occupants.	Traditional Japanese housing relies on a frame of a few large posts topped by a heavy tile roof, with no cross-bracing.	In California, average housing uses cross-braced, stud-wall construction, with plywood shear panels lending lateral strength.
Person readiness	Few people seemed to have personal earthquake preparedness kits with emergency food, water, and medical supplies.	In Japan, there are annual earthquake drills, but fewer than 30% of the people in Tokyo participated in last year's exercises; fewer than 10% in Kobe participated.	For all practical purposes, formal emergency drills in businesses and schools are nonexistent. However, personal earthquake kits are increasingly common in California homes and workplaces.

TABLE 8-3   The Lessons from the Kobe Earthquake (after [Hotz 95]) (*Concluded*)

8.4   Summary

This chapter has only scratched the surface on performance modeling and failure analysis. Each infrastructure type requires its own data and modeling. Theoretical equations, such as stress and fatigue, can help in analysis of the structure, but always fall short in modeling the total life cycle. More detailed modeling information is available for most fields of infrastructure.

Failure analysis adds information to the experience base, particularly in defining end points and critical time elements. Detailed examination histories are available in each field.

8.5   References

[Annandale 96] G. W. Annandale, S. P. Smith, R. Nairnes, and J. S. Jones, "Scour Power," *Civil Engineering*, American Society of Civil Engineers, July 1996, pp. 58–60.

[Begley 95] E. F. Begley, and C. P. Sturrock, "Matching Information Technologies with the Objectives of Materials Data Users," *Computerization and Networking of Materials Databases, ASTM STP 1257*, Vol. 4, American Society for Testing and Materials, Philadelphia, Pa., 1995, pp. 253–280.

[Bodzay 91] S. Bodzay, "Failure Analysis in the Forensic Domain," *Conference Proceedings of the International Conference and Exhibits on Failure Analysis*, Montreal, P.Q., July 1991, pp. 249–252.

[C-SHRP 95] Canadian Strategic Highway Research Program, "Bayesian Modeling: Joint C-SHRP/Agency Applications," *Technical Brief No. 8, C-SHRP,* Transportation Association of Canada, Ott., November 1995.

[Calligiuri 91] R. D. Calligiuri, "Failure Analysis, Prevention, and Testing," *Conference Proceedings of the International Conference and Exhibits on Failure Analysis,* Montreal, P.Q., July 1991, pp. 191–197.

[Ghaboussi 92] J. Ghaboussi, "Potential Application of Neuro-Biological Computational Models in Geotechnical Engineering," In *Numerical Models in Geomechanics,* G. Pande and S. Pietruszczak (eds.), Balkema, Rotterdam, the Netherlands, 1992, pp. 543–555.

[Godwin 93] C. J. Godwin, "Slab-on-Grade Foundations on Expansive Soils," *Building Research Journal,* National Consortium of Housing Research Centers, Vol. 2, No. 2, November 1993, pp. 33–43.

[Haas 94] R. Haas, W. R. Hudson, and J. P. Zaniewski, *Modern Pavement Management,* Krieger Publishing Company, Malabar, Fla., 1994.

[Harris 96] S. Y. Harris, "A Systems Approach to Building Assessment," *Standards for Preservation and Rehabilitation, ASTM STP 1258,* American Society for Testing and Materials, Philadelphia, Pa., 1996, pp. 137–148.

[Hotz 95] R. L. Hotz, and K. Reich, "Lessons of Kobe: Hope, Caution," *Los Angeles Times,* February 12, 1995.

[Hudson 97] W. R. Hudson, R. Haas, and W. Uddin, *Infrastructure Management,* McGraw-Hill, New York, 1997.

[Humplick 94] F. Humplick, and W. D. O. Paterson, "Framework of Performance Indicators for Managing Road Infrastructure and Pavements," *Proceedings, 3rd International Conference on Managing Pavements,* National Research Council, Vol. 1, May 1994, pp. 123–133.

[Krinitzsky 93] E. L. Krinitzsky, "The Hazard in Using Probabilistic Seismic Hazard Analysis," *Civil Engineering,* American Society of Civil Engineers, November 1993, pp. 60–61.

[Larson 69] H. J. Larson, *Introduction to Probability Theory and Statistical Inference,* 2nd ed., John Wiley & Sons, New York, 1969.

[Li 96] N. Li, W. Xie, and R. Haas, "Reliability-Based Processing of Markov Chains for Modeling Pavement Network Deterioration," In *Transportation Research Record 1524,* Transportation Research Board, National Research Council, Washington, DC, 1996, pp. 203–213.

[Madanat 95] S. Madanat, S. Bulusu, and A. Mahmoud, "Estimation of Infrastructure Distress Initiation and Progression Models," *Journal of Infrastructure Systems,* American Society of Civil Engineers, Vol. 1, No. 3, September 1995, pp. 145–150.

[NIST 03] National Institute of Standards and Technology, "NIST Response to the World Trade Center Disaster: World Trade Center Investigation Status," U.S. Department of Commerce, December 2003.

[Roberge 95] P. R. Roberge, "Background and Basis for a Knowledge Elicitation Shell for Lifetime Predictions from Stress Corrosion Cracking Data," *Computerization and Networking of Materials Databases, ASTM STP 1257,* Vol. 4, American Society for Testing and Materials, Philadelphia, Pa., 1995, pp. 136–150.

[SHRP 92] *SHRP Product Catalogue,* Strategic Highway Research Program, National Research Council, Washington, DC, 1992.

[Thornton 88] C. H. Thornton, R. L. Tomasetti, and L. M. Joseph, "Lessons from Schoharie Creek," *Civil Engineering,* American Society of Civil Engineers, May 1988, pp. 46–49.

[Uddin 95] W. Uddin, "Pavement Material Property Databases for Pavement Management Applications," *Computerization and Networking of Materials Databases, ASTM STP 1257,* Vol. 4, American Society for Testing and Materials, Philadelphia, Pa., 1995, pp. 96–109.

[Uddin 09] W. Uddin, B.T. Wodajo, K. Osborne, and M. White, "Expediting Infrastructure Condition Assessment for Disaster Response and Emergency Management Using Remote Sensing Data. Proceedings," *MAIRERAV6 International Conference,* Torino, Italy, July 8–10, 2009.

[Uddin 10] W. Uddin, "Flooding, Ecological Diversity, and Wildlife Habitat Studies by Geospatial Visualization and Remote Sensing Technologies," *Proceedings, International Conference on "Biodiversity is Our Life"*—Center for Biodiversity and Conservation, Shah Abdul Latif University, Khairpur (Mir's), Sindh, Pakistan, December 29–31, 2010, pp. 239–254.

[Urban 84] S. R. Goodwin, and A. E. Peterson, *Guides to Managing Urban Capital Series*, Vol. 2: *Guide to Assessing Capital Stock Condition*, The Urban Institute, Washington, DC, 1984.

Concepts of Total Quality Management

CHAPTER 9

Design for Infrastructure Service Life

9.1 Introduction

In Chapters 1 and 2, the interrelationship of design with infrastructure asset management was defined. Design is a uniquely project-level activity of infrastructure management, along with construction, maintenance, and rehabilitation. Planning, budgeting, and programming of maintenance and rehabilitation can be done at the network level, but design is applicable to a specific unit of infrastructure or project.

The process flows from design to construction, maintenance, and rehabilitation. For the infrastructure to function effectively, it must start with good design. In many types of infrastructure, such as pavements, bridges, and buildings, there is talk of theoretical design versus empirical design. This implies, on the one hand, that the choice of physical elements is totally based on theory, or usable mechanistic theory; on the other hand, that the selection or sizing of elements is based entirely upon empiricism or experience. In truth, all design is somewhat subjective, since factors such as criteria, design life, and level of service all change from location to location and from agency to agency.

This chapter is not a "design guide." It is impossible to provide such a design guide in one book because each type of infrastructure has its own design requirements, and within a given infrastructure type each agency has its own specific design elements and approach. Rather, this chapter covers the design process, and the technology and concepts of design. Certain aspects of design are common to any facility and any type of infrastructure. Concepts of design range from visual, such as the design of an automobile or architecture design, to physical and mechanical, such as the design of an engine, an airplane, or a bridge. Engineering of physical infrastructure is not as free-form as architectural design, but it does offer flexibility and an opportunity for creativity.

9.1.1   Reliability and Design

Design is often practiced in a deterministic way, in which a specific design can be defined that uniquely fits a set of physical circumstances. Such an approach is inadequate because of the true variability that faces all designers in terms of inputs and service life of the facility. Bridges, for example, are designed on the basis of predicted traffic and vehicle loads, yet no one has good control over the actual service loads that will be carried on the bridge. Legal load limits vary with respect to bridge configuration, and vehicle operators do not always adhere to legal limits. Furthermore, load zoning and characteristics are political decisions that can change during the life of a facility.

These uncertainties have generally been dealt with by inserting safety factors into the design process. In structural design, for example, the yield strength of the material is determined by test, but a working strength for design is obtained by dividing the yield strength by a factor of safety. For example, if the factor of safety is 2.0, and the yield strength is 276 MPa (40,000 psi), then the working strength would be 138 MPa (20,000 psi).

Such working-stress and factor-of-safety methods are outmoded for the modern design of infrastructure. Reliability concepts are far superior, and any infrastructure engineer is urged to examine reliability-based methods in detail for design. For example, in 1985 AASHTO went to a reliability-based method for pavement design [AASHTO 85]. Updated in 1993, the AASHTO Guide [AASHTO 93] is now incorporated in the current Mechanistic and Empirical Pavement Design Guide [AASHTO 08].

9.1.2   Design Technology Framework

A general framework can be presented for design that is quite comprehensive. Most infrastructure-design methods do not include all aspects of the framework; however, logical extension of past practices will lead to continuous improvement in the design for individual infrastructure types over time. Historically, design has often been compartmentalized; for example, water-pipe design or sewer design is often considered as the process of sizing a pipe or flow structure, when in reality it also should involve selection of material, optimum location of the physical unit, and other details aside from pipe size alone. This problem has certainly applied to bridges and pavements, where design has often been relegated to selection of layer thickness for pavements, or selection of structural-member size and weight for bridges. In bridges, for example, custom within an agency may dictate the use of prestressed concrete members so that the designer overlooks the possibility of comparing other material types, such as steel, or other forms, such as arches. In pavements, it is not uncommon for the agency to preselect a flexible or rigid pavement, so that the so-called design process merely becomes a function of selecting thickness. Modern technology and computer tools make it completely possible for the designer to consider a fully functional design process, and design education should be broadened in this regard. Since design is the first major cost-related element in infrastructure management at the project level, good economical optimum design is critical to all effective infrastructure management. Although it is possible to optimize the balance of infrastructure management given a suboptimal design, such a process will never be truly optimal if the design is shortchanged.

Figure 9-1 presents a comprehensive diagram of most design practices. It is generic and does not cover the details of any specific method. Common to all methods, however, are inputs of loading (either traffic or use), existing environment, and available materials. These factors are handled in two ways. The most common historical method has been to take the input and compare the design for a set of common alternate designs, check these with design models to see if they satisfy requirements, and alter factors and try again until they do. The other option—or more modern method—is to enter the loading, environment, and available materials factors into a comprehensive design model to predict a reliable design for the required service. In each case, standard practice has often been to adjust other details, such as connections, slab joints, joint spacing, and so forth, for the design obtained. This process then culminates with the preparation of plans, specifications, and estimates—the so-called contract documents. Not considered in this process are the following aspects of maintenance, construction practices, and economics.

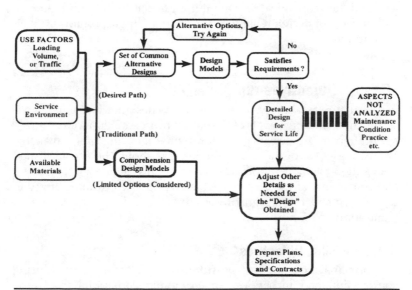

FIGURE 9-1 Description of most design practices.

9.1.3   Evolution of Design Technology

The design process and methods have evolved greatly over the years. Most early designs were necessarily based on experience. These included the use of timber beams developed from local trees near the bridge site or building being constructed, the use of stone pilasters for construction of churches and large historical buildings, the intuitive use of large stone bases in the Roman road system, and the construction of stone aqueducts in ancient Greece and Rome.

However, integrated design is not a totally new development. A recent visit to ancient cities in Greece showed the integration of sewers and storm sewers into the paved-block streets constructed for the main thoroughfare of the old (700 B.C.) city. Fastener holes were designed into some of the paving blocks to provide clean-out access to the sewers underneath.

It can be said that the quality of design, historically, often depends upon the intuition, knowledge, and hard work of the designer. The purpose of good infrastructure management is not to replace these strong characteristics, but to provide tools that enhance the design process so that less-innovative engineers can also become quality designers in the broadest sense.

The general approach used by some agencies for many years has consisted essentially of designating "standard sections" for the various ranges of conditions. These standard sections have historically evolved differently for various environments. The inherent implication in such methods is that performance or service life will be "satisfactory" with the standard design, "unsatisfactory" for anything less, and "uneconomical" for anything greater. In modern methods, the concepts of design reliability have a great impact on this implicit assumption.

9.2   Design Objectives and Constraints

Infrastructure asset management helps the designer focus on the basic function and constraints of the design process. This avoids the rote application of the design procedure while ignoring other available alternatives. One of the first activities that should be accomplished under the systematic approach is to carefully define the objectives and constraints of the problem. Objectives and constraints apply to the facility and the design process, or to the designer's activities. Some apply to both.

9.2.1   Facility Objectives

The principal objectives to be fulfilled by a facility, both during construction and in service, are economic and social in nature.

They include but are not limited to the following:

1. Maximum or reasonable economy, in terms of both agency and user costs
2. Maximum or adequate safety
3. Maximum or reasonable serviceability over the design period (performance)
4. Maximum or adequate capacity for both the magnitude and number of repetitions of demand (loads, volumes, and flaws)
5. Minimum or limited physical deterioration caused by environmental and use influences
6. Minimum or limited noise and air pollution and environmental disturbances during construction
7. Minimum or limited disruption of adjoining land or facility use
8. Maximum or good aesthetics

These objectives pose several major contradictions, which occur with any complex system or facility that attempts to fulfill social and economic needs. Consequently, each unit constructed represents some compromise or trade-off among objectives. The relative influence of the objectives in a particular design situation varies with such factors as rural or urban area, use, and environment.

The first three objectives listed are very important. Economy is a prime requisite because most public agencies feel that they have insufficient funds available for all the desired investments. Safety is another major requirement the facility must fulfill. Serviceability should also be sufficiently high over the design period to ensure the desired level of quality to the user in terms of speed, operating costs, delays, and comfort.

The fourth objective, maximizing capacity, is applicable to both the facility and to the design process. This objective competes directly with the objective of maximizing economy. The fifth objective, minimizing physical deterioration, is related to serviceability in that deterioration is a major factor in causing serviceability loss. However, it is identified as a separate objective because it also competes with the economic objective in terms of maintenance and rehabilitation costs.

The last three objectives are usually of lesser importance, except in buildings. Aesthetics, or appearance, can be a significant objective for certain special applications, or in those cases where patching and repairs are extensive and appear ugly to users or adjacent property owners and observers.

The 21st century's international accords on reduction of anthropogenic CO_2 and public awareness and/or perception of climate change and environmental sustainability concerns have led to many public infrastructure agencies and private entities to consider one more objective—reduce or *minimize carbon footprint*.

9.2.2   Facility Design Objectives

In one sense, the objective of the design process is to design a facility that, when constructed, will accomplish the functional objectives previously defined. However, the specific objectives of the design process can be stated in technical, economic, and social terms:

1. Develop a design strategy of minimum (or reasonable) economy, safety, and serviceability.
2. Consider all possible design strategy alternatives.
3. Recognize and incorporate the variability in the design factors.
4. Maximize the accuracy of predictions of serviceability, safety, and physical deterioration for the alternatives considered.
5. Maximize the accuracy of estimating cost and benefits.
6. Minimize the costs of design, including labor, testing, and computer time.
7. Maximize information transfer and exchange between construction and maintenance personnel.
8. Maximize use of local materials and labor in the design strategies considered.
9. Maximize energy efficient design and minimize CO_2 and other GHG emissions.

Some of the foregoing objectives correspond with the objectives for the facility (or asset) itself, whereas others are peculiar to the design process. Most of the objectives compete with other objectives and require trade-offs.

The first objective simply reflects the need for the design to recognize the objectives of economy, serviceability, and safety. To accomplish this it is necessary, as stated in the second objective, to consider all the possible or feasible design-strategy alternatives. Because these can amount to several hundred combinations for any given situation, the use of computer analysis is necessary for accomplishing this objective.

The third objective, recognition of the stochastic nature of design, construction, and maintenance factors, should be important criteria for any design method. Early contributions to the inclusion of variability in the pavement-design process were presented, for example, in the early 1970s [Darter 71, Darter 72, Darter 73a, Darter 73b], and are now incorporated in the current pavement-design method [AASHTO 08]. There are significant parallel contributions in bridge and foundation design.

The fourth objective, maximizing the accuracy of the design predictions, is related to the third objective, and also to the quality of the input data. Decreased uncertainty in design predictions is important to planning activities so that investment programming can be accomplished more reliably.

The accuracy of the cost and benefit prediction, the fifth objective, is directly related to the accuracy of the design predictions, as well as the ability to estimate the construction, maintenance, and user costs.

Minimizing the costs of the design process in terms of materials testing, personnel time, and preparation of drawings, the sixth objective, is important, but it obviously competes with the other objectives. If the cost of the design process is underfunded, the result can be an incomplete, inadequate, or excessively costly design strategy.

The seventh objective, information exchange between those responsible for construction and maintenance, has unfortunately been neglected to a large extent in design practice. There have been instances in North America of maintenance crews unfamiliar with continuously reinforced concrete pavements using asphalt to fill the numerous hairline cracks that are a design feature of this pavement type and do not require maintenance. The resulting mess might be more readily ascribed to the designers' failure to communicate with the maintenance people than to the maintenance people themselves. It is extremely important that such communication occur, because proper construction and maintenance activities "carry" the design through the life of the facility.

The eighth objective, involving the use of local materials and labor, is perhaps a self-evident desirability for most situations. This objective may complement the first objective of maximum economy, but in some cases local materials may require stabilization or treatment.

The last objective addresses the effort of many cities and public infrastructure agencies to reduce fossil-fuel use and replace that with more renewable energy sources, minimize energy needs during construction and service life, and reduce adverse impacts of built infrastructure on biodiversity, ecosystems, and the environment.

9.2.3 Design Constraints

The major design constraints are economic and physical or technical in nature and include the following:

- Availability of time and funds for conducting the design and the construction
- Minimum level of serviceability allowed before rehabilitation
- Availability of materials
- Minimum and/or maximum dimensions allowed
- Minimum time between successive rehabilitations
- Capabilities of construction and maintenance personnel and equipment
- Testing capabilities

- Capabilities of the structural and economic models available
- Quality and extent of the design information available

The first constraint might involve not only a limit on the funds that can be applied to a particular project but also the opportunity costs of those funds. All the other constraints, which are essentially physical or technological, are relatively self-explanatory. Each has economic implications and is therefore related to the first constraint.

9.3   Design Framework and Components

In an IAMS, the design phase involves several activities. These may be broadly classified as: (1) information needs related to inputs, objectives, and constraints; (2) generation of alternative design strategies; and (3) analysis of structure, flow characteristics, and other aspects of the facility, along with economic evaluation and optimization of these strategies.

9.3.1   Information Needs

The information needs of good infrastructure management are extensive. Detractors may cite this as a reason for not wanting to implement an IAMS, which is a specious argument. The information needs are inherent in the problem not in the IAMS. They only are stated explicitly by the management process. The needs were always inherently there, and any failure to acknowledge and fill those needs means that a suboptimal design will be selected in most cases, or that the life of the infrastructure will suffer.

Much of the needed information can be obtained from network-level management activities, which generally take place prior to project-level design. It is important for the designer to make full use of available information from network-level activities and from associated databases. Failure to do so will result in wasted resources and suboptimal design at best. At worst, it will result in inadequate design and early failure. The designer must not, however, assume that network-level data quality is adequate for design. It merely serves as a starting point. It is essential to supplement network information with data collected specifically for the project being undertaken. Three main categories of data stand out: (1) environment, (2) loading and traffic, and (3) materials characteristics.

9.3.1.1   Environmental Data

Environmental data may be generalized, but a facility often exists in a unique location that requires characterization of the environment with more detail than is available on weather maps, weather databases, or from other general sources. Depending upon the type of environment and the agency involved, experience should govern the way these environmental data are to be handled.

9.3.1.2 Projected Use, Loading, and Traffic

Information on projected use, whether it be volumes of sewage flow, water consumption, average daily vehicular traffic or vehicular axle loads, is a matter of estimating and projecting future needs, rather than a matter of measuring current characteristics. This makes the process more difficult and is an area where problems have often existed in past design. Historical records on similar facilities should be used, but consideration must also be given to potential changes that may occur as the situation evolves. Additionally, such projections should factor in demographic trends, population growth, economic development, land use, and other factors. These generally will be handled in a planning phase at the network level, but sharpening the estimates is required at the design phase.

9.3.1.3 Materials Characteristics

It is impossible to design a facility effectively unless the service characteristics of the materials or elements to be used in the facility are understood. In many cases, the characteristics of the materials will change with environment and use, and in other cases the characteristics change with construction practice, availability of materials sources and other factors beyond the control of the designer. In some cases, the designers have assumed materials characteristics and left it to the construction phase to "provide the materials as designed." This is not a wise practice—materials must be considered effectively in the design process, even though the final function of providing the required materials must be accomplished during construction. Some materials degrade with exposure and/or with time and use because of fatigue. It is essential for the designer to consider any such material degradation.

Ultimately, material properties can only be obtained through appropriate physical testing. Such testing can be expensive and is too often ignored, but all designs depend on adequate knowledge of strength, stiffness, and deformation characteristics.

9.3.1.4 Assessment of Environmental Impacts

In modern design practice, environmental impact assessment is an integral part of infrastructure planning, design, and construction phases with respect to air quality, surface runoff water drainage, archeological site proximity, biodiversity and endangered species, wetlands and other ecosystems, and neighborhood communities and businesses. Additionally, "environmental justice" regulations are enforced on any major construction or renovation project in the United States to ensure that underprivileged communities are not hurt economically and socially. Historical and spatial data on cultural diversity, archeological sites, biodiversity, and ecosystems are needed for environmental impact studies so that design and construction of an infrastructure asset does not have any adverse environmental and social impacts.

9.3.2 Other Information Factors Required

Other factors to be considered in design include: costs, estimated service life, and various details required as input to a specific design method. Items such as inflation, user costs, and related topics may be difficult to obtain; however, this difficulty is not a valid reason for failure to make a reasonable estimate. Some designers, for example, ignore user costs and economic analysis because they say it is impossible to predict inflation and the exact costs to be borne by the user. This fallacy can result in gross error. It is far better, for example, to assume a reasonable interest rate and inflation rate for a combined cost of money of 5 or 6 percent than it is to ignore the cost altogether. Ignoring the factor assumes the rate to be zero, which is obviously erroneous. Such factors can be handled in design through the use of sensitivity analysis for various ranges of user cost, inflation, and interest rates. This approach at least gives the designer a reasonable basis for making a final subjective judgment, or for providing the information to the decision-maker, which may be the administrator or a governing board, such as the city council.

9.3.3 Generating Alternative Design Strategies

Designs of infrastructure involve more than selecting a thickness or a beam dimension; therefore they are termed *strategies* and consist of a combination of elements, such as various configurations and future rehabilitation actions. They also include various material types and sources, expected construction, maintenance, and performance-evaluation policies, as well as construction quality-control and assurance methodologies.

The need for a design alternative to specify materials type and dimensions is apparent. However, without including the expected construction and maintenance-rehabilitation policies as part of the strategy, there may be appreciable error in the predicted performance outputs for that facility structure. Rehabilitation alternatives also become part of any design strategy if the service life of the facility drops to the minimum acceptable or terminal level before the end of the design period. An exception would be a maintenance policy that kept the facility at or above the minimum acceptable level of service to the end of the design period. Alternatively, if financial constraints prevail, maintenance at this level might continue only until funds were available for rehabilitation.

In formulating rehabilitation alternatives there are two major, interrelated aspects that must be considered:

1. Physical or structural aspects with respect to dealing with damage, lack of adequate service, lack of adequate safety, and so forth.

2. Policy aspects with respect to procedures for handling use or traffic during the rehabilitation or remodeling process. In the

case of a building, should the entire building be closed, or should the renovation be handled floor by floor? This decision will depend on how safety aspects are affected, and the type of renovation being undertaken. In a case such as the Pentagon, the headquarters of the U.S. Department of Defense, the facility is so large that it would be physically impossible to close the entire structure for renovation at one time.

In summary, the design phase of infrastructure management will only be adequate if it considers a reasonable number of alternatives; thus, the process of generating alternatives is critical to the process of optimum design.

9.3.4 Analysis, Economic Evaluation, and Optimization

Many design methods still do not include all the activities of economic evaluation and optimization; rather, they include only analysis. However, design methodology has for the past several decades been moving toward incorporating these activities.

The first step in the analysis of any facility alternative is the application of appropriate structural and/or use models. In the case of hydraulic facilities, the models involve flow characteristics and other use factors. In buildings, bridges, and pavements, the primary models are often structural or functional.

In each case, the models must be sufficiently comprehensive to cover the needed areas of concern. The use of incomplete models can have enormous costs. As an example, there was a trend in the 1960s to use thin, lightweight concrete bridge decks to reduce the dead weight of bridge structures, and thus reduce the size of major beam components. The resulting structures, however, deflected more under dynamic loads, and the thinner bridge decks were unable to sustain the additional deflection. As a result the U.S. interstate highway bridges failed earlier than expected and underwent a massive replacement of decks in the 1980s.

A similar example occurred in continuously reinforced concrete pavements (CRCP). Application of structural models showed that a thin slab was sufficient to carry the load; however, no deflection analysis was run in comparison to the underlying subgrade. The large deflection of the thinner slab resulted in permanent deformation of the subgrade in many cases, which then caused loss of support and subsequent failure of the slab. This problem—and the failure to fully understand it—caused the cessation of the construction of CRCP in most of the world, even though it remains a very competitive construction practice in Texas and elsewhere where the deflection criterion has been added.

Current technology cannot adequately predict the type and degree of all distress modes that a facility will be subjected to as a function of time and use. Consequently, composite structural models are often used in an attempt to make a direct prediction of service quality versus age. Limiting values are often placed on other elements. The reader is encouraged to apply these general concepts and principles in reviewing the detailed design in his or her particular infrastructure type. The economic evaluation of a facility alternative should involve the assignment of costs and benefits to the predicted outputs in terms of the user, plus the cost of providing that alternative to the agency in terms of materials construction, maintenance, and rehabilitation. These variables are then incorporated into an economic model to determine the total costs and benefits of the strategy. (See Chapter 14 for a detailed presentation on life-cycle cost and benefit analysis.)

It is desirable to include benefits in the economic evaluation, but this is not currently the case in much of facility-design practice. One reason is that it is difficult, for example, to separate benefits for a pavement or bridge from overall highway benefits. Another is that the difference in user costs between alternative designs can also be considered as benefits but are difficult to determine. Research by the World Bank and subsequently by the U.S. FHWA has successfully related vehicle-operating costs to pavement conditions; other work is applicable to bridges and other types of infrastructure [Zaniewski 82, Chesher 85]. It is strongly urged that a complete economic analysis be made for all design alternatives, treated in detail in Chapter 14.

When all of the alternative design strategies have been analyzed and evaluated, optimization should be used to define the best strategy or strategies for presentation to the decision-maker. Frequently this is accomplished by comparing the discounted total costs for the alternatives, or, when comparing alternatives of different life, an average annual-cost factor can be used to recommend strategy selection.

9.4   Design Effectiveness

Design usually involves a substantial degree of subjectivity, limited by rules of mechanics and physics, but oriented toward optimizing certain features as determined by a client and reflected in the stated criteria of the client. Many additional factors, such as schedule requirements, vendor capabilities, environment, and cost also must be incorporated into the design process. Nonetheless, design is perhaps the most central point of definition for a project in that ideas and information are transcribed to paper in the form of specific and coordinated instructions for the project's construction and documentation. Design effectiveness is related to studies of outputs, or products, of the design process, and is specifically limited to presenting a method for evaluating those outputs.

9.4.1   Measurement of Design Effectiveness

In recent years, increasing attention has been given to productivity and productivity improvement in the industrialized nations. In the

United States, particular interest has focused on productivity in the construction industry, which is often criticized for slow technological growth and escalating costs.

Measuring design productivity is more difficult than measuring construction productivity. More and more people, however, are realizing that the real measures of the effectiveness of the design effort are found in the construction, start-up, and operational phases of a project. Thus, it may be more beneficial to develop a method for evaluating the effectiveness of a design, rather than the productivity during the design activity itself.

In evaluating design effectiveness, it must be recognized that the owner may have as great an influence on the final result as the design engineer. Among the impacts of the owners are proprietary technology inputs, design requirements or constraints, schedule or cost constraints, and changes imposed while the project is in progress. Furthermore, design effectiveness must be measured with recognition of the objectives of the owner, rather than against absolute or independent criteria.

The Construction Industry Institute at the University of Texas at Austin has developed a method for systematically evaluating the effectiveness of a design project [CII 86]. This method has the flexibility to be useful in a wide variety of circumstances, including:

- Widely differing types of projects
- Mixtures of objective and subjective measurements of design effectiveness
- Differing objectives and criteria for design effectiveness
- Measurement of overall design effectiveness regardless of influence source, or measurement of designer performance when influences of designer and owner can be separately identified

Since first conceived and implemented by General Motors in the 1950s and later by the U.S. defense sector [Jackson 06], *Value Engineering* has emerged as a viable approach of looking independently on a "design of infrastructure facility." The *Value Engineering* process identifies one or more design components that can be built using alternative materials and methods on lower costs, and the facility can still meet the intended functions without compromising safety. This topic of enhancing design effectiveness at a lower cost is discussed in detail in Sec. 9.5.

9.4.2   Design Evaluation

Design, including the activities associated with creative thinking, sophisticated engineering calculations, and translation of ideas into drawings, specifications, and procurement of major items for construction, occurs during the early stages of a project. Many contributors are involved during the design phase, including the owner, the various designer groups, vendors, and perhaps construction representatives. It is difficult then to evaluate the performance of a specific party but much easier to evaluate the overall results of all parties.

Any evaluation method evaluates the total design effort, rather than the designer alone. Nonetheless, it can be used to establish goals and benchmarks for tracking selected performance during the design phase of a project and for evaluating performance against prescribed benchmarks.

Most owners, constructors, and designers agree that "design" encompasses the owner and contractor home-office functions of: (1) preliminary design; (2) project management (including cost estimating and control); (3) procurement; and (4) detailed design. None of the existing methods of design evaluation adequately determines the overall effectiveness of these diverse functions.

According to CII [CII 86], many factors are involved in evaluating a design. However, these can be grouped into main components of criteria for evaluation, weights for the various criteria, and a method for combining the various criteria and their weights into a single quantitative number for evaluation. The criteria and their weights are discussed in the following section. Methods for combining the criteria and weights into a single quantitative number are described using a technique called an *objectives matrix* [CII 86].

9.4.3   Design-Evaluation Criteria

Evaluation criteria should be established at the beginning of the design process, understood and accepted by all concerned, and subsequently used to evaluate design effectiveness. The criteria should include only those elements that directly impact attainment of project objectives. Design-performance goals should be reasonable and attainable, and achievements should be dependent primarily on the actions of those involved in the design process. Finally, the criteria should cover all major aspects of design responsibility.

The design-evaluation criteria depend on many variables. The type of project will affect the choice of the design outputs that are selected for measurement. The project schedule also can impact the selection of criteria. Related to the schedule, the type of construction or design contract has a bearing on criteria selection. The design user must be considered when selecting criteria for evaluation of design. The owner, designer, and constructor have different interests in, and uses for, cost-effective design. It is important to keep in mind these interests when selecting criteria to evaluate design effectiveness. All design users, of course, are interested in completing the project on schedule, within budget, and with high quality.

The eight criteria most suited for an initial evaluation of design effectiveness are summarized in Table 9-1. Each criterion is essential to each design user and can be evaluated during or immediately after completion of project construction. Not all of these criteria are easily

Criteria	Quantitative	Subjective
1. Accuracy of design documents	X	
2. Usability of design documents		X
3. Cost of design effort	X	
4. Constructability of design		X
5. Economy of design		X
6. Performance against schedule	X	
7. Ease of start-up	X	
8. Environmental stewardship	X	X

TABLE 9-1   Initial Design Evaluation Criteria

quantified by measurable factors. Table 9-1 also shows which of these eight criteria are quantifiable or must be subjectively rated by personal judgment. Although subjective ratings are valuable, methods should be pursued to develop quantitative evaluations for all criteria.

These eight criteria are not all-inclusive for measuring design effectiveness. Other criteria relating to operability, maintainability, safety, plant operating efficiency, and plant performance are at least of equal importance. These latter criteria, however, require evaluations after a period of plant operation, and by different personnel from those involved in the design construction and start-up of a project. All criteria should be evaluated before the project team disperses and important data become unavailable. Detailed discussions follow for each of the eight criteria.

9.4.3.1   Accuracy of Design Documents
Because specifications and drawings are the most readily identifiable products of a design, they are an important measure of design effectiveness. The accuracy of the design-documents criterion measures the frequency and impact of changes in the drawings and specifications. These changes are related to document revisions and entail extra work in both the design and construction phases.

9.4.3.2   Usability of Design Documents
This criterion determines the ease of use of the design documents by construction forces and relates to the completeness and clarity of the drawings and specifications.

9.4.3.3   Cost of Design
The cost of the design effort is the only criterion that actually occurs totally in the design phase. This can be quantified by the cost-effectiveness of the design activities compared to original (plus approved changes) budgeted amounts and overall project costs.

9.4.3.4   Constructability

Constructability is the optimal use of construction knowledge and experience in planning, engineering, procurement, and field operations to achieve overall project objectives. Implementation of a planned constructability program helps to optimize project costs by successfully integrating construction knowledge into design engineering.

9.4.3.5   Economy of Design

The economy of design criterion relates to overdesign or inefficient design. Overdesign can be reflected by the amount of oversized members and overspecified materials. A poor physical layout of the facility can indicate inefficient design. Context-sensitive design approaches and energy conservation improve design efficiency and aesthetics values.

9.4.3.6   Performance against Schedule

All design users agree that the proper scheduling of design documents and designer specified/procured materials significantly affects a project. The performance-against-schedule criterion reflects the timeliness of design document and materials delivery.

9.4.3.7   Ease of Start-Up

The ease of start-up is a partial indication of the accuracy and efficiency of the design. A measure of the efficiency is obtained by comparing planned to actual start-up time. The number of operators and maintenance personnel required during start-up can also be an indicator of the ease of start-up.

9.4.3.8   Environmental Stewardship

Environmental stewardship in design can be evaluated using environmental impact assessment group of criteria. This group of criteria can evaluate built infrastructure effectiveness in minimizing disruptions to communities, sites of cultural diversity and history, biodiversity including habitats and endangered species, ecosystems including wetlands and forest cover, and air quality and water quality. An example of this type of evaluation is the impact assessment of increased traffic caused by construction of a new shopping mall on increased noise level, groundwater contamination from paved-surface runoff and municipal sewage, harmful vehicle emissions, and GHG emissions.

9.4.4   Criteria Summary

The eight criteria discussed in the previous section are important to all design users, and are relevant regardless of the type of project, construction project activity, or impact of any project variable. Each criterion should be used in any initial evaluation of design effectiveness, but this list should not be construed as absolute or all-encompassing. Because of industry, company, or project differences, some design users may identify additional important criteria.

Not all of these criteria influence or indicate design effectiveness to the same degree. Several of the criteria may be better indicators of an effective design and deserve greater emphasis during an evaluation. Criteria weighting is necessary to properly recognize these variations and evaluations can be combined using an "objectives matrix" technique discussed in the following section.

9.4.5 The Objectives Matrix

Riggs [Riggs 85] and others have suggested the use of a technique called an objectives matrix for productivity evaluation. The same concept can be used to develop an effectiveness measurement for design. An objective matrix consists of four main components: criteria, weights, performance scale, and performance index. The criteria define what is to be measured. The weights express the importance of the relative criteria to each other, and to the overall objective of the measurement. The performance scale compares the measured value of a criterion to a standard or selected benchmark value. Using these three components, the fourth component, the performance index, is calculated, and the result is used to indicate and track performance [Stull 86].

9.4.6 Concluding Remarks on Design-Effectiveness Evaluation

A design-evaluation matrix can be used for any project type or area of a project. The criteria and weights used in the matrix can be modified to fit any project, and the evaluation can adapt to most needs. It can be as simple (subjective ratings) or as sophisticated as desired. Quantitative measures are obviously preferred to subjective ratings as a means of evaluation. Considerable data from many types of projects are needed to develop norms for values of subcriteria.

Given the many complexities and variables of the total design process, no measurement system can yield absolute quantitative results that are applicable without interpretation to all design situations and circumstances. However, the method for evaluating design effectiveness can be used to:

1. Develop a common understanding among the owner, designer, and constructor concerning the criteria by which design effectiveness on a given project will be measured.

2. Compare design effectiveness of similar projects in a systematic and reasonably quantitative manner, highlighting performance trends.

3. Identify opportunities to improve the effectiveness of the entire design process and contributions to the ultimate result of all participants.

9.5   Value-Engineering Applications in Design

Value Engineering (VE) is an effective approach to enhance design and justify sustainable infrastructure management strategies. The VE implementation by FHWA in the United States started in early 1970s after becoming a Congressional law in Title 23 [Jackson 06]:

- The FHWA initiated a VE training program for state highway agencies during the period from 1973 to 1975.

- The National Highway System (NHS) Act of 1995 mandated VE review for all NHS projects costing over $25 million, FHWA's VE regulation-implementing law was published on February 14, 1997.

- The Safe, Accountable, Flexible, Efficient Transportation Equity Act—A Legacy for Users (SAFETEALU) surface transportation legislation (Public Law 109-59) of August 10, 2005 continued environmental stewardship and streamlining efforts. The key provisions of SAFETEALU regulations by FHWA included:

  - States shall provide a VE analysis for each highway project on the federal-aid system with an estimated total cost of $25 million or more.

  - States shall provide a VE analysis for each bridge project with an estimated total cost of $20 million or more.

  - Any other project shall be subjected to a VE analysis if the secretary determines it to be appropriate.

  - Contractors are allowed to propose innovative alternative materials and methods through Value Engineering in Construction Project (VECP) studies.

The total value of FHWA's VE project savings in fiscal year 2005 exceeded U.S.$ 5 billion [Jackson 06]. The VE approach examines all potential cost items for identifying measures that can reduce initial cost and/or life-cycle costs without compromising the function of the item(s) and safety. According to Pareto's law, 20 percent of the items make up 80 percent of the total cost of a product or asset (Figure 9-2) which can help to identify construction items for alternative materials and methods.

In summary, VE methodology adopted by an independent team is described as follows:

- A systematic process of review and analysis of a project after the original design by multidisciplined team of person not involved in the project.

- Recommendations of the VE team are intended for:

  - Providing the needed functions of safety and reliability

  - Achieving the lowest overall cost

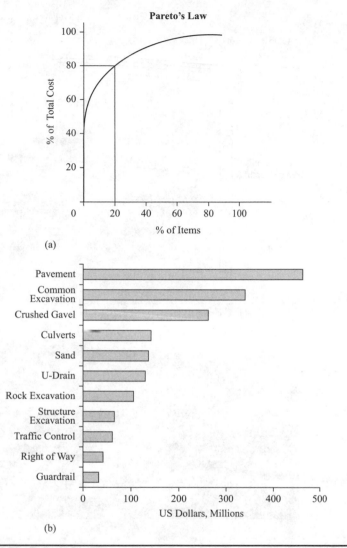

FIGURE 9-2 Illustration of Pareto's law (a) and application to a road project (b) [after Jackson 06].

- Improving the value and quality of the project
- Reducing the time to complete the project
- Combining or eliminating otherwise inefficient use of costly components or using different technologies, materials, or methods so as to achieve the original project objectives.

An example of a VE study is the use of fiber-reinforced plastic (FRP) sheet piles instead of traditional concrete sheet piles by the

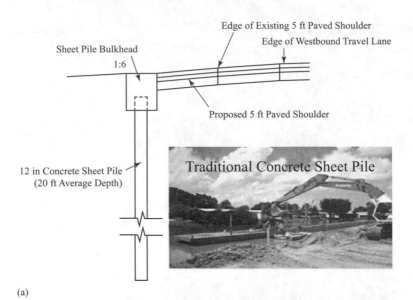

(a)

(b)

Figure 9-3   Traditional concrete sheet piles (a) and FRP sheet pile (b) Installation (credit: Florida DOT) [after Lieblong 06].

Florida DOT to protect coastal roads from hurricane destruction (Figure 9-3). Installation of innovative noncorrosive lightweight polymeric composite FRP sheet pile is easier and 3 to 4 times faster than concrete sheet piles [Lieblong 06]. Based on the results of the Florida VE study the overall initial construction cost was reduced by one-third.

9.6   Summary

Design is a process with a substantial degree of subjectivity, and is influenced by many factors. Nonetheless, it provides a project's central point of translation of ideas into specific instructions.

The method presented herein is for the evaluation of a design, not a designer. It is recognized that many parties and factors influence the final design product. The proposed method is based upon the following eight evaluation criteria:

1. Accuracy of design documents

2. Usability of design documents

3. Cost of design

4. Constructability

5. Economy of design

6. Performance against schedule

7. Ease of start-up

8. Environmental stewardship

Not all of the criteria influence a design to the same degree. A method is presented for combining weights and performance ratings of the eight criteria into a single performance index. The method utilizes an objectives matrix approach and provides the flexibility of using a variety of criteria, weights, and measuring systems to compute the performance index.

The ability to measure design effectiveness using the proposed method represents an important step in a broader effort to improve the total design process. Such an effort encompasses identification of the effect on design effectiveness of: (1) various inputs to the design process; and (2) the systems and techniques employed by the designer.

9.7   References

[AASHTO 85] "AASHTO Guide for the Design of Pavement Structures," Vol. 2, Appendix EE, American Association of State Highways and Transportation Officials, Washington, DC, 1985.

[AASHTO 93] "AASHTO Guide for Design of Pavement Structures," American Association of State Highways and Transportation Officials, Washington, DC, 1993.

[AASHTO 08] "AASHTO Mechanistic Empirical Pavement Design Guide," American Association of State Highways and Transportation Officials, Washington, DC, 2008.

[Chesher 85] A. Chesher, and R. Harrison, *User Benefits from Highway Improvements, Evidence from Developing Countries*, World Bank Report, Johns Hopkins University Press, 1985.

[CII 86] The University of Texas at Austin, "Evaluation of Design Effectiveness," *A Report to The Construction Industry Institute, Publication 8-1*, July 1986.

[Darter 71] M. I. Darter, "Uncertainty Associated with Predicting 18-Kip Equipment Single Axles for Texas Pavement Design Purposes," Center for Highway Research, The University of Texas at Austin, October 1971.

[Darter 72] I. I. Darter, B. F. McCullough, and J. L. Brown, "Reliability Concepts Applied to the Texas Flexible Pavement Design System," *Highway Research Record 406*, Highway Research Board, Washington, DC, 1972.

[Darter 73a] M. I. Darter, W. R. Hudson, and J. L. Brown, "Statistical Variations of Flexible Pavement Properties and Their Consideration in Design," *Proceedings, Association of Asphalt Paving Technologists*, 1973.

[Darter 73b] M. I. Darter, and W. R. Hudson, "Probabilistic Design Concepts Applied to Flexible Pavements System Design," *Research Report 123-18*, Center for Transportation Research, The University of Texas at Austin, 1973.

[Jackson 06] Jackson, Donald R., "Assessment of FHWA Value Engineering (VE) Program," *Presentation at Value Engineering Best Practices Workshop 156*, Transportation Research Board Conference, Washington, DC, January 22, 2006.

[Lieblong 06] Lieblong, K., "Hurricane Case Studies," *Presentation at Value Engineering Best Practices Workshop 156*, TRB Annual Meeting, National Research Council, Washington, DC, January 22, 2006.

[Riggs 85] J. L. Riggs, "What's The Score?" *The Military Engineer*, Vol. 77, No. 50, September-October 1985, pp. 496–499.

[Stull 86] J. O. Stull, and R. L. Tucker, "Objectives Matrix Values for Evaluation of Design Effectiveness," *A Report to the Construction Industry Institute, The University of Texas at Austin*, August 1986.

[Zaniewski 82] J. P. Zaniewski, G. Elkins, B. C. Butler, M. Paggi, G. Cunningham, and R. Machemehl, *Vehicle Operating Costs, Fuel Consumption, and Pavement Type and Condition Factors*, Federal Highway Administration, 1982.

CHAPTER 10

Construction

10.1 Introduction

To fulfill its purpose, an infrastructure asset management system (IAMS) must follow through from the design phase to the implementation phases of construction, maintenance, and rehabilitation, plus data feedback. The transition from design to construction represents one of the most important organizational boundaries in the management system.

Construction converts a design recommendation into a physical reality. Successful construction meets the planning and design objectives within budget and time constraints. It may be sufficient if, after bidding, contract award, construction schedule, materials supply, and processing, actual construction and quality control are conducted in a routine manner. It is not sufficient, however, to perform the design function without thought to construction or maintenance.

If conditions in the field differ from the design inputs or assumptions, changes have to be made. Such changes, at least major ones, should not be made without consulting the design group. Proper management will help ensure that sufficient communication occurs, that new or innovative design or construction solutions are not inhibited, and that as-constructed documentation is complete and understandable. This latter point is most important. Because of the usual hurry and day-to-day problems of construction, it is easy to delay or avoid such documentation as actually built. There have been many experiences in which a search of the files after construction yields incomplete or erroneous information. Proper feedback and plans for rehabilitation depend on reliable and complete as-constructed documentation.

This chapter does not attempt to provide details on construction project delivery protocols, construction practices, construction control, or construction management per se. These are comprehensive subjects that are treated in a variety of construction manuals and guides published by various agencies. Attention in this chapter is focused on: (1) interrelationships of construction with other phases of infrastructure management; (2) constructability of the project;

(3) construction quality control and quality assurance; and (4) documentation or data that construction should produce for construction use and the use of others. If these functions are carried out systematically, then the normal and expected variations in construction methods, equipment, materials, and environment, can be taken into account properly.

10.1.1 Construction Documents

To move the selected design to the construction phase, a set of documents defining the details of the selected design is needed. In some cases, alternative designs may be offered for contract bidding and then the details of each are needed. These documents not only convey details for construction, but also serve as legal documents in procuring the services of a contractor or construction agency.

The traditional documents of design and construction are "plans and specifications." These usually consist of the following:

- A set of drawings that give detailed dimensions and other design aspects
- A set of project specifications that describe in detail the materials to be provided, their arrangement, required characteristics, and so on
- A set of "standards and specifications" that have been previously approved by the agency and are in general use

The first purpose of these implementation documents is to describe the proposed project to the construction group. For example, because a road project is usually built in conjunction with bridges, drainage facilities, and other items, the interrelationships involved also are described. In some agencies, the construction group is equipped to construct the project as designed, and in such cases construction can begin immediately. Usually, however, the construction group has a supervisory or control function and the actual construction is done by an independent agency under a legal contract. In these cases, the documents become highly important because they serve a second purpose as the basis for bidding, pricing, and agreements about payment for the work.

The 21st century has witnessed great strides made in computer-hardware and software technologies, IT infrastructure, Internet sources, and wireless communications. These developments of cyber infrastructure have led to an evolution toward a more paperless world encompassing most sectors of society and including e-mail communication, Internet banking and financial transactions, procurement and inventory processes, organizational message-streaming and statements, supply-chain flow of material and information, and social networking. Construction-management processes and document flow are no exception. Digital data and e-document technologies have been embraced by government agencies, private owners and developers, material suppliers, contractors, and others involved in infrastructure construction.

10.1.2   Construction Management

Construction management involves the use of physical, financial, and personnel resources to convert designs to physical reality. This general concept intersects infrastructure management in the actual building of the facility. The processes are successful when stated planning and design objectives are met and the facility is put into service.

The process of construction management is comprehensive and complex. It involves many considerations, such as estimation, designation, and scheduling of activities; organizational and personnel aspects; legal aspects; finances and cost control; and records of construction documents. There are many published books and manuals available on the subject. Although it is beyond the scope of this book to treat this subject in depth, the levels of construction management as they apply to public agencies are subsequently discussed.

10.2   Construction as Related to Other Phases of Management

An effective IAMS depends on communication and coordination among all phases. General information flows to construction from the other plans of the management system. Details of these other phases have been discussed in previous chapters. Construction receives vital information from all of the phases.

Construction also provides information to these other management phases. The general nature of this set of information flows is given in Figure 10-1. The diagram shows that information from construction is vital to efficient management in the other phases.

Five phases, or subsystems, are considered in Figure 10-1: planning or programming, design, construction, evaluation, and maintenance. On first thought, it might seem that construction should interface only with design and maintenance, but this is an incomplete formulation. Construction is important to the entire management system, and these relationships can be examined in terms of the information flows illustrated in the figure. These information flows occur in both directions—to and from the construction phase.

10.2.1   Planning and Construction

The planning, or programming, phase provides the "what, when, and where" type of information to the construction and other phases. Information on the type of work required and the location of the work can provide direction for both short- and long-term construction planning for the workforce and other resources. Often, this

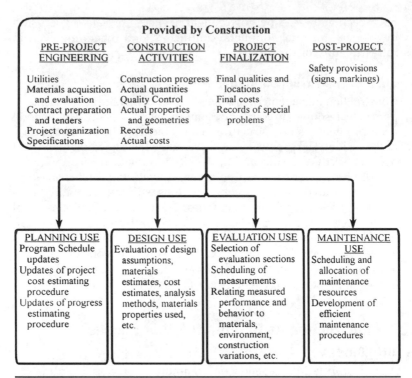

Provided by Construction

PRE-PROJECT ENGINEERING	CONSTRUCTION ACTIVITIES	PROJECT FINALIZATION	POST-PROJECT
Utilities	Construction progress	Final qualities and	Safety provisions (signs, markings)
Materials acquisition and evaluation	Actual quantities	locations	
Contract preparation and tenders	Quality Control	Final costs	
Project organization	Actual properties and geometries	Records of special problems	
Specifications	Records		
	Actual costs		

PLANNING USE	DESIGN USE	EVALUATION USE	MAINTENANCE USE
Program Schedule updates	Evaluation of design assumptions, materials estimates, cost estimates, analysis methods, materials properties used, etc.	Selection of evaluation sections	Scheduling and allocation of maintenance resources
Updates of project cost estimating procedure		Scheduling of measurements	Development of efficient maintenance procedures
Updates of progress estimating procedure		Relating measured performance and behavior to materials, environment, construction variations, etc.	

Figure 10-1 Information provided by construction for potential use by other management phases.

preconstruction information, involving sufficient lead time, can provide better contract prices based on better availability of contractors and less crowding of the work. It is also possible to accomplish the preconstruction activities of surveys, property acquisition, utilities relocations, and so on in a much more timely, economical, and efficient manner. The overall system, or network economic planning and analysis, will provide information for scheduling contract tenders, personnel, and resources and provide for consideration of time restrictions and related items.

In the reverse direction, construction relates to programming in many ways. For example, comparisons of final contract costs and materials quantities with the original programming estimates can help in updating for future programming. Monitoring of construction in progress can point out the need for schedule adjustments. Cost overruns are not always available, and early knowledge can make it possible to reprogram without penalty to other contracts. Finally, completed construction and project openings are directly important to the next round of programming.

10.2.2   Design and Construction

The plans and specifications provide direct design input into construction. This is the most obvious and direct interaction between the two phases. In addition to contract documents and details for construction, however, proper interaction in the early stages can assist construction in the preliminary project field-engineering phase and in setting up preliminary methods for locating materials and establishing quality- and quantity-control techniques.

Feedback from construction to design is equally vital. A seemingly economical design is not effective if it results in unusual or unmanageable construction problems. For example, a material may appear to be economical based on previous data, but a particular job or location may result in short supply and greater expense than expected. On the other hand, new construction techniques may lead to design changes. For example, all major highway departments previously required thin lifts of asphalt concrete in construction to gain adequate compaction. However, with the advent of heavier and variable-tire-pressure pneumatic rollers, and some construction trials, it has been shown that thick lifts may actually give better compaction under many situations. This type of feedback from construction led many agencies to change their specifications. A good management system can help keep new ideas from being discarded because of old methods.

10.2.3   Evaluation and Construction

In its broadest sense, evaluation provides input to the construction phase. Continuous monitoring and evaluations of existing facilities can show that certain construction methods or technology, although acceptable under existing specifications, can lead to premature deterioration or rapid rates of deterioration. For example, aggregate gradation is controlled on a before-compaction basis. Subsequent testing in later evaluations may show that construction has actually degraded the material excessively, which may lead to structural weakness or drainage problems. A continuous and complete data-feedback system obtained by continuous evaluation of a structure and its components can assist in solving many such potential construction problems.

The input from construction to evaluation is direct and important. The as-built construction records provide the initial or zero-age evaluation of the facility. Furthermore, if the construction data are good and effectively recorded, they will form the backbone of all future evaluation because the data will be more detailed than can ever be obtained again. In addition, these records can be used effectively to assist in selecting the initial location of sections to be evaluated periodically. This is important because most evaluation is a sampling process.

During periodic evaluations, unusual problems observed should be checked directly against construction records. Unfortunately, in the past construction records have often been inadequate to fulfill these needs. A properly applied management system will assist in solving this problem.

10.2.4   Maintenance, Rehabilitation, and Construction

The last phase considered in relation to construction is maintenance and rehabilitation. Both routine repairs and work intended to restore or improve a facility are included in the definition of maintenance and rehabilitation. Because this phase is a continuing process, careful consideration of the maintenance required and where it will be done may show some patterns that provide feedback to construction and show up weaknesses in the quality of specifications, construction methods, and/or materials and supervision methods. Because restoration or rehabilitation work is often done by contract, it is considered by many to be a type of construction. In any event, scheduling of the work involved will require consideration and supervision by the construction group. Thus, it requires close cooperation and coordination.

Regarding the information flows from construction to maintenance, effective maintenance techniques and procedures are dependent on the actual materials and construction methods used. They are also related to problems that were encountered during construction.

For example, poor concrete performance can often be shown to be the result of a period of unusually hot weather encountered during construction, perhaps interrupting placement or disturbing finishing operations. In such cases, the as-built construction records form the keystone of the maintenance database. The information on final acceptance, claims, and adjustments can serve as the basis for the preliminary scheduling of maintenance programs and resources. Likewise, knowledge of the actual quantities and qualities of the materials used is essential to future maintenance planning.

10.3   Constructability

Knowing how to build facilities is crucial to optimal planning, designing, and administration of any construction. Know-how is the result of education and experience, and, like any other knowledge, once learned it must be extended and reviewed.

Construction experience is a premeditated learning process whereby a participant in, or observer of, construction operations understands what is taking place and can further describe and communicate to others the events, their purposes, and outcomes.

10.3.1   What Is Constructability?

Constructability has been defined by the Construction Industry Institute (CII) [CTR/CII 87] as the expediency with which a facility can be constructed. Constructability is enhanced by the proper use of construction knowledge and experience in planning, design, procurement, and field operations to achieve overall project objectives [CII 86].

Because constructability enhancement (CE) is by nature multidisciplinary and multifaceted, it means different things to the various participants in a project. To the project owner, constructability affords the opportunity on construction projects to achieve greater efficiency, with resulting lower cost, shortened schedule, or improved quality. To the designer, it is an understanding of the methods and constraints of the actual construction required to execute the design being made. To the contractor, it is a combination of the effort required to implement the design efficiently and the opportunity to minimize his or her effort and resource expenditure.

The CII has identified a number of constructability concepts applicable to the different phases of a project. Briefly, these concepts address project-execution planning, conceptual project planning, specifications, contracting strategies, schedules, and construction methods, including those concerning preassembly, site layouts, design configurations, accessibility, and adverse weather [CII 86]. Constructability improvement has been studied and applied to many segments of the industrial construction industry and to highway construction and public facilities as well [De Vos 89, Fisher 89, O'Connor 86].

Constructability is indeed practiced to some extent by planners and engineers of state Departments of Transportation (DOTs), although it may not have been formally defined and thought of as a primary factor in highway design and construction.

10.3.2   Relationship between Constructability, Value Engineering, and Productivity

Value Engineering (VE), has similarities to constructability, and the differences may not be apparent at first. Innovative construction practices, leading to cost reductions, can be attributed to both constructability and VE, from which it might be concluded that the two are synonymous. There are similarities in objectives and result, but the scope and reach of constructability and VE are different, as shown in Figure 10-2.

VE is defined as a disciplined procedure for analyzing the functional requirements of a product or service for the purpose of achieving the essential functions of the product or service at the lowest total cost. Total cost, in this case, takes into account the owner's cost of planning, design, procurement and contracting, construction, and maintenance over the life cycle of the product or service and may also consider user costs. In the case of a governmental agency responsible for delivering a service, user costs should be considered

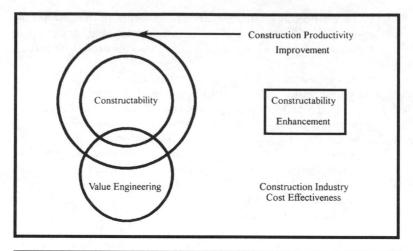

FigurE 10-2   Construction industry cost-effectiveness relationships.

in addition to production costs when the cost-effectiveness of a service is estimated.

Highway constructability, for example, is primarily concerned with optimal construction costs consistent with the function and quality requirements and boundaries set out by the standards or policies of the DOT. Thus, (CE) should be considered, as one of several tools of VE. The effectiveness of both VE programs and CE programs depends upon participants who can contribute and work as members of a team.

From a practical viewpoint, contracted construction costs are by far the largest item in an agency's budget, and it is likely that the most significant improvements in productivity will come about through CE. In the case of an agency which operates more or less on a fixed annual budget, productivity is equated with lower construction costs and/or improved quality, which translates into increased durability and improved operations. One of the functions of CE, therefore, is to provide feedback of construction experience to programming, planning, design, and construction.

The application of VE is similar in many respects to CE in that both concepts are applicable throughout the planning, design, and construction phases, and each is likely to have the most impact during the early stages of project development.

The foregoing discussion underscores the significance of early and effective consideration of constructability as part of an IAMS. This is particularly true during the project planning and design phases, where the value of construction knowledge and experience, applied at the right time, can render the highest dividends. It is clear that once the project advances beyond these phases, investment and

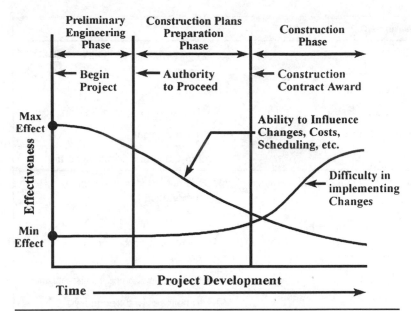

Figure 10 3 Significance of early decisions (adapted from [Azud 69, CTR/CII 89]).

other commitments generally accumulate at rates depicted by the well-known S-curve. The ability to make changes to a project relates strongly to the S-curve. In the same vein, the ability to influence and control costs reduces rapidly, since it is inversely proportional to the same curve. These trends are shown in Figure 10-3.

Thus, in summary, the relationships among productivity, VE, and constructability, all of which are important to effective infrastructure management, are as follows:

- Productivity is a measure of the output/input ratio in constructing a facility.
- Value engineering is concerned with providing the required functions of the facility at the least cost.
- Constructability is a measure of ease or expediency of construction.

10.3.3 Factors Affecting Constructability

During the development of the various elements of a project, constructability is influenced by numerous and diverse factors. A listing of these factors is given in Table 10-1, in which the factors have been grouped into seven categories for subsequent incorporation into a knowledge base.

With the assistance of an agency steering committee, a listing of constructability issues or concerns can be identified and prioritized

Environmental Systems	

• Site
  • Topographical (including accessibility)
  • Geotechnical
  • Hydrological
	• Infrastructural (including vehicular traffic)
	• Political/legal/regulatory
	• Economical/sociological/financial
	• Technological

Project Scope	**Information and Communication**
Operational requirements Facility characteristics • Structural composition • Complexity • Scale Financial and time constraints • Budget and schedule	Documentation/transmission/interpretation • Availability/source/accuracy • Clarity and conciseness/completeness • Consistency/compatibility/ambiguity • Timeliness and frequency • Relevancy
Resources	**Processes/Methods Pertaining to**
Material/manpower/plan and equipment • Availability • Variability/flexibility • Suitability • Intrinsic attributes	Planning, design, specification, and estimate • Procurement/bidding • Construction • Maintenance
Controls	**Innovation**
Quality assurance/quality control; testing and inspection • Cost and financial control • Schedule control • Productivity measurement	Awareness of prompters; recognition of need; stimulation/encouragement • Motivation and freedom to innovate • Capability to innovate • Resources and R&D • Support of champion/innovative leaders

Table 10-1   Factors Affecting Constructability (with Highway as an Example) [CTR/CII 89]

- Planning and design guidelines for enhanced highway construction

- Specification improvements for enhanced highway constructability

- Selection, processing, and management of materials

- Constructability enhancement through innovation

- Facilitating construction under traffic

- Facilitating future expansion and upgrade

- Optimal utilization of plant and equipment

- Optimal risk/responsibility allocation

- Constructability program implementation

TABLE 10-2   Highway Constructability Issues and Concerns

for research and development, as outlined in Table 10-2. Understandably, the list may be long and the topics rather broad in scope.

10.3.4   Constructability Enhancement Programs

To enhance constructability, good objectives must be established. Typically, they are there to.

- Increase productivity
- Reduce project costs
- Reduce project duration
- Reduce delays/meet schedules
- Eliminate unnecessary activity
- Reduce physical job stress
- Promote safety on construction sites
- Reduce conflict
- Increase quality

10.3.5   The Highway Constructability Knowledge Base

As defined by the Texas Department of Transportation (TxDOT) and the Center for Transportation Research [CTR/CII 89], a highway constructability knowledge base (KBS) is a collection of ideas obtained by personal interviews, expert sessions, and literature reviews. A computerized information-retrieval system for detailed treatment of constructability ideas is being developed [Redelinghuys 89]. This will relate the hierarchical diagrams to the various elements of highway projects to be constructed—for example, pavements. The hierarchical diagrams are also related to other aspects, such as the constructability factors in Table 10-1 and the applicable engineering phase.

This approach also offers an efficient structure for continuing analyses and further research. Typical examples of applications or solutions to constructability constraints are given below:

- Acquire right-of-way (ROW) in a timely manner; resolve problematic ROW parcels early.
- Minimize demands for ROW likely to be difficult to acquire; employ techniques for achieving steeper cuts/fills when necessary.
- Reduce delays in utility adjustments; design to avoid utility adjustment by using utility bridges or tunnels.
- Ease or secure the traffic/construction interface; use concrete safety barriers.
- Speed up on-site bridge construction; optimize off-site prefabrication for bridge construction.
- Provide space for contractor accessibility and staging; early consideration of additional ROW needs of constructor for storage and staging, access, and parking.
- Employ tactics to reduce overall project duration; ensure efficient scheduling of construction activities; effective use of liquidated damages and incentives.
- Optimize pavement unit productivity; minimize the number of pavement layers, particularly in intersection design.
- Develop a pavement construction execution plan; develop a comprehensive inspection plan commensurate with the anticipated progress of the project.
- Design earthworks to enhance constructability; allow innovative deep-lift compaction.
- Facilitate future expansion and upgrade; locate utilities to minimize future adjustments; design to take into account future maintenance and expansion.
- Minimize specification-related problems; change unrealistic tolerances; remove references to obsolete methods or materials; ensure consistent interpretation by removing ambiguity and educating and training users that the cost incurred in the design of the utility bridge to carry the weight of the signs and the additional wind load was negligible.

10.3.6 Program Implementation: Barriers and Recommendations

10.3.6.1 Barriers

Although the benefits of a CE program have been demonstrated, barriers to practicing good constructability are common. The CII points out that managers should be aware of these barriers, listed in Table 10-3, and challenge them.

Barriers to Communication and Design-Construct Integration	Barriers to Utilization of Advanced Construction Technologies	Barriers to Innovation
Contract time	Lack of awareness of technologies/inadequate communication	Discouragement of personal initiative, and perceived lack of freedom
Lack of time	Lack of necessary training	Failure to recognize opportunities
Lack of field feedback	Regulatory inhibitors	Lack of personal creative ability
Failure to document and communicate "lessons learned"	Institutional and individual resistance to change	Lack of tools
Lack of construction experience	Reluctance to deviate from current and proven standard operations	Lack of senior support of champions

TABLE 10-3   Barriers to Program Implementation

10.3.6.2   Recommendations

According to the CII, the following recommendations should be considered in implementing a highway constructability program:

1. The commitment of senior management to constructability must be obtained.

2. A strong approach to project management with a single point of responsibility should be pursued.

3. Project-execution plans should be developed for large complex projects during a project-concept conference. Additional planning meetings and design reviews should involve greater participation of involved parties.

4. A proactive approach to constructability needs to be taken. Overreliance on late, reactive design reviews should be avoided.

5. Feedback from the field, if not forthcoming, should be solicited on a periodic basis, prior to, during, and after construction. This feedback should involve department personnel, contractors, and suppliers.

6. Postmortems should be conducted upon completion of all projects. These should be attended by representatives from the owner and the contractor. Other interested parties should also be invited. These meetings should be used to report on

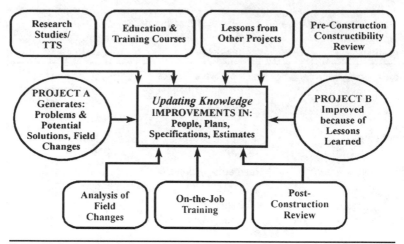

FIGURE 10-4 Improving constructability.

"failures" as well as "successes." Increased opportunities for site visits should be made available.

7. Management training programs that promote communication and integration between design and construction should be conducted. Project "team building" should be initiated on a trial basis and should include exercises for developing team-leadership skills.

8. An accessible and current knowledge base of "lessons learned" should be maintained. Advanced, computerized systems are being developed for storing and retrieving the information.

Figure 10-4 illustrates the foregoing recommendations.

10.4 Construction Quality Control and Quality Assurance

Construction quality assurance is a complex and detailed process to guarantee that the finished facility is built to the standards desired by the owner (more precisely, as defined in the design documents). There are several aspects of quality assurance that are important, including quality control, quality assessment, and testing [Warden 63, Willenbrock 76]. Quality control, quality assurance, and total quality management (TQM) are further discussed in Chapter 11. The discussion in this chapter is directed to construction.

No matter what type of quality assurance is undertaken, all of them start with specifications. There are several types of specifications used in the construction of infrastructure. These are variously termed as:

- Methods and materials specifications
- Recipe specifications
- End-result or end-product specifications
- Performance-based specifications
- Guarantee or warranty specifications for a specified time or utilization period

As of this writing, methods and materials specifications are the most common. Although there is some difference in terminology, recipe specifications are generally the same as method and material specifications. That is, the recipe requires the inclusion of certain ingredients or materials and requires that they be mixed, manufactured, installed, or placed in a specified way.

10.4.1   Recipe Specifications

The most typical specification in use today is the recipe specification. For Portland cement concrete, for example, the construction specification usually specifies aggregate type, and the gradation percent of aggregate, cement, and water that will be blended for the mix plus one or more additives to improve workability and increase set time. A mixing time may also be specified. Such specifications for concrete, however, also generally specify a minimum-acceptable strength; for example, a breaking compressive strength of not less than 20,670 kPa (3,000 psi).

A look through any specification book for a city, county, or state public agency, as well as any major contract for building construction in the private sector, reveals that most of the specifications are of the recipe type. This represents a conflict in that the contractor is told everything to do in the process, and then what minimum strength must be obtained. On the other hand, a liability exists if the recipe is faithfully followed, using the specified materials, but does not meet the strength requirement. In such a case, where does the fault lie?

This has been handled in the past by giving a sufficient factor of safety in the strength requirements, and in reusing good historical materials. Nevertheless, many legal claims must now be settled in court or by arbitration over recipe specifications with conflicting components. It is rather like telling your partner exactly what to include in his or her cake, and then being angry because it does not suit your taste.

10.4.2   End-Result Specifications

There is an increasing use of end-result or end-product specifications to overcome some of the problems with recipe specifications. Rather than specifying the materials and methods to be used, the end-result specification concentrates only on the final product. That is, in our metaphor, the taste of the cake rather than how it was made. End-result specifications are now generally accepted by both contractors (as producers) and public agencies (as purchasers) of infrastructure products [ASTM 78, AGCA 77, Gappinger 85]. There are several reasons for the increasing popularity of end-result or product specifications in contract work. First, end-result specifications help to determine the responsibility for product-quality control and quality assurance. Second, they provide economic benefits to contractors who produce high-quality work and to the purchaser (public agency) in the form of improved products and reduced life-cycle cost. This occurs because end-product specifications allow the contractor or producer the flexibility to use innovation in producing the required end-results.

The use of end-result specifications is, however, a fairly recent phenomenon. In the last decade, for example, concern over ride quality on new pavement surfaces has prompted some state DOTs to develop end-result smoothness specifications [Harrison 89, Hudson 92, SDHPT 82, SDHPT 91].

10.4.3   End-Result Illustration

To apply an end-result specification, it is necessary to have a valid measuring tool that both the producer and owner can accept as understandable and valid and that can be checked and calibrated efficiently as needed.

In terms of pavement smoothness, a proposed specification has been developed by TxDOT [Hudson 92, TxDOT 91]. The main limitation to developing this specification was finding available roughness-measuring equipment that would provide a stable reference for ensuring contractor compliance. Several devices were examined, and a modification of the profilograph, developed in California [AASHTO 87], was selected because it best fulfilled the needs of end-result specifications, even though it was not the most widely used roughness-measuring device available. In this case, a device that produced a visual trace that could be seen by the producer and the buyer, as well as a stable reference platform, were more important than speed of operation or ultimate units used.

Any end-results specification must be applied realistically. That is, the acceptable level must not be set so high that it is unattainable by a good contractor. Conversely, it is undesirable to set the specification acceptance criteria, or level, so low that any contractor can meet it. To follow the principle of setting a realistic inducement, it is necessary to obtain adequate field data on the general performance of the various contractors carrying out the proposed procedures. Usually, the level should be one that 10 to 20 percent of the contractors can typically meet, and 80 to 90 percent will fail to meet. This means that the level is obtainable, but a large number of existing contractors will need to improve their procedures to meet it. Table 10-4 illustrates the specification produced for TxDOT.

Profile Index (PI) Inches/Mile per Each 0.1 Mile Section	Bonus or Deduction Percent of Unit Bid Price
3.0 or less	+5
3.1 thru 4.0	+4
4.1 thru 5.0	+3
5.1 thru 6.0	+2
6.1 thru 7.0	+1
7.1 thru 10.0	+0
10.1 thru 11.0	−2
11.1 thru 12.0	−4
12.1 thru 13.0	−6
13.1 thru 14.0	−8
14.1 thru 15.0	−10
Over 15.0	Corrective work required

TABLE 10-4   Preliminary Smoothness Pay-Adjustment Schedule [Hudson 92]

The specification level also indicates another major principle of good end-result specifications—bonus provisions. Experience shows that contractors subject to being penalized feel much better if they also have the possibility of being rewarded. When such a bonus, or reward, system is included, end-results specifications are much more likely to be successful.

10.4.4   Factors Influencing Evaluation and Enforcement of a Specification

In evaluating pavement smoothness, as with any other end-result specification, it is desirable to have a standard and accurate instrument against which all other instruments can be compared. Unfortunately, many times no standard instrument exists. Lacking an acceptable standard, the next-best thing is to select a particular instrument type and accept it as the standard output device for use in enforcing and developing the specification.

Repeatability of the measurements of compliance must be carefully evaluated before equipment and specification levels are set. The variability introduced by the equipment operator must also be evaluated. It should be remembered that these comparisons must be made under field conditions, that is, the conditions that are expected for acceptance and routine use in the field.

It is also important to know how variability could affect management interpretation by the same reader. That is, might a profilogram, for example, be interpreted differently in the morning (when the reader is alert), than it would be later in the day (after the reader has begun to experience fatigue)? In attempting to answer this important question, the study team in the TxDOT study collected information on variability considered to be realistic to determine what kind of variation can be expected in real-life data collection. This avoids the problem of selecting a band so narrow in the specifications that it is unreasonable to expect the contractor to meet it.

The data for this analysis were obtained by having the same interpreter read several profilograms several times for both high- and low-profile level pavements.

10.4.5 Summary of Example

A new smoothness specification for flexible pavements, based on end-result criteria, was developed and implemented successfully in Texas (see Table 10-5). The development of this specification has led to the following recommendations.

Although most asphalt concrete pavements are relatively smooth, an improved specification shows that they can be made even smoother by paying a bonus to the contractor for high-quality work. End-result smoothness specifications are successful in reducing the small amount of roughness present in new asphalt-concrete pavement contracts. The bonus needs to be sufficiently attractive to induce consistent, high-quality work throughout the entire contract; however, because roughness in new contracts is relatively localized, the total bonus as a percentage of the bid price can be quite small. A highway department, therefore, risks only a small increase in job price for the considerable benefits that accompany a rise in ride quality and serviceability indices. The bonus incentive has an additional feature related to contractor bidding: end-result smoothness specifications make it more difficult for inferior contractors to win bids. This is because experienced contractors familiar with smoothness specifications can discount their bonus into the bid price. In this way, the department does not pay a large total price for the contract. The bid plus bonus becomes very close to original bid-price levels. The Texas experience shows that if the contractor is inferior or inexperienced, problems with ride quality increase. End-result specifications, therefore, reduce the probability of poorer contractors being awarded pavement work.

PI (in/mi)	Percent of Unit Contract Price
PI ≤ 9.0	100
9.0 < PI ≤ 10.9	99
10.9 < PI	Corrective work required

TABLE **10-5** Smoothness Specification for Flexible Pavements

In summary, final acceptance smoothness specifications for newly constructed flexible pavements and newly overlaid flexible pavements are excellent example applications of end-result criteria in highway construction. An end-result smoothness specification improves overall pavement ride quality, gives a better service to the road user, and provides the highway agency with an excellent means of specifying and assuring adequate quality of construction. Total bid plus bonus price for pavement work need not be much higher than traditional bids, and inferior contractors have a lower probability of winning bids. The Texas study suggests that end-result specifications are a highly desirable modification to the traditional smoothness specification [Hudson 92].

10.4.6   Performance-Based Specification

Performance-based specifications are substantially more complicated to develop and use than end-result specifications. The concept is to write a specification for the contractor or the facility provider that will require the entity constructing the facility to provide a given level of performance.

Performance is a time-history function. For example, "a pavement shall retain a serviceability index of greater than 3.0 (on a scale of 5 to 0) for five years or more." Such a specification requires a history basis for defining the performance, and a measure of performance to be used for control or compliance. In some cases, performance is defined in the short term. One example is a high-performance car or boat that is "capable of reaching 96 km/h (60 mph) from a standing start in 6.0 seconds." Another example is a high-altitude airplane that is "capable of performing flight at an altitude of 12,000 m (40,000 ft)." These definitions of performance are not generally applicable to infrastructure management. In this case, performance is time-related, as outlined for the pavement, such that a building, bridge, water line, or sewer "shall serve at an acceptable level of service for a known period of time."

A good illustration of the problem associated with such specifications is the Strategic Highway Research Program (SHRP) in the United States. In 1987, the SHRP was set up to develop performance-based asphalt pavement specifications. The research program set out to undertake all sorts of detailed studies of asphalt aggregates and mixes over a five-year program. It was clear, however, from the start that there was not time within the program to develop new materials, build pavements out of them, and truly observe the performance. Therefore, it was patently impossible to fulfill the stated objectives. The program management suddenly changed the title from performance-*based* specifications to performance-*related* specifications to indicate that the specifications were to include factors that were, in general, at least known to impact the long-term performance of the material. The fact that this was already a part of most existing specifications was perhaps not recognized by the SHRP administrative staff. It should be remembered that performance-based specifications cannot be validated until actual performance has been observed in the field and compared to the performance obtained with different qualities or types of materials.

Infrastructure asset management systems provide the ideal framework for developing performance-based specifications; however, it will take a number of years before such specifications can be developed based upon analysis of feedback data from an IAMS database. Users of IAMS are urged to close the loop between observed performance and the quality and reliability of construction to form a basis of future development of true performance-based specifications.

10.4.7 Guarantee or Warranty Specifications for Specified Period or Utilization

Warranty specifications have not been widely used in infrastructure. They are, however, broadly used in areas such as the automotive industry. Improvements in the quality of automobiles have allowed manufacturers to increase from one-year or 16,000 km (10,000 mi) warranties to five-year 80,000 km (50,000 mi) warranties. Similar mechanical warranties are provided for appliances and other consumer machines, even airplanes and ships. Such specifications and warranties, while slow in coming even in the automobile industry, where millions of vehicles are manufactured and sold annually, are even harder to develop in civil infrastructure. One reason is that the automobile industry's very large size makes it relatively easy to obtain good experience and information on reliability, not only of the automobiles but of the various component subsystems, such as engines, transmissions, and brakes.

These warranties do not come free, and they clearly involve an increase in vehicle price. It is possible to estimate the relative cost of such warranties for automobiles by pricing options made available by the manufacturer for so-called extended warranties. A new automobile purchased by one of the authors came with the option of extending the warranty period for an additional 48,000 km (30,000 mi) by paying $750. The price for extending the warranty to 160,000 km (100,000 mi) would have been $1,500, bumper to bumper, all physical items covered. The relative higher unit cost for the longer extension (which amounted to about 3 to 5 percent of the cost of the automobile) accounts for the added wear and the additional chance of failure over the longer period.

The cost of such warranty specifications for highways, bridges, and buildings would likely be considerably higher than 3 to 5 percent because of the lack of broad-gauged reliability experience and because each construction facility varies from every other facility much more than individual cars vary in the automobile production line.

10.4.7.1   Applications of Warranties to Infrastructure

Since each infrastructure unit, such as a pavement, bridge, building, waterline, or sewer line is unique, the guarantee approach is perhaps a better contracting and pricing mechanism than a direct warranty. This approach is often referred to as "privatization of infrastructure," and it has become somewhat more popular in the last few years. It provides the ability to fix the price of the use of the facility for the owner and/or user, but it does not serve in exactly the same role as a recipe or end-product specification. Nevertheless, it fits into construction-quality assurance in the sense that if warranty specifications are used it is no longer up to the owner to inspect, test, and check the construction process on a day-to-day basis.

With warranty specifications, the type of testing for recipe or performance-based specifications is not necessary. However, some way of evaluating the satisfaction of the owner or the facility user over time is needed. In the case of the automobile this is fairly direct, but in the case of a highway or bridge it is more difficult. It is hard to say if the quality of the road serving the user has declined below an acceptable level unless some testing is done during the performance period. One measure might be the number of complaints from users that the ride on the road is rough, that the water pressure is low in waterline, and so forth for other infrastructure assets.

10.4.7.2   Privatization of Maintenance

One major approach to warranty specifications for existing infrastructure is the so-called "privatization of roadway maintenance," as further discussed in Chapters 11 and 13. This approach has been used in a number of countries, such as Argentina and Australia, for roads. It has involved designation of an entire area or network of roads to be bid on by a consortium of financiers, engineers, and contractors. In the State of New South Wales in Australia, for example, each consortium submits a bid in which they agree to provide necessary maintenance, reconstruction, and rehabilitation for a fixed period (e.g., 5 or 10 years) and to do everything necessary to provide the facility at an acceptable service level for the contract price. All of the bids or tenders are then evaluated and a winning group is selected. The group may not necessarily be the lowest price tender, since there may be differences in the level of service and quality provided within the activity of different groups. These approaches have not been in place long enough to fully understand how well they are functioning. The concept is a good one, allowing for economies of scale, good private management, savings in social cost for government workers, and the benefit from the widely accepted concept of higher productivity obtained from private industry. This is the conventional wisdom, which is not necessarily the opinion of the authors and certainly open to debate.

10.4.7.3   Privatization of New Facilities

New facilities are easier to privatize. In the case of roads, this is being done for toll roads in Mexico, Spain, Canada, and other countries. It involves solicitation of offers saying that a roadway system is needed from point A to point B (usually from one major city to another), and it shall have certain connections, certain speeds, certain capacity, and so forth. Proposals are then received from various groups interested in providing a turnkey facility for a monthly or annual cost basis. In this case, the group also includes a bank or financial institution that arranges the necessary financing. Proposals are evaluated, and one is selected—not always on the basis of minimum cost—based on an overall-effectiveness evaluation that is intended to provide the greatest benefit-cost ratio. In some cases, the costs are set as a specific toll for each vehicle using the facility, and the successful bidders are not allowed to charge more than the bid tolls. In other cases the annual cost is set, and the difference between collected tolls and the cost basis are borne by the owner or tolls are raised to cover the difference. In this case all costs, including interest, administration, maintenance, rehabilitation, and operation, are covered in the pricing.

The intention of these initiatives is to take full advantage of the efficiencies attributed to private-sector economics and entrepreneurship. There is inadequate history, however, to determine whether or not they are an effective mechanism for warranty specifications. In at least one reported case in Spain, the group winning the bid went bankrupt within three years. It is understood that the additional cost fell back on the government, even though there were warranty bonds and other mechanisms intended to cover this contingency.

10.4.7.4   Warranty Specifications in Shopping Centers

The private sector offers a better model of warranty specifications, and one that has proven to be successful. It is the standard of the industry, and it is being widely used by major corporations that are capable of providing their own infrastructure but choose to obtain their infrastructure from construction and management specialists at a fixed bid price. More recently, some very large and successful retail businesses, such as Walmart and Lowe's, have their own exclusive construction-management teams who carry out all construction phases and annual maintenance.

It has been a common practice of many large retail agencies, such as Sears and Albertsons grocery stores, not to construct their own stores, but to become the so-called *anchor tenant* in a major shopping center. In such cases, the developer agrees to provide retail space, parking, utilities, and maintenance, required by the major tenant at a fixed or agreed-upon price, usually with inflation clauses and scales to cover unforeseen increases in utilities. Such contracts specify the quality and type of parking, air conditioning, floor coverings, lighting, and so forth that will be provided for the contract price. In return, the tenant obtains needed infrastructure at a known price without becoming an infrastructure provider. As a result, the tenants can concentrate on their primary business of retailing, without the need to extend themselves into unknown areas of construction, maintenance, and repair.

This approach is not a panacea that solves all problems, and there can still be disagreements as to whether or not the service contracted for is being provided adequately. For example, the authors were involved as consultants in a major lawsuit at a large shopping center in San Antonio, Texas. In this case, key anchor tenants had a strong disagreement with the shopping center developer about the quality of parking provided and about drainage problems that caused water to pool in front of their stores, making it inconvenient for the customers, who complained avidly. The approach, however, is a valid one, and IAMSs can assist in ensuring that the process works effectively. Infrastructure asset management for users and providers should consider this approach as an option.

10.5   New Technologies in Construction Management

The evolution of modern high-performance computer systems, cyber infrastructure, digital data, and wireless communication in the 21st century has greatly impacted construction engineering and management functions of infrastructure management. Many phases of construction engineering and management, including document transmission and construction-project delivery, are more efficient in a near-paperless world. Modern high-resolution airborne- and space-borne-remote sensing and spatial technologies are providing more accurate digital spatial data at competitive prices and helping expedite planning and design of infrastructure. Terrestrial and kinematic laser-scanning equipment are providing building information-model (BIM) data, enhanced construction-progress monitoring, and three-dimensional visualization capabilities [Al-Turk 99, Uddin 08, 11, Haas 01, Turkan 12]. These remote-sensing technologies for infrastructure applications are discussed in detail in Part 2 of this book.

10.6   Summary

Construction converts a design into a physical reality. To do this successfully, the design must be well documented and "constructible." The actual construction must be well managed, and the work itself must undergo rigorous quality control and quality assurance. The latter can involve a variety of specification approaches, including method, recipe, end-result, performance-based, and long-term warranty. Each has advantages and disadvantages, with the recipe approach relying on an extensive base of experience. End-result and end-product specifications are, however, finding increasing use in infrastructure management, and warranty-type specifications also have been included in a number of countries. The 21st-century developments of cyber infrastructure, digital data, and wireless communication have transformed many phases of construction engineering and management. Modern high-resolution remote-sensing and spatial technologies can provide more accurate data to expedite planning and visualization of construction-progress monitoring.

10.7 References

[AASHTO 87] *Survey of Results of the 1984 AASHTO Rideability Survey,* AASHTO, Washington, DC, 1987.

[AGCA 77] *Statistically Oriented End-Result Specifications,* Associated General Contractors of America, August 1977.

[Al-Turk 99] E. Al-Turk, and W. Uddin, "Infrastructure Inventory and Condition Assessment Using Airborne Laser Terrain Mapping and Digital Photography," *Transportation Research Record 1690,* Transportation Research Board, National Research Council, Washington, DC, 1999, pp. 121–125.

[ASTM 78] "Quality Assurance in Pavement Construction," *Special Technical Publication 709,* ASTM, Philadelphia, Pa., 1978.

[Azud 69] G. Azud, "Owner Can Control Cost," *Transactions of the American Association of Cost Engineers,* 1969.

[CII 86] *Constructability: A Primer,* Construction Industry Institute, Austin, Tex., 1986, pp. 2–5.

[CTR/CII 87] *Guidelines for Implementing a Constructability Program,* Center for Transportation Research, Construction Industry Institute, Austin, Tex., 1987, p. 10.

[CTR/CII 89] F. Hugo, J. T. O'Connor, and W. Ward, *Highway Constructability Guide,* Texas State Department of Highways and Public Transportation, The Center for Transportation Research, Austin, Tex., July 1990.

[De Vos 89] J. De Vos, "A Strategy for the Implementation of a Constructability Improvement Program in SASTECH," Master of Engineering thesis, University of Stellenbosch, Republic of South Africa, 1989, p. 5.

[Fisher 89] D. J. Fisher, "Piping Erection Constructability Issues in a Semi-Automated Environment," PhD Thesis, The University of Texas at Austin, 1989, p. 28.

[Gappinger 85] E. P. Gappinger, "End-Product Specifications: State of the Art," *Proceedings, 34th Annual Arizona Conference,* University of Arizona, 1985.

[Haas 01] C. T. Haas, "Rapid Visualization of Geometric Information in a Construction Environment," *IV '01 Proceedings of the Fifth International Conference on Information Visualisation,* IEEE Computer Society Washington, DC, 2001, p. 31.

[Harrison 89] R. Harrison, C. Bertrand, and W. R. Hudson, "Measuring the Smoothness of Newly Constructed Concrete Pavement for Acceptance Specifications," *Proceedings of the 4th International Conference on Concrete Pavement Design and Rehabilitation,* Purdue University, Ind., 1989.

[Hudson 92] W. R. Hudson, T. Dossey, R. Harrison, and G. D. Goulias, *End-Result Smoothness Specifications for Acceptance of Asphalt Concrete Pavements,* Center for Transportation Research, The University of Texas at Austin, January 1992.

[O'Connor 86] J. T. O'Connor, and R. L. Tucker, "Industrial Constructability Improvement," *Journal of Construction Engineering and Management,* Vol. 112, No. 1, 1986, p. 69.

[Redelinghuys 89] J. Redelinghuys, "A Knowledge Base System to Enhance Highway Constructability," Master of Engineering thesis, University of Stellenbosch, Republic of South Africa, 1989.

[SDHPT 82] *Statistical Specifications for Construction of Highways, Streets, and Bridges,* Texas State Department of Highways and Public Transportation (SDHPT), Austin, Tex., 1982.

[SDHPT 91] *Ride Quality for Pavement Surfaces,* Revised 2-26-91, Texas State Department of Highways and Public Transportation, Austin, Tex., 1991.

[TxDOT 91] *Operation of the California Profilograph and Evaluation of Profiles,* Texas State Department of Highways and Public Transportation, Materials and Tests Division, Test Method TEX1000S, Draft, 1991.

[Turkan 12] Y. Turkan, F. Bosche, C. T. Haas, and R. Haas, "Automated Progress Tracking Using 4D schedule and 3D Sensing Technologies Original Research Article," *Automation in Construction,* Planning Future Cities (Editors: Frédéric Bosché and Jan Halatsch), Elsevier B.V., Amsterdam, The Netherlands, Vol. 22, March 2012, pp. 414–421.

[Uddin 08] W. Uddin, "Airborne Laser Terrain Mapping for Expediting Highway Projects: Evaluation of Accuracy and Cost," *Journal of Construction Engineering and Management,* American Society of Civil Engineers, Vol. 134, No. 6, June 2008, pp. 411–420.

[Uddin 11] W. Uddin, "Transportation Management: LiDAR, Satellite Imagery Expedite Infrastructure Planning," *Earth Imaging Journal,* January/February 2011, pp. 24–27.

[Warden 63] W. B. Warden, and L. D. Sandvig, "Tolerance and Variations of Highway Materials from Specification Limits," *Special Technical Publication 362,* American Society of Testing and Materials, Philadelphia, Pa., 1963.

[Willenbrock 76] J. H. Willenbrock, *A Manual for Statistical Quality Control of Highway Construction,* Federal Highway Administration, Washington, DC, 1976.

CHAPTER 11

Maintenance, Rehabilitation, and Reconstruction Strategies, Including Operations

11.1 Introduction

11.1.1 Economic Costs Caused by Delayed or Unplanned Maintenance

U.S. companies collectively spend a trillion dollars a year on freight logistics, nearly 10 percent of the nation's GDP, and on average, 42 tons of freight worth $39,000 was delivered per person in the United States in 2007 [NCFRP 12]. German logistics companies recorded a turnover of about 180 billion euros in 2006 and employed about 2.6 million people with supply chain/logistics being the largest employer [BMWi 08]. These statistics indicate the importance of the transport network to our society. The four transportation modes (shipping port, waterway, aviation, rail, and highway) are owned and operated by different entities in the United States. Unlike federal- and state-funded highway infrastructure, railroads are privately owned. All these modal networks operate within their own policy frameworks and profit motivations. Financing for preserving and upgrading intermodal infrastructure is handled differently among these assets.

A transport infrastructure funding crisis is evident in the United States. Related highlights of these issues follow, which are based on

Internet sources and selected references [BMWi 08, CBI 12, FHWA 09, Gerritsen 09, Uddin 13]:

- The growth in passenger travel, freight demand, and traffic congestion directly contributes to increases in mobility costs, user-operating costs, air pollution and GHG emissions, public health costs, and other societal costs. Congestion also causes fuel waste and stress on commuters.
  - In 2003, traffic congestion in the top-85 U.S. urban areas caused 3.7 billion hours of travel delay and 2.3 billion gallons of wasted fuel, for a total cost of $63 billion. An average of 15 gallons of fuel wasted per person and 25 hours of delay per person per year was contributed to congestion in 2003.
  - Economic costs of congestion in European countries amount to 1 percent of the European Union's GDP. In Germany this is about 20 billion euros annually.
- Based on condition and use limitations much transport infrastructure is aged and in need of repair in the United Kingdom, Germany, and many other European countries.
  - Almost 61 percent of U.K. companies rate U.K. transport infrastructure as below average compared to international standards, and 65 percent of companies say the condition of local roads is declining, with congestion and lack of investment cited as the main concerns.
  - In Germany one in seven bridges on main highways is in critical or unsatisfactory condition. About 83 billion euros is needed just to maintain the current transport infrastructure network.
  - National budget allocation for transport infrastructure expansion is projected at 9 percent GDP in China compared to 1 percent in Germany.

According to a recent U.S. study [Leduc 12], "each dollar of federal highway grants received by a state raises that state's annual economic output by at least two dollars. The 2009 American Recovery and Reinvestment Act (ARRA) allocated $40 billion to the Department of Transportation for spending on the nation's roads and other public infrastructure." In conclusion, maintenance and preservation are extremely important asset-management activities, and this chapter presents guidelines on systematic planning and execution of these phases in the service lives of all infrastructure assets.

11.1.2 Understanding Maintenance Intervention

After construction, the next steps in the management of any type of infrastructure assets are maintenance and operations. In practice, there is a wide divergence in the definition and understanding of maintenance, rehabilitation & reconstruction (M,R&R), particularly in the dividing line between maintenance and rehabilitation. In building infrastructure, for example, terms like *remodeling, restoration,* and *renovation* are commonly used rather than *rehabilitation.*

In highways and bridges, the dividing line between maintenance and rehabilitation is blurred by practice. Often it is divided along organizational lines involving a design division, a maintenance division, and a construction division. Sometimes, the definition depends upon funding sources. Historically, there was no federal funding in the United States for pavement and bridge maintenance. Maintenance was therefore repair work done with local funds while rehabilitation was work done with federal funds.

Some organizations originally defined maintenance and rehabilitation in terms of how the work was accomplished. If it was done by in-house forces, it was called "maintenance." If it was done by contract, it was termed "rehabilitation." These issues have changed, however, over the years. For example, federal funding in the United States is now available for highway and bridge maintenance as well as rehabilitation. Additionally, a growing trend to privatization is blurring the dividing line between maintenance and rehabilitation.

Since there are so many types of infrastructure, this book deals with functional, or global, issues associated with maintenance and rehabilitation definitions. However, concise and relatively precise definitions for maintenance and rehabilitation will be provided.

In one sense, maintenance is perceived to involve routine items like changing the oil, replacing light bulbs, filling potholes, repairing cracks, and associated activities, depending on the type of facility involved.

In another sense, however, maintenance is poorly understood. Remembering the systems concepts and methodologies discussed earlier, two important steps in the process are: (1) problem recognition and (2) problem definition. To truly understand maintenance and be able to apply the understanding to improved infrastructure management, clear recognition and a clear definition of maintenance in a functional sense is essential.

11.2 Definitions

To better define maintenance for infrastructure asset management, it is useful to first look at other industries and at various literature related to civil-engineering infrastructure. In fact a lot of work has been done, particularly in the space and aircraft industries. Areas of civil-engineering infrastructure, such as water and wastewater facilities, that involve both operations and maintenance also have applicable experience.

Examination of the literature from a range of sources shows a well-developed concept in reliability-centered maintenance (RCM). This RCM concept is used in military and civilian aircraft and space activities, where it was originally developed [King 86]. It plays a vital role in the safety of flying, where failure often means death. Later in this chapter, we will discuss RCM in more detail, but the definitions presented here take into consideration that concept.

In the RCM concept, rehabilitation is simply not mentioned. The entire range of maintenance/rehabilitation activities, and even reconstruction and replacement, all fall under a maintenance definition. This is not satisfactory for all infrastructure assets, but the portion of the concept and the definitions that assist in our objectives can be used.

11.2.1 Terms for Defining Maintenance

Terms such as *routine maintenance, corrective maintenance, preventive maintenance, proactive maintenance,* and *reactive maintenance* are commonly used in practice. Other terminology is also used, such as *hard-time replacement, on-condition maintenance, condition monitoring, servicing task, rework task* (repair, overhaul, rebuild), *replacement task,* and *time-directed* (versus condition-directed) *activities.*

Nearly all sources recognize the need for some regular maintenance activities. An example is changing the oil in your car on a regular basis. This is both preventive maintenance and, if done regularly, routine maintenance. The term *routine maintenance,* however, also applies to time-based maintenance; for example, a field crew travels over every bridge within its service area to fill potholes, clean scuppers, and patch approach slabs. Preventive maintenance is done in an effort to forestall trouble and is often done routinely. However, not all routine maintenance is preventive in nature. Corrective maintenance is done to remedy an obvious damage or flaw, and is an after-the-fact activity. Thus, it may also be called *reactive maintenance.*

The term *proactive maintenance* is often applied to work carried out at the discretion of the maintenance crew, or as planned in an infrastructure asset management system (AMS), to prevent impending deterioration or failure. Preventive maintenance and proactive maintenance are similar. After the implementation of the GASB 34 asset-management framework by the USDOT Federal Highway Administration another term, *pavement preservation,* became popular among state highway agencies and asphalt industry and material suppliers. Preservation is primarily another term for preventive maintenance to increase the life of highways.

Condition monitoring leads to on-condition maintenance. In this activity, regular inspections are made to monitor the condition of the infrastructure, and the maintenance is scheduled as a function of that condition. In the aerospace industry, oil changes, tire rotations, rework tasks, such as engine replacement, and so forth, are done as a servicing task on a hard time replacement basis. After several cycles of rework, old components are totally replaced with new ones. These concepts do not generally apply to pavements or bridges, and apply only partially to waterworks, sewage works, and other types of public infrastructure.

We will then limit our definitions to two basic divisions: proactive or preventive maintenance, and reactive or corrective maintenance. Rehabilitation will be defined later in this chapter.

According to the online Merriam-Webster Learner's dictionary, "maintenance is the act of keeping property or equipment in good condition by making repairs, correcting problems, etc." (http://www .learnersdictionary.com/search/maintenance accessed April 4, 2013) Maintenance is defined as "the action of keeping in effective condition, in working order, in repair" [Hudson 97]. A more detailed functional definition suggests that maintenance deals with the specific procedures, tasks, instructions, personnel, qualifications, equipment, and resources needed to satisfy the maintainability requirement within a specific use environment. For our purposes a good, concise definition is as follows: "Maintenance is that set of activities required to keep a component, system, infrastructure asset, or facility functioning as it was originally designed and constructed to function." Within this definition of maintenance, a number of subsets can be defined.

Preventive maintenance (proactive maintenance or *preservation*) is performed to retard or prevent deterioration or failure of a component or system.

Corrective maintenance (reactive maintenance) is performed to repair damage and/or to restore infrastructure facilities to satisfactory operation, or function, after failure.

Routine maintenance is any maintenance done on a regular basis or schedule. It is generally preventative in nature, but may be corrective.

Hard-time replacement (HTR) is replacement after a certain length of time, regardless of whether the component has failed or not. It is therefore a type of routine maintenance, but may also be corrective or preventive.

On-condition maintenance (OCM) is maintenance done in response to condition-monitoring actions indicating impending deterioration or failure. By definition it is a type of preventive maintenance.

Critical maintenance is defined by some as the maintenance that must be done immediately to prevent imminent collapse or functional failure. An example might be a bridge with a fractured steel girder, or an abutment that has been hit and fractured by a passing barge or ship.

The relationships of the above subsets are shown in Figure 11-1. A major distinction between corrective and preventive maintenance is whether or not failure or damage has already occurred. In many cases, no distinct failure point can be identified; thus, the dividing line between the two is shown as a gray area.

11.2.2 Maintenance Standards for Historic Properties

The U.S. Department of Interior has established standards for treating and preserving historic properties that have been used by the

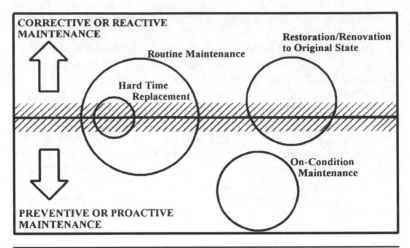

FIGURE 11-1   The relationships between subsets of maintenance.

National Park Service since the mid-1970s. The four treatment approaches based on these standards emphasize the retention and repair of all historic fabric, as briefly described in the following sections [Weeks 96].

11.2.2.1   Preservation
Preservation is defined as "the act or process of applying measures necessary to sustain the existing form, integrity, and materials of a historic property." Key ideas relating to preservation standards include the following:

- Use the property as it was used historically, or find a new use that maximizes retention of distinctive features.
- Preserve the historic character (continuum of property's history).
- Stabilize, consolidate, and conserve existing historic materials.
- Replace minimum amount of fabric necessary and in kind (match materials).

11.2.2.2   Rehabilitation
Rehabilitation is defined as "the act or process of making possible a compatible use for a property through repair, alterations, and additions, while preserving those portions or features that convey its historical, cultural, or architectural values." Key ideas relating to rehabilitation standards include the following:

- Use the property as it was used historically, or find a new use that minimizes retention of distinctive features.
- Preserve the historic character (continuum of property's history). Do not make changes that falsify the historical development.

- Repair deteriorated features. Replace a severely deteriorated feature with a matching feature (substitute materials may be used).

- New additions and alterations should not destroy historic materials or character. New work should be differentiated from the old, yet remain compatible with it.

11.2.2.3   Restoration

Restoration is defined as "the act or process of accurately depicting the form, features, and character of a property as it appeared at a particular period of time." This is accomplished by removing features from other periods in its history and reconstructing missing features from the restoration period. Key ideas relating to restoration standards include the following:

- Use the property as it was historically, or find a new use that reflects the property's restoration period.

- Stabilize, consolidate, and conserve features from the restoration period.

- Replace a severely deteriorated feature with a matching feature (substitute materials may be used).

- Replace missing features from the restoration period based on documentation and physical evidence. Do not make changes that mix periods and falsify history.

- Do not execute a design that was never built.

11.2.2.4   Reconstruction

Reconstruction is defined as "the act or process of depicting, by means of new construction, the form, features, and detailing of a non-surviving site, landscape, building, structure, or object, for the purpose of replicating its appearance at a specific period of time and in its historic location." Key ideas relating to reconstruction standards include the following:

- Do not reconstruct vanished portions of a property unless the reconstruction is essential to public understanding.

- Reconstruct based on documentary and physical evidence.

- Precede reconstruction with a thorough archaeological investigation.

- Preserve any remaining historic features.

- Recreate the appearance of the property (substitute materials may be used).

- Identify the reconstructed property as a contemporary recreation.

- Do not execute a design that was never built.

11.3 Maintainability

Maintainability has become an important concept in infrastructure management. For example, urban freeways and bridges often carry more than 200,000 vehicles per day; as a result it is very difficult to close such a facility, even for a short time at night. Ideal infrastructure management would balance the high user costs during maintenance shutdowns against the additional costs required to keep such facilities operating without closure for maintenance. This analysis involves the costs and benefits of maintainability.

Maintainability is defined in different ways in the cited literature. The basic subjective definitions are simple statements like "the ease with which a facility or system can be maintained" or "the reciprocal of the mean net time required to perform maintenance." Another example is "the capacity to carry out maintenance with ease and minimum expenditure while maintaining the safety of the crew and a desired level of accuracy in the repair work."

Maintainability is also a design element that considers the ease of maintenance in the future. Explicit interaction between maintenance personnel and designers in the design stage can significantly improve the maintainability of the design in terms of minimum costs with minimum environmental impact and a minimum expenditure of resources.

Maintainability often relates to access or "access areas." This involves being able to get to the area for servicing or repair. It is an important consideration for facilities that have mechanical and electrical equipment, such as wastewater and water-treatment plants, and hydroelectric dams where spacing and location of equipment elements are important. It can also relate to bridges. The Connecticut bridge panel that fell out several years ago, for instance, resulted from the corrosion of a pin connection that was covered with a plate, which made it not easily accessible for inspection, maintenance, or replacement.

Maintainability is, therefore, an inherent characteristic of a design. Blanchard et al. [Blanchard 95] define maintainability in a more objective way as:

> A characteristic of design and installation which is expressed as the probability that an item will be retained in or restored to a specified condition within a given period of time, when maintenance is performed in accordance with prescribed procedures and resources.

11.3.1 Measures of Maintainability and Availability

The following points must be considered for maintainability:

- If maintainability is to be an inherent characteristic of design, it must involve the designer's vision of the end product. However, it is obvious that the as-built end product is not completely in the hands of the designer. Therefore, true maintainability characteristics may be different from those perceived in the design.

• A fundamental question is whether maintainability is related to the frequency of maintenance actions performed during a system's life cycle. In one sense, we might say maintainability should be simply expressed in terms of the mean time to repair [Smith 93]. However, this does not capture the probability aspect over a given period of time as defined by Blanchard et al. [Blanchard 95].

The design features on which maintainability depends are so varied, and their interrelationships are so complex, that no single all-encompassing indicator can be determined as representing the desired maintainability characteristics for a given system. The following measures of maintainability are reported by the U.S. Department of Defense report [DOD 66].

11.3.1.1 Qualitative Maintainability
As a requirement, this would be a general statement to be incorporated in a system-design process. Examples are minimizing complexity, accessibility of components and parts, and built-in self-test features. These qualitative statements become maintainability objectives for the design process.

11.3.1.2 Quantitative Maintainability
A quantitative maintainability requirement is a definite statement of the allowable resources or time required to perform a given type of task. The measure of maintainability works in parallel with the availability of the system to perform specific functions, as discussed later in this chapter. The mean time to repair (MTTR) is often selected as a measure of maintainability, as follows. MTTR is the mean of the times required to repair an item or system during a given period. Other important parameters used in maintainability and availability are MDT (mean down time), which is the total inactive time for the equipment during a given period, and MTBF (mean time between failures), which is the mean of the total active time for the component during a given period.

11.3.2 Maintainability Analysis
Maintainability analysis precisely defines a system's maintainability requirements and provides the designers and the organizations responsible for the system's design and development with timely information as to how effectively the established repair-time goals are being met.

The first step in conducting a maintainability analysis is to establish maintainability controls in the form of qualitative and quantitative requirements. To determine the quantitative maintainability requirements for each subsystem, the repair time of the subsystems must be budgeted so that the mean is less than the allowable MTTR for the system.

Since MTTR for a system is its principal maintainability requirements, it is essential that means be developed early in a program for allocating and controlling the MTTR of each subsystem. This allocation is managed by determining the contribution of active downtime for each subsystem to the total system downtime, and evaluating these contributions against the established MTTR for the system.

11.3.3 Availability Concept

Availability is the probability of a system or component being in service when required. Inherent availability is given by the following expression:

$$\text{Availability} = \text{MTBF}/(\text{MTBF} + \text{MTTR}) \tag{11.1}$$

Provision of availability is a major objective of the design process. Thus, the availability goal for a given system specifies certain demands on reliability in terms of MTFB and on maintainability in terms of MTTR. The trade-offs discussed in the next section provide an optimum combination of these design objectives.

11.4 Trade-Offs among Design Objectives in Relation to Maintenance

Trade-off techniques are analytical processes whereby a complex design problem, involving the selection of one of several possible design variants, is broken down into a number of smaller problems. Each problem is studied in the light of all the system parameters, such as reliability, availability, safety, production, and schedule. Finally, an optimum-design solution is established. The overall objective is a life-cycle program considering various other design requirements, or data sources, as shown in Figure 11-2, including reliability and maintainability data.

There is a definite relationship between availability, reliability, and maintainability. Reliability and maintainability contribute in varying proportions to provide certain levels of availability. It is evident that if a certain degree of system availability cannot be economically achieved through emphasis on reliability, then it can be achieved only by incorporating better maintainability in the design.

Maintainability is the most significant factor in the eventual solution of the optimum readiness of system. Efforts to achieve system readiness just by means of greater design reliability have often proved to be economically infeasible. The alternative in such situations is a

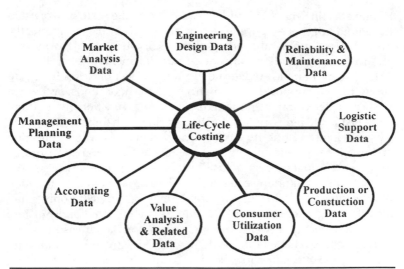

Figure 11-2   Data sources for life-cycle cost analysis [Blanchard 95].

greatly increased emphasis on maintainability. All combinations of maintainability design features, together with the cost and associated MTTR of each, must be considered to find the combinations that best meet the maintainability requirements of the system.

Following is the priority order for design objectives while carrying out trade-off studies [Anderson 90]:

1st order	System effectiveness
2nd order	Operational availability
3rd order	Reliability, maintainability, support
4th order	Parameters for reliability, maintainability, support

11.5   Rehabilitation

As opposed to the aerospace industry and other machine-related activities, civil-engineering infrastructure usually incorporates rehabilitation as a major element of the infrastructure management system (IMS). The dictionary definition of *rehabilitation* is "the action of restoring something to a former condition or status." Maintenance, on the other hand, is seen as the continuous retention of something "in an existing state," as discussed previously. These dictionary definitions imply a difference in timing of the two types of actions. It is obvious that the dividing line between the two will always remain hazy, since it is very difficult to draw a line between continuous retention and stepwise restoration. The dividing line between maintenance and rehabilitation, in practice, is often policy- and rule-dependent, and the factors affecting the dividing line can relate to the scale of work and the amount of money put into a particular activity. For example, the first 20 years of federal funding of the U.S. interstate highway program did not include maintenance, but did include rehabilitation. There is considerable evidence that under these policies some agencies withdrew support for maintenance, for lack of funds, and when the facility deteriorated enough to be defined as functionally deficient by the Federal Highway Administration rules, rehabilitation funding was provided to bring it back to current standards. The implication of "functionally deficient" in this case meant that the road or bridge needed to be upgraded so as to adhere to current standards.

Rehabilitation often seems to be related to a change in function of a facility. An old airport or airbase may be servicing only small aircraft or light loadings, but it could be rehabilitated and put into service as a heavy-duty air-carrier airport or bomber air base. In some infrastructure, rehabilitation is defined as "the process of undertaking remedial measures to solve or correct inadequacies resulting from load or natural (environmental, seismic, or hydrologic) causes." Rehabilitation, therefore, often involves changes in engineering properties, modernization, and, in many cases, changes of size, scope, function, or geometry.

Some authors point out that rehabilitation occurs when repairs and maintenance will no longer solve the problems observed. This is particularly true for a facility involving mechanical and electrical equipment, such as wastewater and water plants. In effect, continuous maintenance is unable to "retain the original state," and "restoration" to this state (or rehabilitation) is required. This also implies a difference in the size or scale of the operation. Even well-maintained infrastructure will continue to deteriorate in its functional performance to the point where it eventually requires rehabilitation or restoration.

Two main themes are apparent in practice. First, rehabilitation is generally seen as involving a larger scale of work than maintenance. Second, it involves modifying or upgrading. To clarify understanding, the following concise definition for rehabilitation is offered: "Rehabilitation is that set of activities involved in restoring a component system or facility to a modified or changed state."

Under this definition, filling potholes, sealing cracks, painting walls, and replacing light bulbs all remain under maintenance. In fact, any actions directed at fixing or returning something to its *original* state should be termed maintenance under this definition. Such a definition fits well with the aircraft industry, where the term rehabilitation is not used (i.e., a 747 aircraft is maintained as a 747). It also fits one of the U.S. Federal Highway Administration's definitions of rehabilitation, in which funding is often only made available when a road or bridge is upgraded, either structurally or functionally. It thus follows that rehabilitation should occur much more often in highways, bridges, and other infrastructure that are being continually upgraded and modified, but not, for example, in the aircraft industry.

The foregoing definition highlights the consequences for infrastructure management: We cannot simply maintain facilities in a single state—we also have to continually upgrade, improve, and modernize them, either to just meet existing demand or to service continually changing demand. Some facilities, therefore, evolve during their overall lifetimes to very different forms or structures. For example, aircraft 747s do not slowly evolve over time into 767s! A stepwise evolution brought about by rehabilitation, in addition to the intermediate periods of continual maintenance on the maintainable parts of facilities, therefore leads to complex and interesting challenges for infrastructure managers.

11.6   Reliability-Centered Maintenance

Reliability is usually defined as "the probability that a component or system will satisfactorily perform its specified function for the specified or required period of time under given or predicted operating conditions." Satisfactory performance over time is a probabilistic situation related to variations in material properties, physical environment, and load.

The reliability of a complex infrastructure facility depends on the reliability of individual components. Satisfactory system performance is achieved when all or most of the elements are performing satisfactorily. Even a slight reduction in component reliability may drastically reduce the overall system reliability.

Data regarding failure rate versus time are usually the basis of reliability predictions for infrastructure components. A system may experience different failure rates, depending upon where it is in its life cycle. It is important to know the details of the failure mechanisms and their causes, so that proper design, maintenance, or operation actions can be taken to achieve the specified reliability.

Six generic trends of failure probability against age, for a variety of electrical and mechanical items, have been reported by Moubray [Moubray 92]. These are shown in Figure 11-3. Type A is the well-known "bathtub curve." It begins with a high incidence of failure, known as infant mortality or burn-in, followed by a relatively constant failure rate, and ends in a wear-out zone. The other trends are variations on this basic theme. Generally, it has been proven that a constant failure probability, indicating that something is just as likely to fail at one time as any other (as shown by pattern E), is more applicable to electrical systems and some mechanical systems. Structural components, and therefore most public infrastructure, generally experience low or insignificant failure rates until near the end of their "structural life," at which time failure rates begin to climb (as shown

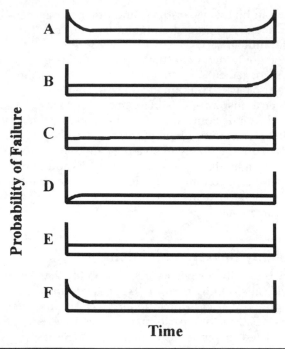

Figure 11-3   Six patterns or types of failures [Moubray 92].

by pattern B). Age is therefore observed as the major determinant of the failure probability of an infrastructure system.

A case study of reliability centered maintenance for railway transit maintenance to achieve optimal performance examined the types of obstacles and patterns experienced by a heavy rail transit agency located in North America [Marten 10]. The study also examined the impact of reliability centered maintenance process outcome on availability, reliability, and safety of rolling stock. Details can be found in the cited reference.

11.6.1   Reliability in Design as Related to Maintenance

Reliability is a design attribute. It means that the system's reliability is established by how well or poorly the design process is accomplished. The design, or more broadly the system definition, which also encompasses how the system must be operated and maintained, is the sole determinant in setting the inherent reliability that can be achieved. Good, sound maintenance procedures are vital to attain "potential" reliability. Maintainability is "built in," but not necessarily reliability—especially in a complex system.

The concept of reliability is a part of some contemporary infrastructure design practices. A reliability factor was introduced in the 1986 version of the AASHTO pavement design guide [AASHTO 86].

The use of safety factors in most structural design codes is an indirect way to achieve reliability in design. However, these factors of safety are basically subjective, and function as human-ignorance factors for complex design problems.

Reliability engineering advocates a comprehensive review of system specifications to ensure that all objectives and supporting requirements are properly included. The design process should recognize the importance of a proper life-cycle maintenance program in attaining the inherent reliability of the system. Unfortunately, this aspect of the design process is often relegated to a secondary priority.

The application of the reliability concept is very useful in infrastructure management, where a significant relationship exists between age and performance. Although such relationships are subject to variability, the benefits in developing reasonable performance prediction models cannot be overestimated. These models tell us about the operational reliability of certain infrastructure in certain conditions. This knowledge can help in developing different construction and maintenance alternatives and allowing the selection of the most reasonable combination of initial design and future maintenance strategies.

Once the infrastructure is constructed, the subsequent management objective is to use the designed-in reliability and maintainability levels to retain the functional performance levels through the life of the infrastructure. RCM offers a solid method to achieve this objective.

In general, RCM methodology sets up a detailed maintenance plan for a particular aircraft, power station, water-treatment plant, and so forth, in a rational and efficient manner. Maintenance may then continue according to this plan either until a modification takes place or throughout the operating life of the system.

The primary objective of RCM is to prioritize maintenance activities according to both the risk and consequence of failure. One of the first and most fundamental steps in setting up a detailed maintenance schedule using the RCM method is to do a failure mode and effects analysis (FMEA). This forces a breakdown of the facility into systems and components to the desired level, which is followed by a systematic assessment of the different modes of failure for each component, the probability of those failures, and the consequences. Once this has been accomplished, various maintenance actions can be assigned that will minimize the risk of serious consequences while balancing the economic and practical viability. In rare cases, the process forces redesign if fatal flaws are found where it would be impossible to reduce risks to acceptable levels. This sort of problem is often solved by designing-in redundant or backup systems, such that failure can occur in the primary system, be detected, and corrective maintenance performed without necessarily affecting the overall performance.

It is not within the scope of this book to exhaustively discuss the intricacies of RCM, but it is felt to have considerable potential for application in infrastructure management.

11.7   Maintenance Management

Maintenance management is an important area of concern in overall infrastructure AMS. In the broad, generic sense, maintenance management means ensuring that proper maintenance is applied at the proper time in the life cycle of the infrastructure. Over the past four decades, however, the term has taken on a more specific meaning and is usually associated with a maintenance management system (MMS). Even the term *maintenance management system* has a large number of meanings. Generally, it is similar in concept to a production control process in manufacturing, and involves setting up a database to record data, particularly cost and volumes of work, related to various maintenance activities. This was done, for example, in the 1960s and 1970s in highway departments across the United States and in other countries [Jorgensen 72, Buttler 75]. Such MMS involved recording costs, volumes of work, labor hours, and related topics associated with the day-to-day operations of highway networks.

MMSs have preceded AMSs by a number of years. In particular, the early AMSs were primarily pavement management systems (PMS) and bridge management systems (BMS). MMSs, however, were not as successful as PMSs. Although they required data collection, inventory information, and condition assessment, inadequate attention was given to inventory location and identification detail. As a result, when attempts were made to use the data to more specifically define, for example, the cost of maintenance of a particular road section, it was found impossible to separate the cost of pavement maintenance from mowing, ditch maintenance, and other maintenance activities within a linear link of highway. It was impossible to obtain data specific enough for any given length of highway—or for the pavement alone—to develop maintenance cost models for use in life-cycle costing.

Although a great deal of attention was given to propagating the systems across the United States and in many other countries, not enough attention was given to training and actual implementation into the daily work of the DOTs involved. As a result, both the inputs and outputs of most of the systems were far too broad to be of true value in an infrastructure management system.

The authors of this book have examined 10-year-old maintenance management systems in which the data were never used during that period. These MMSs were uncovered while trying to get maintenance cost information for use in life-cycle cost analysis for PMS and BMS. It was found that, in many cases, all maintenance costs for a particular mile of highway from fence line to fence line were accumulated in the same file, but there was no way to segregate specific costs of patching and crack filling, from mowing and striping, for example.

And while developing BMSs, the authors found that, in most cases, all costs on a short bridge were lumped together over a given period of time. It was impossible to separate data on patching versus cleaning scuppers versus rail repair. Modern MMSs, such as used in North Carolina and Texas, do a much better job of separating these activities and costs for analysis [Bhargava 12, Hudson 11, Perrone 10, Pilson 10].

In defining an IMS, a MMS should be developed as part of or linked to the database of the basic IMS. In those types of infrastructure where regular operations (that is water-treatment plants, sewage-treatment plants, etc.) are a major element, it might be an operations- and maintenance-management subsystem (OMMS). In roads and bridges, operations usually take on a more passive role, except in situations requiring significant traffic or flow control, for example.

Maintenance-related functions involve condition assessment, preventive-maintenance tasks, corrective-maintenance tasks, and documentation as a minimum. The condition-assessment tasks and detailing of inventory are part of the broader IAMS database.

Figure 11-4 shows an interrelationship between condition assessment, operations, preventive maintenance, and corrective maintenance [Grigg 88]. All are related to the system or network in terms of the inventory.

Corrective maintenance requires a decision: Is the deficiency or damage observed serious enough to warrant entering the planning-programming-budgeting (PPB) cycle, leading to a capital budget request, or is it minor enough to take care of from field-maintenance budgets? To a degree, this depends on the level of cost involved. If the problem is major, the PPB activity incorporates information about

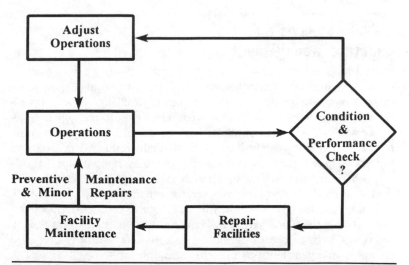

FIGURE 11-4   Linkage of operations and maintenance [Grigg 88].

new standards and growth forecasts, which may put the need into a rehabilitation category. There will then be a linkage between maintenance, rehabilitation, and replacement. Because of the interface in the budgeting and planning process, the same staff should be involved with planning and budgeting for rehabilitation and new facilities.

Ultimately, maintenance management involves detailed activities, such as work orders, time sheets, work-record histories, scheduling, and crew assignments, which are too detailed to be adequately treated in this book. Sixty-two cities were surveyed by the Urban Institute about their practices and experience in maintenance management. The survey results indicate that the principles of good maintenance are well known, but there has been a problem with the lack of their application [Urban 84]. The selection of strategies involves systematic identification of available options. These include the following, as well as their combinations:

- Do crisis maintenance only.
- Maintain the worst facilities first.
- Perform opportunistic maintenance when related work is scheduled.
- Use prespecified maintenance cycles.
- Repair those components with the highest risk of failure.
- Use preventive maintenance.
- Reduce the demand for wear and tear on the facility.
- Compare the economic advantages of maintenance strategies.

Note that a rigorous application of RCM methodology, where risk and consequence of failure are vitally important, would often solve most problems associated with choosing the right strategy.

11.8   Operations as a Part of Infrastructure Management

In the broadest sense of infrastructure management, it is necessary to operate as well as to maintain facilities. Infrastructure management systems as we know them do not attempt to deal directly with operations. In other words, the systems were developed to manage maintenance within the context of the facility operation. It is very important, therefore, to realize that the environment in which the IMS is used, and indeed the reason for its existence, is the operations it must support. This operating environment must obviously be considered when planning bridge or road maintenance, since traffic must be delayed, detoured, or stopped. Other types of infrastructure require even closer cooperation between the management system and the operating environment. Buildings, dams, electrical facilities, as well as airports, air bases, and chemical- and petroleum-manufacturing plants are some examples.

Grigg explains some of the interaction between an IMS and its operating environment and shows that the initial design of the IMS is dependent on the target performance objectives of the agency, which are, in turn, often influenced by the effects of bureaucracy [Grigg 88, 12]. In the vast majority of cases, at least three distinct phases characterize the life of a facility: (1) planning and design; (2) construction; and (3) operation and maintenance. The latter is generally the longest and often the most important. It should always be remembered that the initial design should include a maintenance schedule, which implies that this schedule must take due cognizance of the operations and that flexibility is greatly reduced after construction.

The concepts outlined by Grigg and other authors point out the value of setting goals and objectives and defining methods and rules that will motivate the staff workers dealing directly with the IMS. This is a critical part of an effective infrastructure-management process.

The whole purpose of an operations-management activity is to provide the maximum productivity within the organization. This leads to concepts of performance measurement, effectiveness, and efficiency. It is helpful to think of the output of an organization in terms of production. For those types of infrastructure that have a major operations component, it is important to measure effectiveness in terms of the actual production compared to the theoretical production.

11.8.1   Quality Control and Quality Assurance

Quality control is a vital part of operations, as well as all other aspects of infrastructure management, particularly construction. Quality control has become increasingly important as we learn that high-quality products are not only the best way to compete, they are essential [ASCE 88]. We have also learned that the best way to have high quality is to do things right the first time. There is a lot of activity in infrastructure today associated with quality control (QC), quality assurance (QA), and continuous quality improvement (CQI) or total quality management (TQM).

11.8.1.1   Definitions
The following are definitions of quality control and quality assurance.

QC stands for operational techniques and activities aimed both at monitoring a process and at eliminating causes of unsatisfactory performance. QC refers to technical detail and is the method manufacturers and builders use to ensure that their materials are up to standard and that products are fabricated or built correctly.

QA is defined as "all planned actions necessary to provide adequate confidence that a product or service will satisfy given requirements for quality." QA is the process used to ensure that quality controls and service aspects are carried out, and to monitor and record the appropriate verifications applicable to activities during each phase of a project.

QA is a key consideration in end-result specifications (ERS), based on statistical concepts. The growing use of performance-related criteria in ERS should evolve to long-term performance requirements. This would be a move in the direction of European contracting procedures.

The quality of materials incorporated and work performed directly influences the service life of an infrastructure, maintenance costs, level of service, and user costs. Quality can be defined as "the characteristics of a product (materials and methods of fabrication, for instance) that provide a level of functional (i.e., serviceability) and structural performance and design life." QC ensures that the specified "ingredients" are combined and placed so that the product will have the desired level of performance. QA is all of the activities necessary to verify, audit, and evaluate quality. With the move to ERS, the contractor/subcontractors/suppliers take on QC responsibility (such as process control), and the agency/consultant uses QA for assessing acceptance.

11.8.1.2   Scope of Quality Control

The scope of QC in an ERS framework, from a contractor or supplier's viewpoint, involves the following:

- QC is the method used to ensure the product/work will be accepted.
- Contractors or suppliers impose a QC system of inspection, sampling, and testing (process control) on themselves.
- Without QC, a contractor or supplier does not know whether a proper level of quality is being achieved until the product/work is either accepted, price-adjusted, or rejected by the agency.
- QC may not be specified by the agency but is essential.

11.8.1.3   Scope of Quality Assurance

The scope of QA from an agency (owner) viewpoint can be summarized as follows:

- What do we want? (planning and design)
- How do we order it? (construction plans and specifications)
- Did we get what we ordered? (inspection, sampling, and testing)
- What if we don't get what was ordered?
  1. Accept substandard (marginal or borderline).
  2. Do not accept (repair or reject).
  3. Assess price adjustment (penalty).
- What if we get "better" than ordered? (bonus?)

11.8.2   Total Quality Management

With a move toward ERS or warranty specifications, the need for improving the quality of work increases. Improved quality will have a direct impact on bonus/penalty provisions, and contractors are accordingly seeking assistance through the new philosophy of TQM. Partnering, an integral part of TQM, is rapidly becoming a project strategy for agency-contractor commitment and communications.

TQM incorporates all of the activities associated with the continuous improvements of quality and productivity (service), with an emphasis on customer-supplier relationships, employee involvement in decision making, teamwork, rigorous analysis of work as a process, statistical quality control, and managerial focus on leading. This is essentially the definition of TQM that has been suggested for the engineering and construction industry [FHWA 93].

11.8.2.1   Principles of TQM

According to Deming, the following obligations must be fulfilled by executive management to improve quality, productivity, and competitive position [Deming 86]:

1. Create constancy of purpose toward improvement of service and product to be competitive, stay in business, and provide jobs.

2. Adopt the new philosophy. Management must awaken to the challenge, learn their responsibilities, and accept leadership for change.

3. Create sole dependence on inspection to achieve quality. Eliminate the need for inspection on a mass basis by building quality into the product in the first place.

4. End the practice of awarding business on the basis of a price tag. Instead, minimize total cost. Move toward a single supplier for any one item, on a long-term relationship of loyalty and trust.

5. Improve constantly and forever the system of service and production, to improve quality and productivity and thus constantly decrease costs.

6. Institute training on the job.

7. Institute leadership (management and production workers) to help people and machines do a better job. Leadership of management is in need of overhaul, as well as leadership of production workers.

8. Drive out fear, so that everyone may work effectively for the organization.

9. Break down barriers between departments, so that people work as a team.

10. Eliminate slogans, exhortations, and targets for the workforce, asking for zero defects and new levels of productivity.

11. Eliminate work quotes on the production floor; eliminate management by objective and by numbers (numerical goals). Substitute leadership.

12. Remove barriers that rob hourly workers, management, and engineering staff of their right to pride of workmanship.

13. Institute a vigorous program of education and self-improvement. Start with executive managers.

14. Put everybody in the organization to work to accomplish the transformation.

The following important concepts are involved in the TQM philosophy:

Quality strategy—The most important aspect of TQM is the commitment of the decision-makers and executive managers.

Customer orientation—This is the core of TQM. Customer means both the internal (within organization) as well as the external (end-users and public) customers. (Public agencies have customers, too.)

Employee involvement—Employee involvement in organizational activities and other steps of the process is equally important. People conceive innovations. It means the involvement of all employees in activities relating to quality, including analysis of a particular problem and finding solutions.

Process-centered approach—TQM shifts the emphasis from the earlier concepts of the quality of product or quality of service to the process itself. If the process is of high quality, then the product is bound to be of high quality.

Problem-solving tools—The basic management tools also apply for TQM. These basic tools include statistical techniques, check sheets, histograms, cause-and-effect diagrams, design of experiments, multivariate analysis, stratification, control charts, design review, failure analysis, and analysis of alternative solution strategies.

Continuous improvement—Every organization requires improvement, which is the most important aspect in implementing TQM.

Benchmarking—It is a continuous process of measuring one's own products and services, business practices, and related activities against the leading competitor.

People orientation—It is the mutual trust of the employee and the management in good and bad times that leads to a motivated and dedicated team.

Life-cycle maintenance plan—Life-cycle maintenance plans reduce breakdowns, maximize machine-operating efficiency, increase life and encourage maintenance-free designs, and help to build up a trained and motivated team.

11.8.2.2 TQM's Implementation and Costs

Without an adequate QA system, fluctuating levels of quality may result in extra production costs. The installation of a quality system like ISO 9000-9004 standards may stabilize the quality at a predetermined level. However, TQM provides for constant quality improvement through a strong customer focus, as opposed to mere conformity to specifications.

It is expected that the implementation of a TQM will result in increased profit and less employee turnover while reducing costs, loss of breakdown time, rejections, and complaints. Figure 11-5 shows the concept of reduced quality costs for process checking versus product checking [Hayden 89].

Considering infrastructure management as a process, and implementing TQM concepts, provide numerous benefits, including the possibility of aggressive commitment from executive decision-makers and the participation of the public. The City of Louisville airport expansion project is an excellent example of active participation by customers and the public in the planning and implementation stages [Michael 96].

11.8.3 Role of ISO 9000 Standards

A new series of quality standards called ISO 9000 has been created by the International Organization for Standardization (ISO). Its adoption by the European Economic Community has dramatically increased

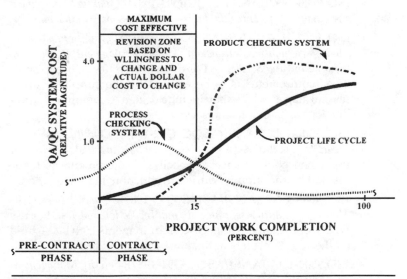

FIGURE 11-5 Process-checking versus product-checking quality systems (after [Hayden 89]).

the use of QA programs. The ISO 9000 Series of Quality Standards is now the worldwide accepted QA standard, but it is still in the evolutionary stage. In essence, it is a stepping-stone toward a full TQM process. Future revisions and supplements will be published to address issues that are raised as more businesses learn and use these standards. There are several categories, including ISO 9001 to 9004. The appropriate QA category should be selected considering the following six factors:

1. Design-process complexity

2. Design maturity

3. Production-process complexity

4. Product/service characteristics

5. Product/service safety

6. Economics

While many companies in the United States are becoming ISO 9000–registered, the quality-system requirements specified in American National Standards Institute (ANSI)/American Society for Quality Control (ASQC) Standards Q91, Q92, Q93, and Q94 are complementary to the technical specified requirements of a product and/or service (http://www.ansi.org/ and http://prdweb.asq.org/). These standards are technically equivalent to the international standards ISO 9001, 9002, 9003, and 9004, respectively [ASQC 87].

Following is a brief list of the applicable areas for ISO and ANSI/ASQC standards.

- **ISO 9000-1987 (ANSI/ASQC Q90-1987),** *Quality Management and Quality Assurance Standards—Guidelines for Selection and Use.*

- **ISO 9001-1987 (ANSI/ASQC Q91-1987),** *Quality Systems— Model for Quality Assurance in Design/Development, Production, Installation, and Servicing.* For use when conformance to specified requirements is to be assured by the supplier during several stages, which may include design/development, production, installation, and service.

- **ISO 9002-1987 (ANSI/ASQC Q92-1987),** *Quality Systems— Model for Quality Assurance in Production and Installation.* For use when conformance to specified requirements is to be assured by the supplier during production and installation.

- **ISO 9003-1987 (ANSI/ASQC Q93-1987),** *Quality Systems— Model for Quality Assurance in Final Inspection and Test.* For use when conformance to specified requirements is to be assured by the supplier solely at final inspection and test.

- **ISO 9004-1987 (ANSI/ASQC Q94-1987),** *Quality Management and Quality Systems Elements—Guidelines.* For design of an effective quality-management system to satisfy customer needs and expectations while serving to protect the company's interests. A well-structured quality system is a valuable management resource in the optimization and control of quality in relation to risk, cost, and benefit considerations.

Among several national efforts in the United States to promote quality, AASHTO has begun to develop a QA/QC guide specification for the highway industry. This guide follows the efforts of numerous states that have developed and used quality-assurance specifications, some since the mid-1960s [FHWA 93]. ISO 9000 reinforces the importance of employee and management participation in the quality-production process. Integrating the TQM principles of employee participation and continuous process improvement with the ISO 9000-9004 standards can give any organization better tools.

11.8.4   Role of Automation, Role of Automation, Workforce Productivity and Management Approach in Infrastructure Construction and Maintenance

The Construction Industry Institute (CII), an organization of over 100 major member corporations acting at the national (U.S.) and international level, has incorporated extensive involvement by various researchers and practitioners. This includes, for example, researchers at the University of Texas at Austin, Georgia Tech, University of Kentucky in the United States, Chang-An in China, and the Universities of Waterloo, Calgary, and Alberta in Canada, and others.

Among the advances attributable to these researchers are the applications of automation in construction, including activity and progress tracking [Ahmed 12, Shahi 13, Turkan 13, Razavi 10]. These advances have made extensive use of sensors and 3D technologies, both photogrammetric and laser scanning, which incorporate sophisticated point-cloud analysis algorithms.

Productivity in construction and maintenance is an ongoing issue for both the industry and public agencies in almost all areas of infrastructure. Several researchers, such as [Goodrum 11, Young 11], have done extensive work in this area and have related workforce productivity to technologies, materials, and supply-chain networks.

Management approaches in construction and maintenance may vary somewhat in applications but basically are in accordance with the framework described in this book. Noteworthy in wastewater collection, for example, is the framework described in [Rehan 13]. As another example, the use of automated earned-value tracking as an activity, and progress tracking using 3D sensing technologies linked to 4D project scheduling has been described in [Turkan 12, 13].

In summary, automation, including sensing technologies, robotics and information technologies, workforce productivity and safety, and management approaches including organizational aspects, life-cycle analysis, optimization, performance modeling, scheduling and execution, are providing leading-edge advancements in construction and maintenance.

11.9 References

[AASHTO 86] "AASHTO Guide for Design of Pavement Structures," American Association of State Highway and Transportation Officials, Washington, DC, 1986.

[Ahmed 12] M. Ahmed, C. T. Haasand, and R. Haas, "Using Digital Photogrammetry for Pipe-Works Progress Tracking," *Canadian Journal of Civil Engineering*, Vol. 39, No. 9, 2012, pp. 1062–1071.

[Anderson 90] R. T. Anderson, and L. Neri, *Reliability Centered Maintenance: Management and Engineering Methods*, Elsevier Applied Science, New York, 1990.

[ASCE 88] American Society of Civil Engineers, *Quality in the Constructed Project*, manual of professional practice, Vol. 1, 1988.

[ASQC 87] American Society for Quality Control, *American National Standard, ANSI/ASQC 094-1987, Quality Management and Quality System Elements— Guidelines*, Milwaukee, Wis., 1987.

[Bhargava 12] A. Bhargava, A. Galenko, and T. Scheinberg, "Asset Management Optimization Models: Model Size Reduction in The Context of Pavement Management System," prepared for presentation at the *92nd Annual Meeting of the Transportation Research Board* and publication in the *Transportation Research Record*, Washington, DC, 2013.

[Blanchard 95] B. S. Blanchard, D. Verma, and E. L. Peterson, *Maintainability: A Key to Effective Serviceability and Maintenance Management*, John Wiley and Sons, New York, 1995.

[BMWi 08] Federal Ministry of Economics and Technology (BMWi), Mobility and Transport Technologies: The Third Transport Research Programme of the German Federal Government," April 2008, www.bmwi.de, accessed on October 20, 2011.

[Buttler 75] B. C. Buttler, and L. G. Byrd, "Maintenance Management," *The Handbook of Highway Engineering*, Section 25, Van Nostrand Reinhold, New York, 1975.

[CBI 12] "Better connected, better business," Report, Confederation of British Industry (CBI) and KPMG LLP, United Kingdom, September 2012, http://www.kpmg.com/UK/en/IssuesAndInsights/ArticlesPublications/NewsReleases/Pages/Positive-outlook-for-digital-infrastructure-according-to-CBI-KPMG-Infrastructure-Survey.aspx, accessed October 15, 2012.

[Deming 86] W. E. Deming, *Out of the Crisis*, Massachusetts Institute of Technology, Cambridge, Mass., 1986.

[DOD 66] "R and D Material: Maintainability Engineering," *Publication 705-1*, U.S. Department of Defense, Washington, DC, 1966.

[FHWA 93] Federal Highway Administration, "Construction Quality Management for Managers," *Publication No. FHWA-SA-93-071*, Demonstration Project 89, Office of Engineering and Office of Technology Applications, Washington, DC, 1993.

[FHWA 09] FHWA, Transit and Congestion Pricing, A Primer, FHWA-HOP-09-015, Office of Innovative Program Delivery, Federal Highway Administration (FHWA), U.S. Department of Transportation, Washington, DC. April 2009.

[Gerritsen 09] E. J. Gerritsen, "White Paper: The Global Infrastructure Boom of 2009–2015," Commentary, *The Journal of Commerce Online*, May 19, 2009, http://www.joc.com/commentary, accessed August 29, 2011.

[Goodrum 11] P. Goodrum, C. Haas, C. Caldas, D. Zhai, J. Yeiser, and D. Homm, "Model to Predict the Impact of a Technology on Construction Productivity,"

ASCE Journal of Construction Engineering and Management, Vol. 137, No. 9, 2011.

[Grigg 88] N. S. Grigg, *Infrastructure Engineering and Management*, John Wiley and Sons, New York, 1988.

[Grigg 12] N. S. Grigg, *Water, Wastewater and Stormwater Infrastructure Management*, Second Edition, CRC Press, Boca Raton, Fla., 2012.

[Hayden 89] W. M. Hayden, *The Effective A/E Quality Management Program: How To Do It*, short course notebook, American Society of Civil Engineers, New York, 1989.

[Hudson 97] W. R. Hudson, R. Haas, and W. Uddin, *Infrastructure Management*, McGraw-Hill, New York, 1997.

[Hudson 11] S. W. Hudson, K. Strauss, et al. "Improving PMS by Simultaneous Integration of MMS," *Presented at the 8th International Conference on Managing Pavement Assets*, Santiago, Chile, November 15–19, 2011.

[Jorgensen 72] Roy Jorgensen Associates, "Performance Budgeting System for Highway Maintenance Management," *NCHRP Report 131*, Transportation Research Board, National Research Council, Washington, DC, 1972.

[King 86] F. H. King, *Aviation Maintenance Management*, Southern Illinois University Press, Carbondale, Ill., 1986.

[Leduc 12] S. Leduc, and D. Wilson, "Highway Grants: Roads to Prosperity?" FRBSF Economic Letter, Federal Reserve Bank of San Francisco, November 26, 2012, http://www.frbsf.org/publications/economics/letter/2012/el2012-35 .html, accessed November 30, 2012.

[Marten 10] F. A. Marten, Jr., "Reliability Centered Maintenance: A Case Study of Railway Transit Maintenance to Achieve Optimal Performance," *MTI Report 10-06*, Mineta Transportation Institute, San José State University, San José, California, December 2010.

[Michael 96] R. S. Michael, "Building Support for Airport Expansion," keynote address, 24th International Air Transportation Congress, Louisville, Ky., June, 1996.

[Moubray 92] J. Moubray, *Reliability Centered Maintenance*, Industrial Press, New York, 1992.

[NCFRP 12] NCFRP, Guidebook for Understanding Urban Goods Movement, Report 14, National Cooperative Freight Research Program (NCFRP), Transportation Research Board, The National Academies, Washington, DC, 2012.

[Perrone 10] E. Perrone, "Bootcamp Training—Pavement Management," ITD MAPS Project, Version 1.0, conducted by AgileAssets, Inc. for Idaho DOT, June 2010.

[Pilson 10] C. Pilson, "Bootcamp Training Manual for Maintenance Management," ITD MAPS Project, Version 1.0, conducted by AgileAssets, Inc. for Idaho DOT, updated January 2010.

[Razavi 10] S. Razavi, and C. Haas, "Multisensor Data Fusion for On-Site Material Tracking in Construction," *Automation in Construction*, Vol. 19, No. 8, 2010, pp. 1037–1046.

[Rehan 13] R. Rehan, Unger, M.A. Knight, and C. Haas, "Financially Sustainable Management Strategies for Urban Wastewater Collection Infrastructure–Implementation of a System Dynamics Model," *Tunnelling and Underground Space Technology*, online available January 2013. in press

[Shahi 13] A. Shahi, J. West, and C. Haas, "Onsite 3D Marking for Construction Activity Tracking," *Automation in Construction*, Vol. 30, March 2013, pp. 136–143.

[Smith 93] A. M. Smith, *Reliability Centered Maintenance*, McGraw-Hill, New York, 1993.

[Turkan 12] Y. Turkan, F. Bosche, C. Haas, and R. Haas, "Automated Progress Tracking Using 4D Schedule and 3D Sensing Technologies," *Automation in Construction*, Vol. 22, March 2012, pp. 414–421.

[Turkan 13] Y. Turkan, F. Bosche, C. Haas, and R. Haas, "Towards Automated Earned Value Tracking Using 3D imaging Tools," *ASCE Journal of Construction Engineering and Management*, Vol. 139, No. 4, April 2013, pp. 423–433.

[Uddin 13] W. Uddin, "Value Engineering Applications for Managing Sustainable Intermodal Transportation Infrastructure Assets," *Management and Production Engineering Review, MPER*, Vol. 4, No. 1, March 2013, pp. 74–84.

[Urban 84] H. P. Hatry and B. G. Steinthal, *Guides to Managing Urban Capital Series*, Vol. 4: *Guide to Selecting Maintenance Strategies for Capital Facilities*, The Urban Institute, Washington, DC, 1984.

[Weeks 96] K. D. Weeks, and H. W. Jandl, " The Secretary of the Interior's Standards for the Treatment of Historic Properties: A Philosophical and Ethical Framework for Making Treatment Decisions," *Standards for Preservation and Rehabilitation, ASTM STP 1258*, American Society for Testing and Materials, Philadelphia, Pa., 1996, pp. 7–22.

[Young 11] D. Young, C. Haas, P. Goodrum, and C. Caldas, "Improving Construction Supply Network Visibility by Using Automated Materials Locating and Tracking Technology," *ASCE Journal of Construction Engineering and Management*, Vol. 137, No. 11, pp. 976–984.

CHAPTER 12

Dealing with New or Alternate Concepts

12.1   Introduction

For many years, civil engineering has dealt primarily with groups of standard materials, including conventional steel, Portland cement concrete, asphalt concrete, vitrified clay (brick and pipe), glass, wood, and naturally occurring aggregates. Other materials are also used, and occasionally new materials are introduced. Traditional design, maintenance, and construction procedures have all been built around these conventional materials and their combinations, such as reinforced concrete for structures and jointed concrete pavements. Historically, structural design has been associated with the design of conventional structural steel and conventionally reinforced concrete. Several generations of engineers have been trained using these materials, and textbooks have been written covering the concepts related to such materials.

Design guides and design manuals are also formulated around these conventional materials. Examples include the Portland Cement Association's pavement design manual [PCA 84], the various design manuals produced by the Asphalt Institute [AI 87, 91], the American Institute of Steel Construction design handbook [AISC 94], and the American Concrete Institute handbooks of various types [ACI 72, 84, 92].

Of course, a variety of new materials have been tried in the last century. However, the frequency with which they have been used successfully, and the difficulty of adapting new materials, has often hampered new developments.

A good example is the California Bearing Ratio (CBR) design method for asphalt pavements, which was widely used by the U.S. Army Corps of Engineers and the Asphalt Institute prior to 1965, and is still being used around the world [USACE 58, 61]. The CBR method is an empirical method based on the use of conventional crushed-stone materials covered by a relatively thin layer of conventional asphalt concrete. In 1962, the AASHO Road Test was completed [HRB 62]. This study showed that the value of a material for carrying repeated traffic loads depended on the quality of the material itself, not just the thickness. The CBR method does not take this concept into account and has now been generally replaced in use. In a similar fashion, various aspects of Portland cement concrete-pavement design were shown by the AASHO Road Test to need adjustment and were modified in the AASHTO Pavement Design Guides [AASHTO 93], which has been updated in the 2000s [AASHTO 08].

Similar examples can be cited for the building industry, as empirical methods developed by Ferguson et al. [Ferguson 88] improved on the empirical knowledge about the use of steel reinforcement in the design of reinforced concrete structures and building elements. This development has continued with the advent of higher-strength concrete, finer-grind Portland cements, and high-tensile-strength steel. Bridges are built with high-strength steel and reinforced concrete to ensure long-term structural integrity and uncompromised safety.

A recent example of the use of construction materials is the New Port Mann Bridge, the widest bridge in the world, which was opened on December 1, 2012 in both directions in British Columbia, Canada. The 10-lane 850-m cable-stayed bridge and part of Highway 1 project was constructed at the cost of $3.3 billion. The following materials were consumed in the construction of this iconic structure: 13,000 tons of structural steel, 157,000 m^3 of concrete, 28,000 tons of rebars, 288 cables (45 km total length), 251 piles (16 km total length), 108 caissons, 100,000 m^2 of retaining walls, 25,000 tons of asphalt used for new bridge deck, and 3.1 million m^3 of earthworks (Twitter tweet, December 7, 2012, via @portmannbridge2).

Many new and innovative materials, like fiber-reinforced plastics (FRP), were not readily accepted in the 1980s and 1990s by the specification and design communities. The most significant reason was the time lag between the development and evidence of long-term performance. Another reason is that the use of existing design and fabrication standards with new material technology may not be the best way to implement the technology; it may be more costly and even fail. These points are obvious if one considers bridge technology. Bridges have always been a permanent record of mankind's achievements in the use of materials [Bridge 95]. After each new material was developed, it took 30 to 40 years for it to gain the acceptance and confidence of engineers before they would use it for a long-span bridge design. For example, the first iron bridge was built in 1779, very soon after the invention of wrought iron, but the material was used in the first world-record span in 1816.

As illustrated in Figure 12-1, increases in bridge span have occurred in three jumps: one in middle 1800s when wrought iron was sufficiently developed; one in the late 1800s when reliable steel was available; and the largest of all in 1930 when high-strength steel wire

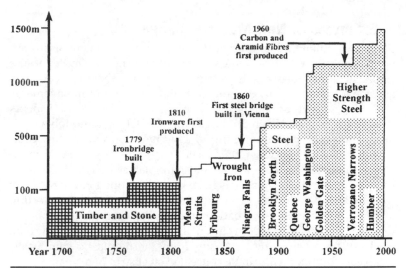

FIGURE 12-1 Historical trend in the world's longest-bridge-span construction [Bridge 95].

was used in a major suspension bridge. The trend indicates that the maximum span has risen by a factor of 3.5 in each 100 years since 1770 [Bridge 95].

Steel is always one of the first choices for structures of the future because of improvements in welding and joint technology and higher yield strength. Concrete technology has also improved rapidly, resulting in very high strength in a shorter time. However, the advanced composites and FRP materials provide new concepts in structural forms, resistance against corrosion, and higher strength-to-weight ratio, as shown in Figure 12-2 [after Jacobs 94].

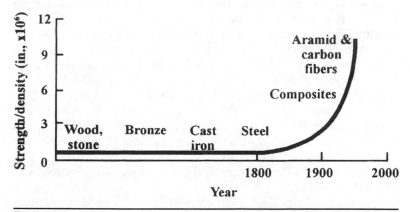

FIGURE 12-2 Historical development of engineering materials (after [Jacobs 94]).

12.2   Examples of New Material Usage

There are numerous examples of attempts to use new materials in conventional infrastructure—some successful, some not. Examples of successes include the use of high-strength steel and high-strength concrete, and the introduction of continuous steel reinforcement in pavements with the advent of continuously reinforced concrete pavements (CRCP) in the 1950s. Even this last example of CRCP, although widely used, shows the difficulty in introducing new concepts as the result of a design-only process. Other recent examples of successful innovative materials are composites and nanomaterial technology. New innovative composite materials have been developed in the past few decades and introduced in practice after meticulous laboratory testing and field performance-evaluation studies. For example, bumpers of cars and pickup trucks conventionally manufactured using chrome steel are now being made mostly from thermoplastic composites (with each bumper featuring a concealed steel strip of a smaller cross section). Additionally, this approach is a solution for minimizing chrome-related environmental hazards and the long-term benefit of reduced use of more precious metals.

12.2.1   CRCP Pavements

CRCP was first introduced in reasonable quantities in the state of Texas in the 1950s [McCullough 60, 68]. From early success it appeared to be a panacea because of its very smooth riding quality and its apparent ease of construction without joints. However, problems soon developed in the as-constructed product.

The original design was based upon strength characteristics of the steel and concrete in which the steel was placed at the mid-depth of the concrete in sufficient quantity to hold cracks tightly together, but the concrete itself cracked to relieve concentrated tensile stresses. This part of the design worked well; however, the design did not account for the significant increase in deflection that occurred with the thinner reinforced slabs with no joints. The resulting deflection increase later caused failure in the subbase and subgrade under the pavement within a period as short as two to five years. This loss of support then resulted in even higher deflections in the pavement and subsequent punch-outs and failure. The Texas Department of Transportation (Texas DOT) immediately spotted these problems, subsequently modified their designs, and have continued to build the pavements successfully. The revised design produces pavements, in general, 51 to 102 mm (2 to 4 in) thicker than the original 203-mm (8-in) thick design. This improvement was the result of the interaction of design, construction, maintenance, and data feedback.

Another problem associated with CRCP was the need for different construction techniques. When CRCP was first developed, most highway contractors had conventional form-type paving equipment. However, CRCP lends itself to formless paving because of the continuous nature of the steel. When pavement contractors learned how to build the new pavement, a rapid reduction in construction costs resulted. Unfortunately, such success did not occur on major projects in Florida and Indiana, where construction, inspection, and supervision was reportedly inadequate. These projects resulted in devastating early failure, which caused the FHWA to cancel recommendations for constructing this type of pavement. As of 1996, this recommended support is still suspended. If an effective infrastructure management system had been in effect as this new concept was being developed, it may well have provided codified technology for dealing with not only the design concept changes but also the construction and maintenance requirement changes.

12.2.2 CORTEN Steel-Building Construction

Some new materials may be valuable in the proper setting but misused. One such example involves introducing a new building material on the campus of the University of Texas at Austin. The concept of CORTEN steel [Bradford 93] is that it is "designed to oxidize or rust." Such oxidation is supposed to occur quickly and uniformly to produce a thin, hard, impervious layer, which presumably prevents further rusting and deterioration of the steel. In 1970, an architect accepted the recommendation of steel manufacturers to use some of the new steel as a facing material on the Communication Building at the University of Texas in Austin. This material rusted as planned and formed a reddish-brown color. However, it did not stop rusting as predicted, and subsequently rusted completely through within a period of five to six years. The result was that many sections had to be replaced, and eventually a complete new surface cladding had to be installed on the building at tremendous cost. This material has been used successfully as a structural material but was not acceptable in thin layers.

12.2.3 Chemcrete Asphalt-Concrete Mixes

Chemcrete is another example of premature acceptance of a new concept without adequate evaluation. In the late 1970s, a major company undertook the development of a "new" material called Chemcrete. The material was purported to be a panacea for asphalt-concrete construction. It produced very stiff pavements, which its proponents labeled "a stronger pavement." Unfortunately those involved did not adequately recognize that asphalt-concrete mixes must fulfill several purposes to perform successfully. These include not only stability, but also strength, flexibility, and durability. The company involved enlisted several well-known professionals in the pavement field to help promote their product. The laboratory tests and the field trials, however, were observed for only very short periods of time, so the durability aspect of the material was not properly evaluated. Subsequent pavements using Chemcrete cracked badly and rapidly, and resulted in extremely early failure. The product ceased to be marketed a few years later. Proper application of an infrastructure management system, or in this case a pavement management system (PMS), would likely have shown the deficiencies at a very early date during the test period.

12.2.4   The Use of Geotextiles and Thermoplastic Grids

Another example of materials use that has both positive and negative aspects involves fabric or geotextiles and thermoplastic grids in pavement construction.

Since 1960, various manufacturers of nonwoven fabrics (mostly using polypropylene fibers) have been seeking to broaden the markets for their products. Pavements appeared an attractive target, and they began vigorous efforts to market their product for use in pavement rehabilitation. These materials have been marketed as a cure-all that will waterproof (because of the application of a tack coat on the existing, old pavement) and prevent reflection cracking in overlaid asphalt-concrete pavements. Television and trade-magazine advertisements have appeared depicting the apparent speed and quality with which various city engineers were supposed to be able to repair their streets using these products. The ads never mention that the cracks do not disappear but are only delayed in returning by one or two years. They also do not refer to the fact that the use of a fabric is the equivalent in cost to one or more additional inches of asphaltic concrete, which would show better pavement performance.

On the other hand, there is a great deal of evidence of successful and effective use of fabric as a construction technique when placed as a separating medium on wet or soft subgrades. The use of the fabric or geotextile provides a platform for placing granular material that will not punch into the soft subgrade, and which saves considerable money, time, and materials over waiting for drier conditions or compacting the aggregate into the soft subgrade. This use of fabrics can be valuable and meets the test of economic expediency. Fabrics are also successfully used as drainage layers and for confining drainable trenches in some cases, although clay particles can tend to clog the fabric and prevent the desired drainage.

In the late 1990s and 2000s, thermoplastic grids have been promoted to minimize and delay reflection cracking in pavement construction. Due to the use of high-strength glass and carbon fibers in proprietary layouts, these woven products are theoretically more durable and resistant to stresses caused by reflection-cracking mechanisms and likely more effective in rehabilitation of rutted and distressed asphalt-surfaced pavements. Most published literature shows successful installation locally with only one-year or less assessment. Again, the long-term performance studies are important before dependable standards can be developed to support suppliers' claims.

12.2.5 Applications of Polymers in Construction

12.2.5.1 Polymers

The benefits of polymer-material technology are evident in many areas of life, including sports gear, auto parts and autobody repairs, buttons, the aerospace industry, bikes, musical instruments, and home furnishings. Most items are made of cast-polymer or fiber-reinforced composites.

12.2.5.2 Cast-Polymer Products

The use of polymers and particulate composites is not new in construction applications. Cast-polymer products have been used for everything from home furnishings, such as kitchen countertops and bathrooms, to rapid repair of bomb damage [Dawson 95].

Cast polymer is a nonreinforced particulate composite that combines polymers, fillers, and additives to meet specific application requirements. Fillers account for up to 90 percent of the material volume. After the material is mixed, it polymerizes and becomes rigid [Dawson 95].

Methyl methacrylate acrylic (MMA) and unsaturated polyesters resins (UPR), both orthophthalic and isophthalic, dominate the gel-coated cultured marble and solid surfaces markets, which include home furnishings for kitchen and bathrooms, and whirlpool spas. Solid-surface densified and nonporous products made from high-quality resin can be cut, sanded, and seamlessly joined. Architects and specifiers can take full advantage of these properties, as well as resistance to corrosion and chemicals, for creative architectural and interior designs.

Orthopolyester and polyurethane are commonly used for decorative castings and flexible moldings. Polyester, epoxy, and vinyl ester are the predominant matrix resins in precast polymer concrete for products such as utility vault doors and electrolytic mining cells. They also are used in field applications, such as the polymer-concrete composite overlay used for the Golden Gate Bridge approach [Dawson 95]. A polymer-concrete overlay can be opened to traffic within a few hours compared to the 14 to 28 days of curing time required for conventional Portland cement concrete.

12.2.5.3 Polymer-Modified Asphalt

Asphalt is used with sand and selected sizes of aggregates as hot-mix, emulsion and cold mix, and seal coats for paving parking lots, streets, highways, and airport pavements. On heavily trafficked roads, rutting has become a major distress and reason for poor performance. Polymer-modified asphalt is becoming increasingly popular to improve the thermo-viscoelastic behavior and in-service performance of asphalt pavements. Mixes paved with asphalt or polymer-modified asphalt are examples of particulate composites, and their stress-strain behavior can be analyzed using the same micromechanical analysis used for advanced polymeric-matrix fiber-reinforced composites [Plaxico 95, 96]. Polymer-modified asphalt is widely used in pavement construction because of its superior performance over virgin-asphalt binder [Uddin 03].

12.2.6 Fiber-Reinforced Plastics for Structural Applications

12.2.6.1 Overview

For over three decades, polymeric fiber-reinforced plastics (FRPs), broadly known as *advanced composites*, have been used for grates, steps, ladders, pipes and tanks, and numerous other applications in petrochemical plants and chemical industries where resistance to corrosive chemicals is a primary concern as compared to structural performance. In these applications, composites outperform steel, concrete, and wood. The aerospace industry has used these advanced composites in nonstructural components because of their light weight, higher strength-to-weight ratio, and superior resistance to temperature changes and fire [Technology 95]. Metal composites produced primarily from titanium and high-strength steel materials have been traditionally used in the aircraft industry. But in the quest to reduce the overall empty weight of large commercial aircraft to gain in fuel efficiency, polymeric composites such as FRP materials are being increasingly used in the 21st century.

Fifty percent of Boeing's B787 (popularly known as the Dreamliner), including the entire fuselage and most of the wings, is built from carbon FRP. This is about four times more FRP than used in the B777, which contains 12 percent carbon composites [Forbes 12]. This makes the B787 far lighter than any plane of its size and allows it to fly at the same 0.85 Mach top speed like the B777 but consuming 20 percent less jet fuel and producing 20 percent less emissions [Forbes 12]. That means less travel time and fewer connections for international passengers. The increased range of the B787 allows nonstop flight from Houston to Auckland for which a connection in Los Angeles is required flying aboard the B777.

The application of polymeric advanced-composite materials has been rather slow in civil-engineering structural market and construction industries. IAMS can show owners, operators, and designers the importance of low maintenance, longer life, and improved structural performance offered by advanced high-performance construction materials. The application of advanced composites is being sought in all areas of construction, particularly for waterfront and off-shore structures and seismic retrofitting of bridges. Another application area is the rehabilitation of reinforced concrete structures, where the major problem is reduced life caused by the corrosion of steel reinforcement in a harsh environment. Use of FRP products could significantly increase the life and performance of these structures. The light weight of FRP products makes handling easy and reduces assembly costs. The FRP materials made from polymeric resin matrix and fibers are becoming popular, as the trend toward achieving low-cost manufacturing methods increases.

12.2.6.2 Selection of FRP Material and Application Process

The proper use of FRP material and the application process is the key to successful and longer-lasting applications in structural strengthening and retrofit of infrastructure assets. FRP composites are anisotropic in nature, and this should be used to advantage in innovative designs. Three primary limitations of FRP are long-term loss of strength caused by creep effects; debonding and fiber material degradation caused by the likelihood of water absorption by the resin matrix; and durability problems in an alkaline environment. These should be appropriately considered in design applications. The following examples illustrate innovative applications of FRP materials.

Naval FRP piers. The U.S. Naval Facilities Engineering Service Center (NFESC) installed an all-composite catwalk at its Advanced Water Technology Test Site (AWTTS) at the Port Hueneme site, California, in May 1995 to test composites in marine structures. The total length of the test pier is 48 m (160 ft), consisting of several spans. The composite portion of the new pier consists of two bays: a fiberglass-composite deck and a concrete deck prestressed with carbon-fiber/epoxy rods. The decks are the result of a cooperative Construction Productivity Advancement Research (CPAR) project among the Composites Institute, the Army Corps of Engineers CERL, and the South Dakota School of Mines and Technology [Technology 95, Iyer 95]. The deck was designed to carry pedestrian loads, withstand seawater exposure and even total submersion during extreme high tides, and be subjected to cyclic loading from wave action during tide charges.

Seismic retrofit of bridge columns. Seible and Innamorato [Seible 95] identified the possible failure modes of concrete bridge columns based on a study sponsored by the California Department of Transportation (Caltrans). For these cases, the three primary failure modes are: (1) column shear failure; (2) concrete confinement failure of the flexural plastic-hinge region; and (3) debonding of lap splices in the column reinforcement.

A reliable retrofit measure should consider all three possible failure modes. Such retrofit methods include steel-column casing to jacket bridge columns. For conventional columns with a uniform circular cross section, the steel jacket can be fabricated in two halves, bolted together, and filled with grout. For nonuniform cross sections or other shapes, steel casings are difficult to fabricate. For strengthening and rehabilitation of bridge columns, a composite wrapping and winding system is an efficient alternative, from a functional viewpoint, to provide hoop constraint. Seible and Innamorato [Seible 95] outline the principles of an automated continuous carbon-fiber wrap system that addresses the transverse reinforcement deficiency by wrapping prepregnated carbon tows in the horizontal, or 90-degree, direction to the column axis to provide the required transverse confinement, clamping, and buckling restraints. This innovative composite retrofit system ensures anchorage of the wound carbon tows by the continuity of the fiber wrap for the entire column jacket, and provides close control and monitoring of lay-up thicknesses with an automated winding system.

12.2.6.3   Responsibility of Specifying and Approving Agencies

Development of acceptable test standards and specifications for emerging technologies, such as FRP composites, is a major concern for the specifying and approving agencies. In particular, agencies are interested in the time-critical emergency situations that warrant structural applications in life-line infrastructure facilities, such as seismic retrofitting of bridges and repair and strengthening of structural members. Caltrans [Sultan 95] has already initiated an active program in this area for performance testing of FRP to verify:

- Strength and performance of the fiber itself
- Strength and performance of the resin
- Strength and performance of the composite
- Behavior of the system as a structural member
- Validity of the design approach

Durability of FRP composites should be checked for resistance to water, ultraviolet rays, high temperature, salt water, alkaline environment, freeze-thaw, and fire. Table 12-1 outlines a list of important issues that an approving agency should examine to qualify and implement any proposed FRP material-fabrication-installation process [Sultan 95].

As discussed in Chapter 9 on value engineering applications, noncorrosive lightweight FRP-sheet piles are included in the materials specifications and design standards of several government agencies in Florida and other states in place of traditional reinforced concrete-sheet piles for waterfront protection of infrastructure assets.

1. Product documentation
2. Understanding of the process
3. Type of materials
4. Material physical properties
5. Long-term durability (chemical and physical) testing of composite
6. Quality control in manufacturing, mixing, and applying
7. Fiber content, voids, resin ratio
8. Design guideline for the specific composite
9. Safety factors
10. Damage and failure modes
11. Adequate specifications
12. Repeatability and consistency
13. Acceptable field erection methods
14. Effect of fatigue on bond behavior
15. Performance under dynamic load
16. Testing under sustained loading
17. Qualifications of suppliers and product designers
18. Cure temperature
19. Transportation and handling
20. Maintenance issues

TABLE 12-1   Checklist for Qualifying and Implementing FRP Technology [Sultan 95]

12.3   Handling Diminishing Resources

During the 1940s, 1950s, 1960s, and 1970s, the United States and many other areas of the world were on a veritable building binge. Notable construction included the interstate highway system, the rapid development of Toronto, New York, Chicago, Los Angeles, and other large cities, including buildings, sewage facilities, water supply network, dams, and related infrastructures. This construction grew out of the completion of the greatest war ever fought, World War II. Techniques developed during the war for rapid construction of facilities and equipment at any cost continued to be used in the creation of the new national infrastructure. Although this approach was successful in getting construction done, it resulted in a massive depletion of the world's natural resources.

The construction of the U.S. interstate highway system alone consumed a large quantity of good natural aggregates in many sections of the United States. An outgrowth has been the creation of tighter and tighter specifications requiring higher and higher quality materials, which have become more difficult for the manufacturers and producers to provide. In some cases, this has made costs prohibitive.

An infrastructure management system can properly consider both the benefits and costs of the materials, with their limitations, and provide an effective and economical way for using some of the so-called marginal materials.

Since the early 1980s, the world, and in particular North America, has become concerned with environmental issues and the *reusing, reclaiming, recycling,* and *reducing* consumption of certain materials. This problem has two aspects: (1) the required change of material usage because of environmental issues; and (2) the potential successful use of reclaimed, recycled, reused, or otherwise waste materials.

In the 1980s, steel from cheaper export and "dumping" practices hurt steel industries in the United States so much that steel production from raw materials almost went out of the manufacturing business. However, in the late 1990s the U.S. steel industry developed new technologies for efficient manufacturing of steel from recycled steel waste. Currently, about 90 percent of steel for domestic use is produced in the United States from recycling [Mittal 12].

12.3.1 Environmental Impacts

In the last few decades, much has been learned about the environmental impact of certain materials used in infrastructure development. A well-known example is the use of asbestos. For about 50 years prior to 1970, asbestos was used as an insulation material in buildings, particularly around heating ducts and heating pipes, and as a fire-resistant material in the form of sprayed-on insulation around structural steel and reinforced-concrete beams. It was also used as a siding and roofing material in the form of pressed asbestos sheeting and corrugated paneling, and even as a filler material in asphalt concrete to replace certain types of fine aggregate. Environmental tests subsequently showed asbestos to be carcinogenic. Its use in construction was suspended, and efforts began to remove it from many construction locations. Two aspects of this problem could be handled by an appropriate infrastructure management system: (1) the development of new materials to replace asbestos; and (2) maintenance and removal procedures for dealing with the material in place. Unfortunately, the basic and very costly remedial approach has been to remove the material. However, a thorough evaluation of the costs and benefits might conceivably show that the benefits of removing the material are minimal, and that the damage done by disturbing the material and removing it is greater than would ever be caused by sealing it and leaving it in place.

This is an example where external factors—that is, public fear and the profit motive for certain industries—has overridden good systems analysis. This is not to say that the environmental aspects should be ignored, but merely that they should be thoroughly evaluated in a realistic sense along with the total costs and benefits.

12.3.2   Environmentally Safe Alternatives

The asphalt industry provides examples of existing materials that proved to be environmentally questionable, and the environmentally safe alternatives that were developed in their place. Specifically, cutback asphalts, those asphalt binders that are diluted with gasoline and diesel fuel, were widely used in the United States and around the world prior to 1970 for constructing low-volume roads and for priming subgrade and base layers on high-quality construction. The volatile component of the cutback asphalt, that is, gasoline, kerosene, and diesel, evaporates into the atmosphere, thus adding to the problems of air quality, particularly in urban areas. Alternate methods of changing the viscosity and transportability of asphalt are: (1) to heat the material; and (2) to dilute it with water in the form of an emulsion. Considerable efforts have been made in the past 30 years to improve the quality of emulsified asphalts to replace asphalt cutbacks.

The use of polymer fibers and waste tire rubber in pavement construction as a modifier has been tried in different forms. The fibers started dislodging from concrete pavement within six months after a field trial on the I-20 highway near Vicksburg, Mississippi. This was a failure example of a new product without adequate pilot tests. On the other hand, the use of very fine waste tire rubber (80 microns or less in size) in hot-mix asphalt provided the best field performance on a test section of I-55 highway in northern Mississippi against rutting, compared to the performance of Superpave-grade virgin asphalt and other polymer-modified asphalt binders [Uddin 03].

Environmentally safe handling and disposition of waste materials during construction and demolition is another aspect of and partial solution for diminishing natural sources of aggregate and steel. Pollution-prevention practices were demonstrated at the successfully completed $250 million relocation and reconstruction "design-build" project of the U.S. Naval Sea Systems Command (NAVSEA) headquarters in Washington, DC. Some salient construction and environmental stewardship features of the project included [Gonzalez 05]:

- One million sq ft (about 100,000 m^2) complex, housing over 5,000 employees.

- Eight-story parking garage for over 1,500 vehicle spaces.

- Monthly billing of over $10 million and 600 construction workers on site.

- Pollution-prevention program included: (1) removal of cooking-oil fryers from new naval ships, (2) new fluorescent-light fixtures for aircraft hanger bays, and (3) plastic-waste compaction and disposal system.

- National Environmental Protection Act (NEPA) compliance program included: (1) environmental friendly surface-runoff drainage and disposal facilities, (2) uniform national-discharge standards, and (3) improved sonar soundings for monitoring and protecting marine life and ecosystem.

- Sustainability considerations: contractor's removal of all old demolished brick, concrete, and steel materials for recycling or applications on other sites.

Other examples of environmentally safe alternatives for waste-disposal infrastructure construction are the design and construction of landfills for rubbish waste and disposal of hazard materials. Government agencies must plan environmentally safe waste-disposal infrastructure or enforce these regulations on corporations and private entities. Selected case studies follow:

- In aftermath of Hurricane Katrina's landfall along the Mississippi Gulf coast on August 29, 2005, more than 90 percent of structures within a half mile of the coastline were destroyed; the storm affected 250 miles of coastline and more than 100 miles inland, destroying 150,000 properties. Hurricanes Katrina, Wilma, and Rita in 2005 left 87,000 sq mi (222,720 sq km) of debris in parts of Mississippi, Louisiana, Alabama, Arkansas, Florida, and Texas. In 2005, there were 60 landfill sites in Mississippi, which increased to 69 sites in 2006. One reason for the development of new landfill sites was to handle millions of tons of debris from the hurricane destruction [Uddin 10a].

- New York City has already begun implementing measures that contribute to reducing the city government's GHG emissions. These measures include, among others, landfill methane recovery at a 75 percent capture rate and solid waste recycling. Since the closure of Fresh Kills Landfill in 2001, all solid waste generated in New York City is exported to managed landfills or waste-to-energy facilities out of the city [NYC 07].

- Low-radiation nuclear-waste storage at each nuclear power plant site in the United States is the most critical issue facing the nuclear power-plant disaster after the earthquake and tsunami that struck northeast Japan coast in March 2011. Furthermore, efforts to transport the nuclear waste to dedicated depository locations are met with resistance from local communities. This issue is faced by nuclear power-plant operators worldwide because of security concerns and hazards of community exposure from disaster events. Research efforts are needed for the development of improved methods and materials to handle nuclear waste.

12.4   Considering New Methods and Materials for Infrastructure Use

How does an agency involved in the provision of infrastructure go about considering new materials, methods, and designs? No agency wants to be blind to new innovations, but neither do they want to fall prey to wild marketing claims and promises. The best approach to this problem is a formalized IAMS. During the interim, however, general systems-management concepts can be of assistance. The first step would be to look at the standard or typical procedures that are currently used. The second step would be to define the alternatives that should be compared to the standard. The third step is then to list all of the criteria that must be used in making the comparison. Performance, strength, stability, and service life, of course, immediately come to mind, but in some cases the more important criteria may be maintainability, recycling potential, use of local materials, environmental impact, energy impact, and political necessity. It is important to list these factors a priori, that is, before comparisons are made.

A thorough and complete listing of factors and costs (not only economic costs) involved in any issue is critical. If a preliminary comparison is made that does not consider all pertinent factors, it is very difficult to readjust the final solution based on the addition of other factors. It is always tempting to be biased to the initial solution, or to add those factors that point to the solution you prefer if the "first solution" does not seem suitable. Such an approach is not objective, and is far less likely to give meaningful results than a truly objective approach.

A well-defined infrastructure management system will provide the listings and approaches necessary to include the majority of factors from the outset. However, no system is perfect, and the judgment of the infrastructure management team is critical in setting up the problem. It will be preferable, and a better solution will evolve, if an interdisciplinary team is used to assist in formulating the problem, defining the criteria and parameters of a solution, considering the alternatives, and defining the best potential solution. These should all be done before fixing a mindset. Even a computerized IAMS does not make decisions, but rather provides an organized, effective method of processing the large amount of data and information that will be critical in making decisions.

12.5   Dealing with Shrinking Natural-Aggregate Sources

For nearly 100 years, the United States has been constructing major civil-engineering works. Most of these civil works use natural aggregates. Aggregates make up 90 to 95 percent of almost any civil-engineering infrastructure project, either alone as an aggregates base or as the major component of asphalt concrete, Portland cement concrete, or vitrified clay. They also form part of any stabilized materials option. In many places in the world, the natural aggregates are almost depleted, and there is a strong need to replace natural high-quality aggregate sources with either manmade materials, waste materials, recycled materials, or lower-quality aggregates that heretofore have been rejected under existing specifications.

Historically, civil engineers have incorporated high quality in specifications. Such specifications were presumed to have little or no potential for failure in an absolute sense, but few if any comparisons were made. It has now become necessary to modify standard specifications to accept aggregates that do not meet the extremely high standards used in the past. Such materials can serve well and may cost less in the long run. A methodology for dealing with this concept has been developed by Saeed [Saeed 96].

12.5.1   An IAMS Subsystem for Evaluating Materials Including Waste and Reclaimed Materials

Not all materials are equal in quality, and quality itself has a wide variety of criteria, ranging from stiffness and initial strength to durability and long-term performance. Energy consumption in the production of new materials and shrinking supplies are all related to the problem. To deal with new materials, as well as waste and reclaimed materials (WRMs), an IAMS subsystem has been developed by Saeed [Saeed 96].

Significant pressure is building to use a wider range of materials in infrastructure construction, maintenance, and rehabilitation. Primary reasons are: (1) shrinking funds; (2) limited supplies of high-quality materials near the point of utilization; (3) difficulty in opening new quarries because of zoning laws; (4) increased haul distances and transportation costs; (5) societal concerns and environmental pollution; (6) increased cost of land utilization; and (7) an urgent need to reduce the amount of energy consumed in the production of natural aggregate.

No known method used in design prior to 1996 provided a method for considering all of the diverse factors associated with selecting materials. The stand-alone materials-evaluation system [Saeed 95], called the WRM Evaluation System, was developed in a research project for the Texas DOT. It can, however, be applied to all types of infrastructure. The method considers four categories of attributes: (1) technical; (2) economic; (3) societal; and (4) environmental. These come into play after an initial screening process based on local information. The screening process ensures that the proposed material possesses certain minimal fundamental physical, chemical, mechanical, and thermal properties before it is run through the selection subsystem for possible use. The method considers trade-offs between the various strength and performance categories against

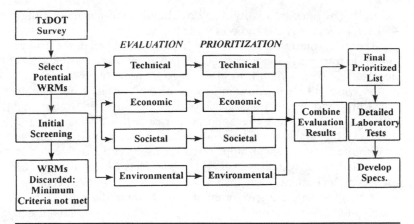

FIGURE 12-3 Outline of method to evaluate factors in selecting materials [Saeed 95].

economic, environmental, and societal issues. While it is difficult to evaluate the societal and environmental issues economically, Saeed provides a method for considering expert opinion and public opinion in this regard. The outline of the method is presented in Figure 12-3.

Enormous amounts of new waste materials are generated in the United States daily. For some agencies and persons, using these waste materials has become a responsibility, one that can assist in preserving the world's natural resources. Using waste materials can also reduce the world's dependence on natural aggregate [Saeed 95, TGLO 93].

In the technical literature pertaining to use of waste materials, the term *recycling* is often used. Recycling, in essence, is taking an item, after it has been through its useful life, and remaking it into a new, useful product instead of discarding it. *Reusing,* on the other hand, refers to using an item over and over again in its current form without converting it into something new. *Reclaiming,* as defined by Wood et al., is the process during which a waste material is removed from its original location for reuse or recycling at some other location [Wood 89].

Construction, maintenance, and rehabilitation (M&R) of transportation infrastructure is costly and uses large quantities of material. Utilization of WRMs in these activities is expensive but at times has advantages over the use of conventional materials. Prominent among the major benefits are the conservation of energy and natural resources, as well as preservation of the environment and existing highway geometrics [NCHRP 90].

The typical approach to characterizing WRMs has been to evaluate them technically in the laboratory, and compare the findings with standard specifications for standard natural materials. This is not an effective method because these materials generally do not equal natural aggregates in quality. They do, however, often have a high societal, environmental, and economic value.

There are many reasons for an agency to consider the use of WRMs in transportation infrastructure construction and M&R activities [Han 95]. These include local shortage of natural aggregates, high cost of waste disposal, commitment to the environment, availability of natural and waste materials, political pressure, environmental safety, and others. Cost-effectiveness, performance, availability, and the prevailing political climate are the four issues fundamental to determining if a material is appropriate to use in construction.

An IAMS subsystem is needed to objectively evaluate meaningful recommendations. Conclusions can be categorized as technical, economic, societal, and environmental in nature in the subsystem. A simple methodology is needed to evaluate WRMs for potential use in infrastructure construction and M&R activities.

12.5.2   Evaluation Subsystem

Any WRM evaluation method must consider the technical, economic, societal, and environmental aspects of their use, after an initial screening process. A technical evaluation unit must evaluate WRMs based on the physical, chemical, mechanical, and thermal properties required of a material for transportation project use. Technical properties determined to be of importance will be investigated in laboratory studies to technically evaluate WRMs.

The economic evaluation unit must identify waste materials that are most viable in an economic sense. This unit must consider all the costs related to the production of construction aggregate from WRMs and compare these to costs incurred in the production of natural aggregate and the disposal of waste material.

Societal and ecological implications often cannot be measured directly in dollar terms. However, the social evaluation unit should consider the interests or desires expressed by society in general, as well as the government and the private sector.

The environmental impact of using WRMs includes some factors that are related, directly or indirectly, to technical, economic, and social evaluation. The environmental evaluation unit must measure the actual impacts as opposed to how people feel about them.

Objective data is required for any WRM evaluation method in all four evaluation units. This objective data should be supplemented with subjective data in cases where objective information does not exist or is difficult to quantify, such as environmental impact and social values.

12.5.3   Current Issues and 4R Practices

A recent news headline shows "China to flatten 700 mountains to build a city" (Twitter tweet, December 7, 2012, via @TIMENewsFeed, http://time/RKgEmn). The megaproject will cost $3.3 billion to blast mountains and make space and $11 billion to build on the outskirts of Lanzhou, the capital of Gansu province in northwestern China. This mountain blowup will be the source of aggregate needed for construction of this metropolis in China. Other countries in north America and Europe need to deal with shrinking aggregate sources and find creative ways to minimize waste by adapting 4R (4R: Reclaim, Reuse, Recycle, Reduce waste) policies. Infrastructure applications of 4R practices are being considered by specifying agencies and accepted on increasing number of construction projects as agencies embrace the importance of sustainability.

A great deal of literature material has been developed by the FHWA and National Center for Asphalt Technology (NCAT) on pavement recycling. A synthesis study by Tighe and Gransberg made an extensive literature survey on recycled, alternative, and renewable highway materials, as listed in Table 1 of their referenced study [NCHRP 11]. The study shows the following survey results of 42 state DOT agencies in the United States and seven Canadian provincial ministries of transportation:

- Recycled roof shingles and tire rubber used by many agencies in asphalt pavements
- Recycled glass and foundry sand used only by a few agencies in both asphalt and concrete pavements
- Fly ash (a waste by product of coal-fired electric power plants) used mostly for concrete pavements

Different recycling techniques are available to address specific pavement distress and/or pavement structural requirement. The following discussion provides an overview of benefits and types of common recycling and other 4R practices [Uddin 08]:

- Benefits of recycling include: reduced cost of construction, conservation of aggregate and binders, preservation of existing pavement geometrics, preservation of environment, conservation of energy, efficient use of natural resources, and less user-delay costs.
- Concrete-pavement recycling includes: cracking and seating (with overlay), rubblization (with overlay), and crushing for use as coarse aggregates. Pervious or porous concrete is an environmental friendly surface-runoff application for concrete paving.
- Recycling methods for asphalt pavements are: hot-mix recycling, hot in-place recycling, cold in-place recycling, full-depth reclamation, and reclaimed-asphalt pavement (RAP).

Most state DOT specifications include the use of RAP for paving asphalt mixtures. The Mississippi DOT requires 10 percent RAP on all highway jobs. It is estimated that 10 percent of RAP results in 7 percent saving in total construction costs. Some state DOT agencies even require higher percentages of RAP in asphalt-paving projects. Caution must be exercised to avoid inferior quality and early failure while using recycling options, such as the presence of stripping in existing asphalt pavement should limit its use on shoulders or secondary roads. Another example is the use of an upper limit of percentage RAP on heavily trafficked highways that needs to be thoroughly evaluated to avoid early failures.

12.6   Energy-Related Issues

Energy consumption is sometimes a separate and more important issue in selecting infrastructure design, construction, and maintenance techniques than economics alone. In oil-importing countries, such as Chile, while the economics of asphalt pavements may be acceptable, the strategic issues and support of the local Portland cement industry may suggest the application of a systems concept. Even in the United States, which is relatively energy-rich, the balance of payments strongly suggests the need to examine the energy efficiency of a system as well as its overall performance and economy [Solminihac 92, Jones 93].

IAMSs offer methods for managing roadway infrastructure to reduce maintenance problems, rehabilitation delays, and lane-closure times. Thus, energy can be saved through improved infrastructure management. A coordinated methodology for planning, design, construction, and maintenance of transportation infrastructure minimizes vehicle delays and congestion and saves millions of gallons of gasoline and diesel fuel.

Technically speaking, there is no difference between an IAMS with and without energy savings taken into account. The difference lies in the variables used in the systems. In general, more attention should be paid to decision variables related to current and future user cost, fuel consumption, and so forth, to calculate energy savings. Although variables related to energy saving may be difficult to quantify, they can be used indirectly in decision making. Just like other issues, energy consumption can be considered in all the activities from planning, design, and construction to maintenance.

Regarding pavements, attention should be paid to better network-level planning design, material use, construction, and maintenance to save energy over the long run. Better design increases pavement life, thus decreasing the number of rehabilitation cycles and saving fuel because of reduced delays during pavement construction. Another method of saving energy is to recycle existing material. Recycling existing materials saves the energy that would be required to make new materials to replace them. It further reduces the need for energy and the costs associated with the disposal of old materials.

The Urban Roadway Management System (URMS) developed by Chen et al. at the University of Texas [Sohail 95] was one of the first comprehensive pavement management systems for application in small- to medium-size cities. It provides managers and engineers in public works departments with a computer-based tool to assist in all their activities, including planning, design, construction, and maintenance. For agencies desiring to evaluate associated "energy costs," the URMS can be adopted to minimize energy use as well.

During the first decade of the 21st century, emphasis on sustainability and reduction in energy use and emissions produced led the development of *warm* asphalt that can be produced and compacted at lower temperature than traditional hot-mix asphalt [Hutschenreuther 10]. The use of titanium oxide–treated brick/paver block (on sidewalks and parking areas) is shown to absorb NO_2, a precursor of tropospheric ozone, from the air and reduce smog in summer months that is flushed away in drains after rain [Uddin 10b]. The long-term performance of some of these new paving materials has not been fully assessed.

The 4R practices are beneficial for environmental sustainability and energy conservation by reducing processing of raw materials. Aggregate processing and cement manufacturing consume significant energy and produce considerable CO_2 emissions. It is estimated that CO_2 emissions from the cement clinker production process is the largest source of non-combustion-related CO_2 emissions, contributing 4 percent to the global total in 2010. In 2011, this source of CO_2 emissions increased globally by 6 percent, mainly due to an 11 percent increase in China [Olivier 12].

A liquid natural gas (LNG) trucking-corridor network has been under construction in the United States, and so far 30 LNG truck-fueling stations are complete and providing fuel service to LNG-powered 18-wheelers (http://www.cnn.com/2012/12/07/us/trucking-gas-future; accessed December 7, 2012). LNG-fueling infrastructure needs to be expanded throughout the United States to take advantage of 20 percent lower CO_2 emissions from burning of natural gas and the abundance of natural gas reserves in North America. Many parts of the United States are dotted with other no-fossil renewable energy sources (wind and solar). However, unless these infrastructure facilities are networked with national electric-power grids, they are not effective to boost the electric-power capacity of a country. Obviously, this vital lifeline infrastructure needs to adapt IAMS examples presented in this book.

12.7 Summary

Infrastructure asset managers continue to be faced with complex issues in providing the best infrastructure benefits to the public, as well societal benefits of energy conservation and emission reduction. As outlined in this chapter, dealing with new concepts and materials are on the list. Several examples have been given, although comprehensive details are not included. It is hoped that the examples given here will prompt readers, both managers and operational-level people involved in optimizing infrastructure resources, to use an IAMS as a tool to assist in their work. Examples of several working systems, presented in Part 5 of this book, can serve to assist readers.

12.8   References

[AASHTO 93] *AASHTO Guide for the Design of Pavement Structures,* American Association of State Highway and Transportation Officials, Washington, DC, 1993.

[AASHTO 08] "AASHTO Mechanistic Empirical Pavement Design Guide," American Association of State Highways and Transportation Officials, Washington, DC, 2008.

[ACI 72] "A Design Procedure for Continuously Reinforced Concrete Pavements for Highways," *ACI Journal,* American Concrete Institute, Detroit, Mich., 1972.

[ACI 84] *Design Handbook: In Accordance with the Strength Design Method of ACI 318–83,* American Concrete Institute, Detroit, Mich., 1984.

[ACI 92] *Concrete Slabs on Grade: Design, Specifications, Construction, and Problem Solving,* American Concrete Institute, Detroit, Mich., 1992.

[AI 87] "Thickness Design-Asphalt Pavements for Air Carrier Airports," *Manual Series No. 11,* Asphalt Institute, College Park, Md., 1987.

[AI 91] "Thickness Design-Asphalt Pavements for Highways and Streets," *Manual Series No. 1,* Asphalt Institute, College Park, Md., 1991.

[AISC 94] *Load and Resistance Factor Design: Manual of Steel Construction,* American Institute of Steel Construction, Chicago, Ill., 1994.

[Bradford 93] S. A. Bradford, *Corrosion Control,* Van Nostrand Reinhold, New York, 1993.

[Bridge 95] "Keynote," *Bridge Design and Engineering,* Oct. 1995, pp. 19–20.

[Dawson 95] D. K. Dawson, "Cast Polymers Capture Sophisticated Markets," *Composites Technology,* Vol. 1, No. 3, September/October 1995, pp. 17–20.

[Ferguson 88] P. M. Ferguson, J. E. Breen, and J. O. Jirsa, *Reinforced Concrete Fundamentals,* John Wiley and Sons, New York, 1988.

[Forbes 12] G. Walther, "Fly The Dreamy Skies: Will Boeing's Long Awaited Dreamliner Change the Game for International Travelers?" *Forbes,* March 12, 2012, pp. 104–106.

[Gonzalez 05] H. S. Gonzalez, "Construction and Pollution Prevention Program: Naval Sea Systems Command Head Quarter Relocation Project," ITE Lecture, April 21, 2005, *Construction Engineering and Management Lecture Notebook* (Ed. W. Uddin), 2009, pp. 177–184 (unpublished).

[Han 95] C. Han and A. M. Johnson, "Waste Products in Highway Constructions," *Proceedings, Sixth International Conference on Low-Volume Roads,* Transportation Research Board, Washington, DC, 1995.

[HRB 62] "The AASHO Road Test Pavement Studies," *HRB Special Report 61D,* Highway Research Board, National Academy of Science, Washington, DC, 1962.

[Hutschenreuther 10] Jürgen Hutschenreuther, "New Pavement Technology Developed by the Use of Warm Asphalt Technologies." Presented at *2010 International Conference on Transport Infrastructure,* Sao Paulo, Brazil, August 4–6, 2010.

[Iyer 95] S. L. Iyer, "Design and Construction of FRP Cable Prestressed Navy Pier at Port Hueneme, California (CPAR Program)," *Proceedings, Fiber Reinforced Plastics Workshop,* sponsored by the Federal Highway Administration and TRB Committee A2C07, January 1995.

[Jacobs 94] J. A. Jacobs, and T. F. Kilduff, *Engineering Materials Technology,* Instructor's Manual, 2nd ed., Prentice-Hall, Englewood Cliffs, N. J., 1994.

[Jones 93] M. A. Jones, "Cost Differences between Expediting and Standard Urban Freeway Construction," Master's Thesis, the University of Texas at Austin, December 1993.

[McCullough 60] B. F. McCullough, and W. B. Ledbetter, "LTS Design of Continuously Reinforced Concrete Pavements," *ASCE Journal Highway Division*, Vol 86. No. HW4, December 1960.

[McCullough 68] B. F. McCullough, and H. J. Treybig, "A State Deflection Study of Continuously Reinforced Concrete Pavement in Texas," *Highway Research Record 239*, Highway Research Board, National Research Council, Washington, DC, 1968.

[Mittal 12] *Design and Construction of Steel Sheet Piling Structures Course Lecture*, Arcelor Mittal, Skylinesteel, Nucor-Yamato Steel, San Francisco, Ca., March 2012.

[NCHRP 90] J. A. Epps, *Synthesis of Highway Practice Report 160: Cold Recycled Bituminous Concrete Using Bituminous Material*, National Cooperative Highway Research Program, Transportation Research Board, National Research Council, Washington, DC, 1990.

[NCHRP 11] National Cooperative Highway Research Program, Sustainable Pavement Maintenance Practices," *Research Results Digest 365*, Transportation Research Board, Washington, DC, December 2011.

[NYC 07] New York City (NYC) Mayor's Office, "Inventory of New York City Greenhouse Gas Emissions," New York City Mayor's Office of Long-Term Planning and Sustainability, April 2007, http://www.nyc.gov/planyc2030, accessed September 15, 2010.

[Olivier 12] Jos G. J. Olivier, G. Janssens-Maenhout, and J.A.H.W. Peters, "Trends in Global CO_2 Emissions," *2012 Report—Background Studies*, PBL, Netherlands Environmental Assessment Agency, Ispra: European Commission's Joint Research Centre, The Hague, 2012.

[PCA 84] *Thickness Design for Concrete Highway and Street Pavements*, Portland Cement Association, Skokie, Ill., 1984.

[Plaxico 95] C. A. Plaxico, R. M. Hackett, and W. Uddin, "A Micromechanical Analysis of the Viscoelastic Response of Resin Matrix Composites," *Proceedings, International Conference on Fibre Reinforced Structural Plastics in Civil Engineering*, Madras, India, December 1995, pp. 99–108.

[Plaxico 96] C. A. Plaxico, W. Uddin, and R. M. Hackett, "A Micromechanical Model for Asphalt Materials," *Proceedings, Fourth Materials Congress, American Society for Civil Engineers*, Washington, DC, November 1996, Vol. 1, pp. 761–770.

[Saeed 95] A. Saeed, W. R. Hudson, and P. Anaejionu, "Location and Availability of Waste and Recycled Materials in Texas and Evaluation of their Utilization Potential in Roadbase," *Research Report 1348-1*, Center for Transportation Research, the University of Texas at Austin, Austin, Tex., October 1995.

[Saeed 96] A. Saeed, and W. R. Hudson, "Evaluation and Use of Waste and Reclaimed Materials in Roadbase Construction," *Research Report 1348-2F*, Center for Transportation Research, the University of Texas at Austin, Austin, Tex., August 1996.

[Seible 95] F. Seible, and D. Innamorato, "Earthquake Retrofit of Bridge Columns with Continuous Carbon Fiber Jackets," *Report No. ACTT-95/08*, Vol. II: *Design Guidelines, Report to Caltrans*, Advanced Composites Technology Transfer Consortium, University of California, San Diego, Calif., August 1995.

[Sohail 95] F. Sohail, "Implementation of the Urban Roadway Management System," Master's Thesis, the University of Texas at Austin, August 1995.

[Solminihac 92] H. de Solminihac, "System Analysis for Expediting Urban Highway Construction," Ph.D. thesis, the University of Texas at Austin, May 1992.

[Sultan 95] M. Sultan, "Caltrans Program for Performance Testing of Fiber Reinforced Plastics for Seismic Retrofit and Rehabilitation of Structures," *Proceedings, Fiber Reinforced Plastics Workshop*, sponsored by Federal Highway Administration and TRB Committee A2C07, January 1995.

[Technology 95] "Naval Test Pier Gets Composite Catwalk," *Composites Technology*, Vol. 1, No. 3, September/October 1995, pp. 5–6.

[TGLO 93] G. Mauro, *Texas Recycles: Marketing Our Neglected Resources*, Texas General Land Office, Austin, Tex., 1993.

[Uddin 03] W. Uddin, "Viscoelastic Characterization of Polymer-Modified Asphalt Binder for Pavement Applications." *Applied Rheology Journal*, Vol. 13, No. 4, 2003, pp. 191–199.

[Uddin 08] W. Uddin, "Pavement Recycling Practice in the U.S." *Recycling Symposium, 2008 ICTI, First International Conference on Sustainable Transport Infrastructure*, Beijing, China, April 24–26, 2008.

[Uddin 10a] W. Uddin, "Flooding, Ecological Diversity, and Wildlife Habitat Studies by Geospatial Visualization and Remote Sensing Technologies." *Proceedings, International Conference on "Biodiversity is Our Life"*—Center for Biodiversity and Conservation, Shah Abdul Latif University, Khairpur (Mir's), Sindh, Pakistan, December 29–31, 2010, pp. 239–254.

[Uddin 10b] W. Uddin, "Environmental, Energy, and Sustainability Considerations for Life-Cycle Analysis of Pavement Systems." *Proceedings, International Workshop on Energy and Environment in the Development of Sustainable Asphalt Pavements*, US-China Workshop, Xian, China, (sponsored by U.S. NSF and China NSF), June 7–9, 2010.

[USACE 58] "Engineering and Design-Flexible Pavements," *EM-1110-45-302*, Waterways Experimental Station, U.S. Army Corps of Engineers, 1958.

[USACE 61] "Revised Method of Thickness Design of Flexible Highway Pavements at Military Installations," *Technical Report No. 3-582*, Waterways Experimental Station, U.S. Army Corps of Engineers, August 1961.

[Wood 89] L. E. Wood, T. D. White, and T. B. Nelson, "Current Practice of Cold in-Place Recycling of Asphalt Pavements," *National Seminar on Asphalt Pavement Recycling*, Transportation Research Board, National Research Council, Washington, DC, 1989.

PART 4

Economics, Life-Cycle Analysis, and M,R&R Programming

CHAPTER 13

Maintenance, Rehabilitation, and Reconstruction Policies and Treatment Alternatives

13.1   Introduction

The strategic aspect of maintenance, rehabilitation/renovation, and replacement/reconstruction (M,R&R) was presented in Chapter 11. In essence, the term *strategic* refers to a planned approach. This includes a definition of what is meant by maintenance, rehabilitation, renovation, and reconstruction, and a description of the major elements of such a strategic approach or direction. Included are the concepts of maintainability, explicit consideration of future maintenance and rehabilitation at the design stage, reliability-centered maintenance, maintenance management systems, and ongoing operations.

By comparison, the term *policy* involves the set of rules and guidelines required to realize the strategy set out by an agency in regard to the way work will actually be done. This includes the following elements:

- Creating an operational form of the maintenance, rehabilitation, or overall infrastructure asset management system
- Determining whether the work should be contracted out or done by in-house forces
- Determining or specifying an appropriate policy, such as preservation type of policy (i.e., preventive maintenance and timely rehabilitation) or a reactive type of policy (i.e., dealing only with corrective actions or problems as they arise)

- Identifying the M,R&R treatment alternatives to be considered and, in turn, identifying those that are feasible for a given set of conditions
- Defining procedures for evaluating the effectiveness of M,R&R treatments
- Defining the collecting and reporting of M,R&R data
- Explicitly recognizing the importance of environmental impacts and policies

13.2   Maintenance Management Systems

An overall framework for infrastructure management is outlined in Chapter 1 (see Figure 1-7). It can apply to any infrastructure asset. More particularly, it is useful to review the operational form of management systems within the overall framework. One approach, which has existed for several decades, is the maintenance management system (MMS) (see also Sec. 11.7). In effect, most maintenance management systems are a form of the production control process, especially in the roads and highway sector of infrastructure management. Figure 13-1 presents

Figure 13-1   Basic operational form of most maintenance management systems.

the basic operational form or functional extent of most such maintenance management systems.

The functional extent of maintenance management systems represented in Figure 13-1 has existed relatively unchanged for 25 years or more. A study by the U.S. National Cooperative Highway Research Program [NCHRP 94] has noted, however, the following:

> The current maintenance management systems were right for their time, helping in the planning, budgeting, monitoring, and evaluation of maintenance work, and fostering standard methods and productivity guidelines. Many changes have occurred since these systems were first developed, however; in this light, several improvements are needed in current systems.

The NCHRP study goes on to point out that most maintenance management systems use a one- to two-year outlook rather than multiyear planning. They employ a centralized approach to planning, scheduling, and control that can be burdensome for field personnel, and they are not flexible enough to adjust work plans and schedules to reflect changing conditions. They do not incorporate distinct measures of levels of service or quality standards, and they do not have the capability to adjust resource requirements based on the degree of outsourcing to be used. Finally, NCHRP reports that current MMS are not able to generate timely reports to various management levels. All of these deficiencies should be addressed in future MMMs, preferably as a synchronous part of an IMS. In fact, the NCHRP report specifically recommends that the next generation should build upon the experience gained from the integration of engineering, economic, and management principles in PMSs.

Another source of information relevant to the operational form that maintenance and/or other infrastructure management systems might consider are in the American Public Works Association book, *Public Works Management Practices*. It recommends criteria and procedures for public works management services [APWA 91].

13.3 Private Contracts for Maintenance

It has been widespread practice in the United States, Canada, and other countries to contract out rehabilitation and reconstruction work in almost all areas of infrastructure. By comparison, however, normal practice has been to use in-house forces for routine maintenance of roads, water, sewer systems, buildings, parks, and utilities. The general rationale for the latter approach has been historical (i.e., things have always been done that way). The perception was that permanent teams could be more responsive to short-term, immediate needs. Although some contended that it was more cost-effective, in many cases it has been a convenient mechanism for awarding political jobs and thus gaining votes at the local level.

In recent years there has been a growing shift to the privatization of maintenance, particularly in the highway sector but also involving other areas of infrastructure, such as building maintenance. A noteworthy and very ambitious initiative comes from Australia, where a 10-year total road maintenance contract was awarded in late 1995 involving 2,000 lane-km in the northeastern region of Sydney [TEACC 96]. It covers pavements, nearly 200 bridges, over 600 traffic signals, guardrails, curbs and gutters, signs, and drainage systems (i.e., everything within the right-of-way of each road). Maintenance, in this case, has a very broad definition, ranging from routine, reactive type of work, such as pothole repairs, to preventive and corrective work to reconstruction. The contract is performance-based in that certain long-term performance indicators and asset preservation targets have to be met. It represents a true partnership between the contractor and the Road Transport Authority (RTA). The contract price is $170 million (Australian dollars), with periodic lump-sum payments to the contractor.

13.4   Identifying M,R&R Alternative Treatment Policies

Pursuant to discussions in Chapter 11, M,R&R policies need three ingredients: (1) M,R&R intervention-level criteria; (2) M,R&R alternative strategies with unit cost and production data; and (3) life estimates of each strategy or an appropriate performance model. As discussed in Chapter 12, sustainability and environmental impacts should be considered, such as the 4Rs (reuse, reclaim, recycle, and reduce) for M,R&R work. Innovations in construction materials and methods for 4R work can provide economic benefits of reduced material costs, as well as additional societal benefits of reduced energy consumption and emissions. The selection of an appropriate policy depends on a number of technical and economic factors. Broadly speaking, the following categories of policies are included.

13.4.1   Do-Nothing Policy

A do-nothing policy implies that no significant maintenance is done to the facility, and the condition deterioration continues until the time at which the facility is abandoned, reconstructed, or replaced. This may be a viable policy if a new facility is planned to replace the existing one (to satisfy a growing need or demand), or if budgeted rehabilitation/restoration methods cannot extend the life to the desired number of years. A do-nothing policy can be used as a base case to calculate the benefits of other alternatives. It also works well if replacement is funded within three years.

13.4.2   Routine Maintenance Policy

Routine maintenance policy generally involves periodic actions to preserve the condition and service levels. It is usually programmed as an annual lump-sum expenditure. Example actions are periodic mowing; general cleanup of common areas, parks, and roads; and regular maintenance and cleanup of electrical and mechanical equipment, such as elevators, pumps, and air-conditioning units. This is sometimes termed *preventive maintenance.*

13.4.3   Critical Maintenance Policy

Critical maintenance involves a policy of immediate action(s) to avoid imminent failures and emergency situations resulting from accidents and natural disasters. It also includes actions responding to users' or the public's complaints. Emergency repairs are performed in the case of structural damage and failures caused by accidents (for example, derailment of railroad cars, spillage of hazardous materials, gas pipeline blowups, and failure of underground water pipes) and natural disasters (e.g., collapse of bridges and buildings during earthquakes, landslides, floods, tornadoes, and hurricanes).

13.4.4   Scheduled M,R&R Policy

Scheduled M,R&R include both timed maintenance and replacement/ reconstruction. This approach is applicable based on available past experience on, for example, painting of steel bridges, pavement markings, exterior painting of a building, replacement of gutters on residential buildings, and scheduled maintenance of pumps and auxiliary equipment of a water-supply system, etc. This type of policy would also include: (1) hard-time replacement (e.g., the shingle roof of a residential building might be replaced after 20 years because of the known average life of this roofing material) and (2) cost-effective capital replacement/reconstruction strategies (for example, asphalt resurfacing of a bridge deck every 15 years). This differs from preventive maintenance since it is longer term.

13.4.5   Condition-Responsive M,R&R Policy

Condition-responsive M,R&R policy is based on preselected threshold values of appropriate performance indicators that will trigger major rehabilitation/renovation and replacement/reconstruction actions. It is important to carefully identify these performance indicators. Examples would be frequency of breaks, amount of leakage in water pipelines, or level of scour near a bridge pier. However, good predictive performance models are needed for timely budgeting.

13.4.6   Database of M,R&R Actions

A database of all M,R&R actions taken under the foregoing policies, or for any reason, is essential for updating the maintenance programs and evaluating their effectiveness. At a minimum, the database should include M,R&R activity codes, timing of the work, units of measure, unit costs, expected life, production rates, and condition improvement after implementation. Additionally, it is desirable to evaluate condition before and after each action. The historical database also is valuable for performance-modeling purposes (see Chapter 8).

13.4.7   Economic Analysis of M,R&R Policies

Economic analysis of competing M,R&R strategies on a life-cycle basis is discussed in detail in Chapter 14. It is important to compare the overall economics of each policy. Such comparisons can be based on the more detailed analysis of competing strategies and may reveal, for example, whether a scheduled or condition-responsive M,R&R policy is more cost-effective for given conditions.

13.5   Example M,R&R Treatment Alternatives for Roads

Haas, Hudson, and Zaniewski [Haas 94] present detailed examples of M,R&R strategies and analysis methods for pavements. Maintenance treatments depend on the pavement-surface type and its use, predominant distress, and deterioration mechanisms [Haas 94, Roberts 91]. Long-term performance of various maintenance strategies is being evaluated in the Strategic Highway Research Program (SHRP) and is periodically reported in the literature [SHRP 99]. For flexible pavements, surface treatments, thin-asphalt overlays, and crack sealing are generally used as global preventive maintenance treatments, whereas patching is used as a localized, corrective maintenance treatment. Major rehabilitation alternatives include hot-mix asphalt overlays, concrete overlays, surface milling and recycling, and reconstruction. For rigid pavements, cleaning and rehabilitation of joints, slab replacements, patching, and thin-asphalt overlays are generally used as preventive maintenance treatments. Rehabilitation alternatives include hot-mix asphalt overlays, concrete overlays, and reconstruction.

Noise generation on heavily trafficked roads can force the use of special treatments, such as sound-barrier walls erected on both sides of the right-of-way. Innovations in thermoplastics are providing new materials that could prove to be maintenance-free, show relatively higher sound-absorption characteristics (as compared to conventional concrete materials), and reduce the wall height. An example is the application of polycarbonate material to reduce the sound-wall heights on the I-5 freeway in Los Angeles [Gharabegian 96].

A 2011 study of maintenance practices [NCHRP 11] shows 4R interventions are common in maintenance, such as recycling and use of waste tire rubber. However, the study shows that only 21 percent of the 49 highway-agency survey respondents in the United States and Canada believed that noise pollution is an important issue in their maintenance programs. Similarly, the respondents to the survey were not concerned about the water-quality issue or environmental sustainability. These study results show that at least among the surveyed agencies the environmental/sustainability concerns of noise and water quality are not on their radar during their maintenance work programs. This agency perspective is different from the sustainability awareness shown by the general public who are primary stakeholders of transportation infrastructure assets and primarily affected by the extra noise and poor water quality.

13.6 Example M,R&R Alternatives for Water Mains

An analysis of the deterioration and maintenance problems of aging water-supply systems has identified three key areas of M,R&R planning [O'Day 84]: (1) leakage, (2) main breaks, and (3) risk analysis. Utilities must recognize the risk of a main failing, and the risk of damage and loss from the failure. Such damage can result in floods, disruption of traffic, fire, and substantial losses to affected businesses. In general, a break in a 122-cm (48-in) main will cause more loss and damage than several breaks in a 15-cm (6-in) main. There are three basic types of commonly used M,R&R alternatives or actions: general maintenance (main cleaning, equipment repair, valve service, treatment-facility cleaning), repair (main lining, main repair), and replacement (main relaying). Table 13-1 lists the major deterioration problems and maintenance actions reported for water systems.

Problems/Deterioration	M,R&R Action
Water quality affected by tuberculation and internal pipe corrosion	Main cleaning and lining, cathodic protection, facility maintenance (preventive and corrective maintenance)
Low pressure and high-head loss caused by main tuberculation	Main cleaning and lining (preventive maintenance)
Valves and hydrants in poor condition or inoperable because of neglect and/or unavailable replacement parts	Equipment maintenance (routine and scheduled maintenance)
Distribution system leakage caused by joint leaks, main breaks, and service leaks	Leakage survey and repair, service leak inspection and removal of abandoned service (condition-responsive/preventive maintenance)
Main breaks caused by main deterioration from internal and/or external corrosion	Repair of replacement of main breaks, main relaying

TABLE 13-1 Water-Distribution System Problems and Maintenance Actions [O'Day 84]

Water-main repair or replacement decisions become critical when mains deteriorate structurally and break frequently. In-service evaluation of water mains is discussed in Chapter 7. The following evaluation criteria can be used to make replacement decisions [Stacha 68]:

- Projected future costs of continued repair
- Initial cost of main replacement
- Public risk from future breaks
- Frequency and severity of inconveniences to the public from future breaks
- Hydraulic adequacy of the main
- Future plans for streets and/or utility construction that could affect break patterns and alter costs through joint construction
- Severity of disruptions caused by main-replacement construction activity

An innovative preventive maintenance treatment is the use of in situ lining to improve leakage control and repair existing mains, such as the Insituform process [News 94]. This lining employs thermoplastic technology for rebuilding deteriorated pressure pipes, without relying on the structural properties of the pipe. Economic analyses of leak control and repair versus replacement of main breaks are discussed in Chapter 14.

13.7 Example M,R&R Alternatives for Sewer Mains

Urban water-supply and wastewater systems are regulated in the United States by the Clean Water Act. Sewers collect wastewater from residential communities and businesses. Control of pollution is achieved by the proper treatment of wastewater. Maintenance of filter systems in drainage basins, trenches, and wells is necessary on a periodic basis. Catch basins should be inspected and cleaned after major storms. Corrugated steel pipes for sewer lines and culverts have been used for 100 years [AISI 90]. The life of these pipes can be extended by 25 percent with external bituminous coating, and 15 to 35 years of service life can be added if invert bituminous or concrete paving is applied to the bottom 25 percent of the pipe circumference. The most frequently used equipment for infiltration maintenance includes vacuum pumps to clean sumps and pipes, compressed air jets to clean wells, and flushing. Methods of rehabilitating deteriorated steel pipes include installation of a concrete invert for large diameters, relining with smaller-diameter pipe, shotcrete relining, and cement-mortar relining [AISI 90].

Like water mains, sewer mains can also be lined with new thermoplastic materials to provide an acid-alkali-resistant and structurally durable rehabilitation/repair alternative [News 94]. The Washington

Type of Job	Excavation/ Replacement Cost per Foot	Slip Lining Estimated Cost per Foot	Insituform Actual Cost per Foot
Sewer line, 25.4-cm (10-in) diameter and 88.5-m (295-ft) long; 100% of the work involved paving replacement, and five service laterals were reconnected.	$222	$150	$52
Sewer line, 15-cm (6-in) diameter and 96-m (320-ft) long; 100% of the work involved paving replacement, and nine service laterals were reconnected.	$80	$150	$59
Sewer line, 25.4-cm (10-in) diameter and 684-m (2280-ft) long; 27% of the work involved paving replacement, and no service laterals were involved.	$69	$60	$52
Sewer line, 25.4-cm (10-in) diameter and 428-m (1425-ft) long; 80% of the work involved paving replacement, and six service laterals were reconnected.	$84	$80	$52

TABLE 13-2   WSSC Cost Comparison of Sewer Reconstruction Techniques [Urban 84]

Suburban Sanitary Commission (WSSC), responsible for providing water and sewer services to Montgomery and Prince George counties in Maryland, compared the costs of replacement, slip lining, and Insituform for 15-cm (6-in) and 25.4-cm (10-in) sewers in a number of projects. The result was significant savings using the newer "trenchless" technology of Insituform, which involved no excavation [Thomasson 82, Urban 84]. These comparisons are summarized in Table 13-2.

13.8   Example M,R&R Alternatives for Buildings

Buildings generally have more components than other types of infrastructure. The effective life of a building will end or be greatly reduced if one or more of these components (structure, foundation, mechanical, roof) fails. The finishes, fixtures, and furnishings may be critical for the owner or user because of the visual, decorative aspects; however, the architectural and engineering-condition assessment is usually focused on functional and structural integrity. Harris defines six functional subsystems of a building that account for the service provided, the individual failure of any of which can lead to overall functional failure [Harris 96]. These six subsystems are listed as follows:

1. *Structure*—to minimize deformation and prevent cracking. This includes foundations, compacted ground, earth-retaining structures, bearing walls, columns, beams, girders, trusses, and braces.

2. *Vertical closure*—to provide weather barrier. This includes nonstructural components of exterior walls, siding, windows, doors, and sills.

3. *Horizontal closure*—to collect and transport precipitation. This includes roofing materials, gutters, downspouts, drains, retention ponds, sump pits and pumps, and sewers.

4. *Climate stabilization*—to maintain desirable levels of heat, humidity, and oxygen. This includes heating, ventilation, air conditioning, ducts, controls, fireplaces, insulation, fans, pumps, motors, and other necessary equipment.

5. *Hydraulic*—to distribute water and collect and dispose of sewer. This includes water mains, water-distribution systems, plumbing, grease traps, water heaters, wells, pumps, storage tanks, drains, valves, vents, sanitary sewers, septic tanks, and hydrant and fire protection.

6. *Energy distribution*—to distribute power and collect by-products. This includes fuels and energy of all types and return of wastes, electrical installation, wiring, meters, gas lines, oil tanks and pumps, and pressure regulators and controls.

Obviously, there are many potential M,R&R alternatives to address failure or deterioration of the above functional subsystems. The reader may consult various industry and public-agency documents for details, particularly the publications of the Building Research Board, National Academy of Science, Washington, DC.

13.9   Evaluating the Effectiveness of M,R&R Alternatives

The evaluation of effectiveness of M,R&R treatment alternatives can be carried out in one or more of the following ways:

- Economic effectiveness
- Technical effectiveness
- Visual, or aesthetic, effectiveness
- Public or owner satisfaction/acceptance
- Environmental impact

Economic effectiveness is usually evaluated on a life-cycle basis in terms of uniform annual costs, which will be described in

Chapter 14. Technical effectiveness is usually evaluated in terms of performance and service to the users (see Chapter 8 on performance modeling) and is thus an integral part of any life-cycle-based economic-effectiveness evaluation.

It is sometimes important to evaluate the visual, or aesthetic, effectiveness of roads, bridges, retaining walls, parks, buildings, and various other types of infrastructure that are observed by users or owners. The simplest approach might consist of a quality and appearance scale, such as: (1) excellent; (2) good; (3) fair; (4) poor; and (5) unacceptable. More sophisticated approaches could also be used, but the determining factor would be the degree of importance that visual/aesthetic effectiveness plays in the overall effectiveness evaluation. This value could range from essentially zero for underground services to high for a retaining wall. Clearly, park- and public-service facilities, such as museums, find this important, as do schools.

A similar situation exists for public or owner satisfaction. This is quite a general category, and may include economic, technical, and visual effectiveness, depending on the owner. Very few IAMSs currently evaluate such overall satisfaction.

Environmental-impact effectiveness is usually addressed through an environmental-impact assessment prepared according to applicable federal, state, or local regulations. If the standards incorporated in the regulations are met, then the infrastructure is generally environmentally acceptable. In other words, it is acceptable or not acceptable. Of course, if acceptable, there can be a range of effectiveness within that category that can be considered.

13.10   Collection and Reporting of M,R&R Data

Section 13.1 noted that a comprehensive policy approach includes not only the necessity of collecting and reporting M,R&R data, but also how this is carried out. Chapters 4, 5, and 6 have addressed the details of the data needs, data analysis, database management, and procedures for collecting and processing data. It is not intended herein to repeat these details. Rather, it is intended to emphasize the importance of ensuring that an agency policy would include the capability of determining the following fundamental information (see also Sec. 13.4):

- The M,R&R actions or treatments that were carried out, in what locations, when, and to what degree/quality/extent
- The costs, problems encountered, and expected life of the infrastructure

In addition to the above, a variety of data may be required to assess effectiveness and develop performance models, as discussed in Chapters 4, 5, 6, and 8.

13.11   Recognizing Environmental Stewardship and Sustainability

Most M,R&R actions must conform to environmental regulations or policies that require that the impacts of each action be considered. Exceptions would be minor repairs or renovations.

Usually, the environmental impact assessments include possible disturbances to the natural environment, creation of bad by-products and emissions, disturbances to manmade works, noise creation, effects on public health, and environmental justice to minimize disruptions to the affected communities. As previously indicated, it is essential to check with the relevant environmental office (federal, state, local) to determine which environmental requirements have to be involved. Generally, most governmental agencies have a regular channel of environmental coordination.

Traditionally, environmental impact studies are time consuming due to detailed manual inspections involved, permitting protocols with numerous agencies, and the overall manual workflow process. However, a comprehensive five-year study sponsored by the U.S. DOT in cooperation with the National Aeronautics and Space Administration (NASA) during 2000 to 2005 implemented airborne and spaceborne remote-sensing and spatial technologies to expedite environmental impact assessment process based on the 1969 National Environmental Policy Act (NEPA) guidelines [DOT 02, Uddin 04, Xiong 04]. Readers can find details on these technologies and case studies in these references.

13.12   Summary

All M,R&R treatment possibilities for all types of infrastructure cannot be covered in this chapter. Instead, this chapter illustrates the need to consider a range of possibilities, including new technology innovations of materials and construction methods which can minimize disruptions to communities and adverse impacts on the environment. One should not get stuck in a rut of outdated methods; approaches should be evaluated and assessed as outlined for the specific type of infrastructure involved. In this way, a more complete and effective IAMS can be developed to provide the maximum benefits for available funds.

13.13   References

[AISI 90] *Modern Sewer Design,* American Iron and Steel Institute (AISI), Washington, DC, 1990.
[APWA 91] *Public Works Management Practices,* American Public Works Association (APWA), Chicago, Ill., 1991.

[DOT 02] Department of Transportation, "Achievements of the DOT-NASA Joint Program on Remote Sensing and Spatial Information Technologies: Application to Multimodal Transportation, 2000-2002," U.S. DOT Research and Special Program Administration, Washington, DC, April 2002.

[Gharabegian 96] A. Gharabegian, "Soundwall Extension Quiets Noise along Los Angeles Freeway," *Roads and Bridges*, April 1996, pp. 71–72.

[Haas 94] R. Haas, W. R. Hudson, and J. P. Zaniewski, *Modern Pavement Management*, Krieger Publishing Company, Malabar, Fla., 1994.

[Harris 96] S. Y. Harris, "A Systems Approach to Building Assessment," *Standards for Preservation and Rehabilitation, ASTM STP 1258,* American Society for Testing and Materials, Philadelphia, Pa., 1996, pp. 137–148.

[NCHRP 94] National Cooperative Highway Research Program, "Role of Highway Maintenance in Integrated Management Systems," *NCHRP Report 363,* Transportation Research Board, National Research Council, Washington, DC, 1994.

[NCHRP 11] National Cooperative Highway Research Program, "Sustainable Pavement Maintenance Practices," *Research Results Digest 365,* Transportation Research Board, Washington, DC, December 2011.

[News 94] "Buyer's Guide," *Civil Engineering News,* Vol. 6, No. 2, March 1994, p. 43.

[O'Day 84] D. K. O'Day, "Aging Water Supply Systems: Repair or Replace," *Infrastructure—Maintenance and Repair of Public Works, Annual of the New York Academy of Science,* Vol. 431, December 1984, pp. 241–258.

[Roberts 91] F. L. Roberts, P. S. Kandhal, E. R. Brown, D.-Y. Lee, and T. W. Kennedy, *Hot Mix Asphalt Materials, Mixture Design, and Construction,* NAPA Education Foundation, Lanham, Md., 1991.

[SHRP 99] Strategic Highway Research Program (SHRP) Project H-106, "LTPP Pavement Maintenance Materials: SHRP Crack Treatment Experiment," Final Report for Oct 93-Jun 99, DOT project DTFH61-93-C-00051 (Authors: K. L. Smith and A. R. Romine), ERES Consultants, Inc., Champaign, IL., September 1999.

[Stacha 68] J. H. Stacha, "Criteria for Pipeline Replacement," *Journal of American Water Works Association,* May 1968.

[TEACC 96] "The New Force in Road Maintenance," *The Earth Mover and Civil Contractor,* August 1996, pp. 31–37.

[Thomasson 82] R. O. Thomasson, "In-Place Sewer Reconstruction Proves Cost-Effective," *American City and County,* February 1982.

[Uddin 04] W. Uddin, "Air Quality Project—Remote Sensing Tunable Laser Measurements of Air Pollution," *Technology Guide NCRSTE_TG004,* Center for Advanced Infrastructure Technology, The University of Mississippi, The National Consortium on Remote Sensing in Transportation—Environmental Assessments (NCRST-E), U.S. DOT Research and Special Program Administration, Washington, DC, June 2004.

[Urban 84] H. P. Hatry, and B. G. Steinthal, *Guides to Managing Urban Capital Series,* Vol. 4: *Guide to Selecting Maintenance Strategies for Capital Facilities,* The Urban Institute, Washington, DC, 1984.

[Xiong 04] D. Xiong, R. Lee, J. B. Saulsbury, E. L. Lanzer, and A. Perez, "Guidance on Using Remote Sensing Applications for Environmental Analysis in Transportation Planning," *Report WA-RD 593-2,* Oak Ridge National Laboratory/Washington State Department of Transportation Environmental Office, Olympia, Washington, August 30, 2004.

CHAPTER **14**

Life-Cycle Cost and Benefit Analysis

14.1   Introduction

Applying the principles of economic analysis to infrastructure projects occurs at two basic levels. First, the overall economic viability and timing of a project must be determined. This is most often carried out in comparison to and/or in competition with other projects. Second, there is the requirement to achieve maximum economy for a project once it has been selected, programmed, and budgeted. Such within-project economic analysis is achieved by calculating the costs and benefits of the various alternatives that satisfy overall project requirements. In terms of economic analysis, the major difference between these two levels of infrastructure management is the amount of detail or information needed. Otherwise, the basic principles are the same.

This chapter describes both the economic principles and their incorporation into methods of economic analysis. Such analysis is a vital component of any infrastructure management system [Army 86, DOI 83, TRB 85, Hudson 97]. Assessment of costs and benefits over the entire *life cycle* (also called *whole of life,* in some countries like Australia) is integral to economic analysis. Examples of life-cycle benefit and cost analysis, involving several case studies, are provided to illustrate the principles and the methodology. More complete analysis details can be obtained in any good engineering economics text.

14.2   Basic Principles

Considerable information exists on the principles of engineering economy and the methods of economic analysis. The principles applicable to infrastructure management include the following:

- The level at which the analysis is to be made must be clearly identified, that is, expenditure programming for a number of projects or networks, or within-project optimization.

- The economic analysis provides support for a management decision but does not alone represent a decision.

- Criteria, rules, or guides for such decisions must be separately formulated before the economic analysis, even though such criteria may be straightforward and simple (i.e., minimization of life-cycle costs, maximization of internal rate of return or maximization of benefit-cost ratio).

- The economic analysis itself has no relationship to the financing of a project. Such financing considerations can either limit the number of feasible projects (on a network-level basis) or limit the amount available for a particular project. They do not affect the methodology or principles controlling the economic analysis per se.

- An economic analysis should consider as many feasible alternatives as possible, within the constraints of time and other resources.

- All alternatives should be compared over the same life cycle or time period. The time period should be chosen so that it is possible to forecast the factors involved in the analysis with a reasonable degree of reliability. Uncertainties can be considered in making the decision to choose the best alternative.

- The economic analysis of infrastructure projects should include life-cycle assessment (LCA) of agency costs, user costs and benefits, and societal costs and benefits, if possible.

- Life-cycle assessment of costs and benefits can help to justify innovative long-lasting alternatives, which may not be selected based on initial costs alone.

- Value-engineering applications of design and construction projects, discussed in Sec. 9.5, also require LCA of costs and benefits over the analysis period.

14.3   Cost and Benefit Factors

14.3.1   Agency Costs

Agency costs are those costs directly represented by the budget or out-of-pocket costs paid by the owner. External or exogenous costs, such as those associated with user delays during infrastructure work and public health costs caused by air pollution (usually termed *user costs*), environmental damage to biodiversity and ecosystems, and the like (usually termed *societal costs*), may be combined with agency costs in a total economic analysis. These exogenous costs may or may not affect agency's decisions; they do not usually appear in an agency's budget.

The major initial and recurring costs over the life cycle that a public agency may consider in the economic analysis of infrastructure project alternatives include the following:

- Initial capital costs of construction or major M,R&R intervention
- Cost of maintenance and protection of traffic
- Future costs of M,R&R
- Salvage return or residual value at the end of the period (which may be a "negative cost")
- Engineering and administration
- Costs of borrowing (if projects are not financed from current revenue)

14.3.2   Nonagency Costs

Nonagency costs can involve the user of the infrastructure or facility, or they can be incurred by nonusers, and they can include costs to the society. They include the following:

- User costs
  - Occupancy time in or on the facility and time delays
  - User costs related to service interruptions
  - Operating costs (vehicles, tires, etc.)
  - Accidents/crashes and property damage
  - Excessive time delay caused by M,R&R activities
- Nonuser costs
  - Neighborhood disruptions and accidents caused by construction activities
  - Social costs caused by reduced valuation of property
  - Noise pollution, public health costs caused by emissions, and other impacts (visual, aesthetics, etc.)
- Societal costs
  - Costs linked to water pollution, soil contamination, and regional air pollution
  - Costs associated with environmental damage to biodiversity and ecosystems
  - CO_2 capture and sequestration costs related to energy, transportation, and deforestation

Some of the above costs are difficult to quantify. In the transport field, however, considerable progress has been made on quantifying user time and time-delay costs, crash-related costs, and operating costs [TRB 85, Uddin 93]. For nonuser costs, most work in the transport field has focused on emissions and noise.

In recent years reasonable societal cost models have been developed for public health costs and loss of productivity attributed to air pollution and smog. Monetary values of vehicle emissions in terms of medical costs and other related societal costs have been established and the measurable costs of air pollution are found high enough to justify substantial expenditures on reducing vehicle emission rates [Small 95]. According to Murphy and Delucchi [Murphy 98] the total societal cost is about 33 percent of the full transportation costs and about one-third of that is air pollution cost. Other societal costs not considered include: loss of wetland and agriculture land, change in property values, damage to historical property, equity, and urban sprawl. If any of the urban areas in the United States, monitored by the EPA, exceeds the established air pollution threshold and is declared a "nonattainment area," it may lose federal funding for construction of transportation projects. Societal public health costs can therefore be a significant part of the total life-cycle cost.

The societal public health costs are calculated as a function of vehicle miles traveled, average speed, traffic volume, point sources of emissions (industries, airports, power plants, etc.), and climatic data [Uddin 05, 06]. For example, a U.S. study of a rural town, Oxford, Mississippi, showed the following life-cycle user and societal costs [Uddin 06, 11]:

1. The vehicle-operating cost for roads in good to poor condition is 18 to 29 cents per vehicle-mile.

2. The crash- and safety-related cost is 55 cents per vehicle-mile.

3. The public health cost related to vehicle emissions and air pollution is 13 cents per vehicle-mile. (More explanation of this cost calculation is provided in Sec. 14.12.)

4. The total societal cost is the sum of crash and public health costs or 68 cents per vehicle-mile.

14.3.3   Benefits

The benefits of an infrastructure project can accrue from direct or indirect cost reductions, and from gains in business, land use and values, aesthetics, and community activities in general. Benefits would, however, accrue primarily from direct reductions in user costs. Benefits include:

- Additional taxes or total income generated by a project
- User savings because of reductions in operating costs (vehicles, tires, etc.)
- Reduction in service interruption costs and time delays
- Reduction in fuel costs caused by traffic congestion
- Reduction in emissions and air pollution–related medical costs and lost productivity
- Reduction in CO_2 and other GHG emissions

To measure or calculate benefits, it is necessary to define those facility or infrastructure characteristics that will affect the previously noted user costs of time, accidents, operating costs, and so forth. These could include level of service, appearance, or aesthetics. However, the first factor is level of serviceability, which can be measured in a number of different ways, depending on the infrastructure component involved, and would have a major influence.

In assigning benefits for the purpose of project evaluation, the question of whether generated or transferred usage of the facility should be included must be considered. Usually, this question arises with respect to the overall project.

This question can relate to the public-policy viewpoint adopted [Winfrey 69]. One viewpoint involves a "sales concept," in which all usage is considered regardless of sources; the other involves a concept of true savings, in which only existing plus normal usage growth over the analysis period is considered. It is usual in transport projects to use the first concept. Thus, including total expected usage in the calculations for costs and benefits for infrastructure-improvement projects is not unreasonable.

14.4 Analysis or Life-Cycle Period

A general guideline for selecting the length of an analysis period is that it should not extend beyond the period of reliable forecasts. For short-term or high-uncertainty projects, 10 years could be an upper limit. For other long-term projects, such as bridges or structures, 50 years or more would be reasonable; however, the present worth of costs or benefits at future times beyond 50 years may not be significant, depending on the discount rate used. Most transport studies use a range of 20 to 30 years. An analysis period of 50 years is used for sewer pipeline materials. The analysis period can also be based on service-life estimates. Examples of service life are provided in Chapter 3 (Tables 3-1 and 3-2) and Chapter 8 (Table 8-1). The particular period chosen is basically a policy decision for the agency concerned, and can vary.

14.5 Discount Rate, Interest Rate, and Inflation

Discount rate, or time value of money, must be used to relate estimated future costs and/or benefits to present-worth terms. It provides a convenient and understandable means of comparing alternative uses of funds. However, discount rate may not be the same as interest rate, which is associated with actually borrowing or investing money (functional costs).

Inflation is of concern to everyone. But whether and how it should be taken into account in an economic analysis is a complicated issue. Basically, the difference between an interest rate and an inflation rate represents the real return on an investment; consequently, it is reasonable to use that difference as a discount rate, which is the policy of many agencies. For example, in 1996, interest rates in the United States and Canada were at about 5.5 to 6 percent, with inflation at about 1.5 to 2 percent, giving a difference or average discount rate of about 4 percent. This rate is also the approximate long-term growth of the North American economy. Of course an agency may decide, as a matter of policy, to use a higher- or lower-discount rate. As well, some agencies look at a range of discount rates to determine if this has any effect on the comparative rankings of investment alternatives.

For normal IAMS economic analyses, inflation should only be used in setting a real discount rate but not directly as a separate factor in economic analysis for the following reasons:

- Inflation is difficult to forecast and merely introduces another uncertainty into the evaluation. If it were considered, benefits as well as costs would be increased by inflation so that their relative magnitude would still be the same.

- The purpose of economic analysis is to provide a basis for decision making. Inserting a factor for inflation is no guarantee that the results will be more reliable, or that a better selection of alternatives will result.

- The uninflated value of the alternatives in real constant dollars is a better tool for economic analysis than inflated values. It also has more intrinsic meaning to current decision-makers.

Some engineers argue that it is necessary to include inflation in an economic analysis, since ignoring it leads to underestimating costs, and proposed budgets may be incorrect. This argument misunderstands the objective of an economic analysis, which is to provide a tool for the selection of a specific option from a set of alternatives. Once the option is selected, a separate budget and financial analysis is required to determine the cash-flow requirements. The budget analysis generally includes inflation, as well as actual sources of funds, such as taxes, federal grants, bonds, and financial analysis.

14.6 Salvage or Residual Value

Salvage, or residual, value is used by some agencies in economic analysis. It can be significant in the case of buildings, roads, and pavement structures because it involves the real worth of reusable materials, or additional remaining service life at the end of the life-cycle period. With depleting resources, material values will become increasingly important in the future, especially when used in a new application by reworking or reprocessing.

Salvage value of a material or structural component depends on several factors, such as volume and location, degree of contamination, age, durability, and anticipated use at the end of the life-cycle period.

It can be represented as a percentage of the original cost of the new material or component itself, not including the incremental cost of placement and processing.

14.7   Methods of Economic Analysis

A number of methods of economic analysis are applicable to the evaluation of alternative infrastructure investment strategies. They can be categorized as follows:

- Equivalent uniform annual-cost method, or simply the annual-cost method

- Present worth method for

  - Costs, or

  - Benefits, or

  - Benefits minus costs, usually termed "the net present-value method"

- Rate-of-return method

- Benefit-cost-ratio method

- Cost-effectiveness method

In addition to or as part of the foregoing methods, multifactor costs and benefits need to be considered in a trade-off or cross-optimization situation; for example, a bridge versus a pavement investment. This can be handled in a number of ways from a simple ranking, based on the results of any of the foregoing methods, to the use of multiattribute utility theory, which is discussed further in Chapter 15.

14.7.1   Equivalent Uniform Annual-Cost Method

This method combines all initial capital costs and all recurring future expenditures into equal annual payments over the analysis period. In functional form, the method may be expressed as follows:

$$AC_{x1,n} = crf_{i,n}(ICC)_{x1} + (AAMO)_{x1} + (AAUC)_{x1} - sff_{i,n}(SV)_{x1,n} \quad (14.1)$$

where $AC_{x1,n}$ = equivalent uniform annual cost for alternative $x1$, for a service life or analysis period of n years

$crf_{i,n}$ = capital recovery factor for interest rate i and n years = $i(1 + i)^n / [(1 + i)^n - 1]$

$(ICC)_{x1}$ = initial capital costs of construction

$(AAMO)_{x1}$ = average annual maintenance, plus operation costs for alternative $x1$

$(AAUC)_{x1}$ = average annual user costs for alternative $x1$, if applicable

$sff_{i,n}$ = sinking fund factor for interest rate i and n years = $i / [(1 + i)^n - 1]$

$SV_{x1,n}$ = salvage value, if any, for alternative $x1$ at the end of n years

The basic appeal of the equivalent uniform annual-cost method is its simplicity and ease of understanding for public officials. It also permits the comparison of alternatives that do not have the same overall expected life. However, it cannot be used, except intuitively, to determine whether or not a project is economically justifiable because it does not include benefits in the evaluation. Consequently, comparisons among alternatives must be on the basis of costs alone, with the inherent assumption that they have equal benefits. A bridge to "serve" the public can fit into this category.

The present worth of costs method is directly comparable to the equivalent uniform annual-cost method.

14.7.2   Present-Worth Method

The present-worth method can consider costs only, benefits only, or costs and benefits together. It involves the discounting of all future sums to the present, using an appropriate discount rate. The factor for discounting either costs or benefits is:

$$pwf_{i,n} = 1/(1+i)^n \tag{14.2}$$

where $pwf_{i,n}$ = present worth factor for a particular i and n
$\quad\quad i$ = discount rate
$\quad\quad n$ = number of years to when the sum will be expended or saved

The present-worth method for costs can be expressed in terms of the following equation:

$$TPWC_{x1,n} = (ICC)_{x1} + \sum_{t=0}^{n} \left\{ pwf_{i,t} \left[(CC)_{x1,t} + (MO)_{x1,t} + (UC)_{x1,t} \right] \right\}$$

$$- pwf_{i,n}(SV)_{x1,n} \tag{14.3}$$

where $TPWC_{x1,n}$ = total present worth of costs for alternative $x1$, for analysis period of n years
$\quad\quad (ICC)_{x1}$ = initial capital costs of construction, etc., for alternative $x1$
$\quad\quad (CC)_{x1,t}$ = capital costs of construction, etc., for alternative $x1$, in year t, where $t < n$
$\quad\quad pwf_{i,t}$ = present worth factor for discount rate, i, for t years = $1/(1+i)^t$
$\quad\quad (MO)_{x1,t}$ = maintenance plus operation costs for alternative $x1$ in year t
$\quad\quad (UC)_{x1,t}$ = user costs, if applicable for alternative $x1$ in year t
$\quad\quad (SV)_{x1,n}$ = salvage value, if any, for alternative $x1$, at the end of the analysis period, n years

The present worth of costs is used in the equivalent uniform annual cost method when additional capital expenditures occur before the end of the analysis period; that is, where the initial service life is less than the analysis period, and future maintenance, renovation, reconstruction, or rehabilitation (M,R&R) is needed. The equation for this situation, including user costs, is:

$$AC_{x1,n} = [Eq.(14.1)] + R_1 pwf_{i,a1} + R_2 pwf_{i,a2} + \cdots + R_j pwf_{i,aj}$$
$$- (1 - y/L)(R_1, R_2, \ldots, R_j) pwf_{i,a1,a2,\ldots,aj} \qquad (14.4)$$

where $AC_{x1,n}$ = equivalent uniform annual cost for alternative $x1$, for
an analysis period of n years
$R_1, R_2, \ldots, R_j$ = costs of first, second, ..., jth M,R&R treatments
$a_1, a_2, \ldots, a_j$ = ages at which the first, second, ..., jth M,R&R treatments
occur, respectively
y = number of years from time of last M,R&R treatment to
end of analysis period
L = estimated life in years of last M,R&R treatment

The present worth of benefits can be calculated in the same manner as the present worth of costs, using the following equation:

$$TPWB_{x1,n} = \sum_{t=0}^{n} pwf_{i,t} \left[(DUB)_{x1,t} + (IUB)_{x1,t} + (NUB)_{x1,t} \right] \qquad (14.5)$$

where $TPWB_{x1,n}$ = total present worth of benefits for alternative for an
analysis period of n years
$(DUB)_{x1,t}$ = direct user benefits accruing from alternative $x1$ in
year t = indirect user benefits, if applicable, accruing
from alternative $x1$ in year t
$(IUB)_{x1,t}$ = indirect user benefits, if applicable, accruing from
alternative $x1$ in year t
$(NUB)_{x1,t}$ = nonuser benefits, if applicable, accruing from
alternative $x1$ in year t

The net-present-value method follows from the foregoing methods—it is simply the difference between the present worth of benefits and the present worth of costs. Obviously, benefits must exceed costs if a project is to be justified on economic grounds. The equation for net present value is as follows:

$$NPV_{x1} = TPWB_{x1,n} - TPWC_{x1,n} \qquad (14.6)$$

where NPV_{x1} is the net present value of alternative $x1$; and $TPWB_{x1,n}$ and $TPWC_{x1,n}$ are as defined by Eqs. (14.5) and (14.3), respectively.

However, for most infrastructure project alternatives, $x1$ in Eq. (14.6) is not applicable directly to $x1$ itself, but rather to the difference between it and some other suitable alternative, say $x0$. Considering only direct user benefits, $(DUB)_{x1,t}$, these are then calculated as the user savings (resulting from lower operating costs, lower occupancy time costs, lower accident costs, etc.) realized by $x1$ over $x0$.

For some infrastructure situations, alternative $x0$ may be a base case of no capital expenditures (when increased maintenance and operating costs are required to keep the facility in service). The net present-value method may then be expressed as:

$$NPV_{x1} = TPWC_{x0,n} - TPWC_{x1,n} \qquad (14.7)$$

where NPV_{x1} = net present value of alternative

$TPWC_{x0,n}$ = total present worth of costs, for alternative (where $x0$ can be the standard or base alternative, or any feasible mutually exclusive alternative $x1, x2, \ldots, xn$) for an analysis period of years

$TPWC_{x1,n}$ = as defined in Eq. (14.3)

There are a number of advantages inherent in the present-worth method that make it particularly applicable to the infrastructure area compared to the "traditional" annual cost and benefit-cost methods. These advantages include the following:

1. The benefits and costs of a project are related and expressed as a single value.

2. Projects of different service lives, and with stage development, are directly and easily comparable.

3. All monetary costs and benefits are expressed in equivalent present-day terms.

4. Nonmonetary benefits (or costs) can be evaluated subjectively and handled with a cost-effectiveness evaluation.

5. The answer is expressed as a total payoff for the project.

6. The method is computationally simple and straightforward.

There are, however, several disadvantages to the present-worth method, including the following:

1. It cannot be applied to single alternatives when the benefits of those single alternatives cannot be estimated other than to simply calculate the present worth of costs.

2. The results, in terms of a lump sum, may not be as easily understandable to some people as a rate of return or annual cost. In some cases, the summation of costs in this form can act as a deterrent to investment because the lump sum seems so large compared to budgets.

14.7.3   Rate-of-Return Method

The rate-of-return method (sometimes called "internal rate of return") considers both costs and benefits, and determines the discount rate at which the costs and benefits for a project are equal. It can be in terms of the rate at which the equivalent uniform annual cost is exactly equal to the equivalent uniform annual benefit, that is:

$$AC_{x1,n} = AB_{x1,n} \qquad (14.8)$$

where $AC_{x1,n}$ = equivalent uniform annual cost for alternative $x1$ for an analysis period of years

$AB_{x1,n}$ = equivalent uniform annual benefit for alternative $x1$ for an analysis period in years

Alternatively, the rate of return can be expressed in terms of the rate at which the present worth of costs is exactly equal to the present worth of benefits. This can be done by setting Eqs. (14.3) and (14.5) equal to each other:

$$TPWC_{x1,n} = TPWB_{x1,n} \qquad (14.9)$$

In applying the rate-of-return method, each alternative is first compared with the standard or base alternative to establish the difference in benefits. Then using Eq. (14.9) or (14.8), the rate of return can be calculated for alternatives $x1, x2, \ldots, xn$. However, this is only a comparison with the standard. It is necessary to calculate the rate of return or comparison with the standard. It is also necessary to calculate the rate of return for comparisons between alternatives $x1, x2, \ldots, xn$. This is done on the increase in costs between alternatives with successively higher first costs. Proceeding on the basis of such paired comparisons will eliminate all but one alternative—that one having the highest rate of return.

This alternative may or may not be economically attractive, depending on the decision for a minimum attractive rate of return. For example, if it were decided that an investment must have a minimum rate of return of 15 percent to be economically viable, any alternative yielding a lesser return would be rejected.

The internal rate of return method is particularly appropriate to Public-Private-Partnerships (PPPs). A comprehensive example of the application of this method to a toll road project is provided in [Haas 11].

The rate-of-return method has a major advantage in that the results are well understood by most people. It is easy to comprehend a return on investment because of the association and familiarity with normal business terms. However, it must be remembered this is not a "real" interest rate in financial terms.

14.7.4   Benefit-Cost-Ratio Method

The benefit-cost-ratio method has been widely used in the infrastructure area, particularly for large projects such as dams and causeways. It involves expressing the ratio of the present worth of benefits of an alternative to the present worth of costs, or the ratio of the equivalent uniform annual benefits to the equivalent uniform annual costs. The benefits are established by a comparison of alternatives. Using the present-worth formulation, which is preferred by most engineering economists, the benefit-cost ratio may be expressed as follows:

$$BCR_{xj,xk,n} = (TPWB_{xj} - TPWB_{xk})/(TPWC_{xj} - TPWC_{xk}) \quad (14.10)$$

where $BCR_{xj,xk,n}$ = benefit-cost ratio of alternative xj compared to alternative xk (where xk yields the greater benefits, and represents the larger investment), over an analysis period of n years

$TPWB_{xj,n}$, $TPWC_{xj,n}$ = total present worth of benefits, and of costs, respectively, for alternative xj

$TPWB_{xk,n}$, $TPWC_{xk,n}$ = total present worth of benefits, and of costs, respectively, for alternative xk

The calculation of benefit-cost ratios for a set of proposed alternatives is first done by comparison with the standard or base alternative, using Eq. (14.10). Alternatives exhibiting a ratio greater than 1.0 are compared on an incremental basis. This involves calculation of the benefit-cost ratio on the increments of expenditures for successively higher-cost alternatives. Proceeding on the basis of such paired comparisons, using Eq. (14.10), will reveal the alternative of highest economic desirability.

There are some disadvantages to the benefit-cost ratio method. A major drawback is the abstract nature of the ratio, which is difficult for some people to comprehend. Another disadvantage is the possible confusion over whether maintenance cost reductions should be in the numerator or the denominator, and whether cost reductions are "benefits" or "negative costs."

14.7.5 Cost-Effectiveness Method

The cost-effectiveness method can be used to compare alternatives if significant, appropriate measures of effectiveness can be established. This method is extensively used in the pavement field, where effectiveness of an alternative is calculated as the area under the performance curve (i.e., serviceability versus age) multiplied by traffic volume and length of road section. In this form it becomes a surrogate for vehicle-operating costs, which are difficult to determine directly. It also becomes a convenient way to compare alternatives that have different performance curves.

Cost-effectiveness, then, is the ratio of effectiveness divided by the present worth of costs summarized over the life of the facility. It is the inverse of what is termed "cost-effectiveness," but it is used in this way because increasing ratios represent more attractive alternatives, as do increasing benefit-cost ratios. In the cost-effectiveness method, the actual numbers in the ratio are useful only in a comparative sense, because of the mixture of units. They are truly indexes of cost-effectiveness.

As a simple example, consider a facility where effectiveness is related to the number of service interruptions over time, as shown in Figure 14-1. Alternative X_1 involves an initial capital cost of $250,000 with a rehabilitation cost of $80,000 at year seven. It is estimated to have a residual value of $25,000 at the end of the analysis period.

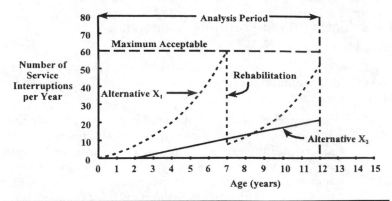

Figure 14-1   Level of service versus age relationship.

Alternative X_2 requires an initial capital expenditure of \$400,000. It is estimated to have a residual value of \$65,000 at the end of the analysis period. The floor areas of X_1 and X_2 are 165 m^2 and 265 m^2, respectively. Assume that maintenance costs are relatively small for each alternative, and that the traffic processed by each is approximately the same.

Calculation of effectiveness, E, for these two alternatives can then be carried out as follows:

$$E = (\text{area under curve}) \times (\text{floor area}) \tag{14.11a}$$

$$E_{x_1} = 442 \times 165 = 72,930 \tag{14.11b}$$

$$E_{x_2} = 620 \times 265 = 164,300 \tag{14.11c}$$

Present worth of costs, C, for the alternatives using a discount rate of 4 percent and analysis period of 12 years, is calculated as follows:

$$C_{x_1} = \text{initial cost} + (\text{rehabilitation cost at year 7}) \times (PWF_{y=7})$$
$$- (\text{residual value at year 12}) \times (PWF_{y=12}) \tag{14.12a}$$

$$C_{x_1} = 250,000 + 80,000 \times \frac{1}{(1+0.04)^7} - 25,000 \times \frac{1}{(1+0.04)^{12}} = 295,178$$
$$\tag{14.12b}$$

$$C_{x_2} = 400,000 - 65,000 \times \frac{1}{(1+0.04)^{12}} = 359,401 \tag{14.12c}$$

Cost-effectiveness ratios, E/C, would be equal to:

$$E_{x_1}/C_{x_1} = 72,930/295,178 = 0.25 \tag{14.13a}$$

$$E_{x_1}/C_{x_2} = 164,300/359,401 = 0.46 \tag{14.13b}$$

On a comparative basis, over the analysis period or life cycle, alternative X_2 is the more attractive alternative, even though it has a significantly higher initial cost.

14.8   Selecting an Appropriate Economic Analysis Method

There are several basic considerations in selecting the most appropriate method for economic evaluation of alternatives. They include the following factors:

- How important is the initial capital expenditure in comparison to future expected expenditures? Often, public officials and private interests are concerned primarily with initial costs. An economic analysis may indicate, for example, that a low capital expenditure today can result in excessive future costs for a particular alternative. Yet the low capital expenditure may be the most relevant consideration to decision-makers, especially in the face of limited funding.

- What method of analysis is most understandable to the decision-maker? The method used may not be technically the best for the situation at hand, but it provides the necessary decision support and understanding, and thus has a high likelihood of being used.

- What method suits the requirements of the particular agency involved? Although the present-worth method may be preferable for many public agencies, an annual cost method or internal rate of return method might be more suitable for a privately owned facility, or a Public-Private-Partnerships.

- Are measures of benefits or effectiveness to be included in the analysis? Any method that does not consider the differences in benefits or effectiveness between alternatives is basically incomplete for use by a public agency. However, for a private-sector situation, an implicit assumption of equal benefits for various alternatives may be satisfactory.

14.9   Effect of Discount Rate

It was noted in Sec. 14.5 that the discount rate selected by most agencies is a policy decision, but usually it is the difference between the interest rate for borrowing money and the inflation rate. In the example given in Sec. 14.7.5, a discount rate of 4 percent was used.

Often, however, it is desired to assess the sensitivity of the economic analysis result to discount rates higher and lower than the discount rate calculated. A convenient way to illustrate the effect of such discount-rate changes is provided in Figure 14-2 [AISI 90]. It represents the use of Eq. (14.2) to show the present worth of $1 expended at sometime in the future for three discount rates. The effect shown has greater significance to future spending at a low discount rate, and less significance at a high discount rate. For example, $1 expended 25 years in the future has a NPV of 48¢ at 3 percent, 23¢ at 6 percent, and only 12¢ at 9 percent. Note that when expended at 75 years, the NPV is nil for both 6 percent and 9 percent, and about 11¢ at 3 percent.

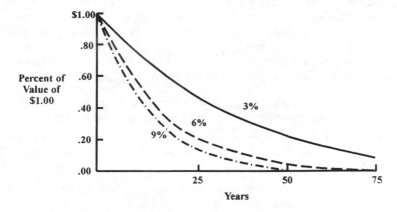

	Discount Rate		
Year	3%	6%	9%
0	1.00	1.00	1.00
25	.48	.23	.12
50	.23	.05	.01
75	.11	.01	.01

FIGURE 14-2   Present value of $1 expended at various intervals and discount rates [AISI 90].

14.10   Application to Transportation and Built Infrastructure

14.10.1   Highways, Roads, Bridges, and Other Transportation Assets

Haas, Hudson, and Zaniewski [Haas 94] present detailed examples of LCA for road and highway pavements. Vehicle-operating cost (VOC) is an important component of road-user cost, which depends upon the vehicle type, speed, pavement geometry and grade, and pavement condition. Reduction in VOCs because of improved pavement condition and/or reduced congestion can be considered as a user benefit. The Federal Highway Administration and the World Bank have developed VOC models applicable to pavement LCA [Zaniewski 82, Watanatada 87, Uddin 87, 93, Paterson 92]. Reportedly, these VOC models have been updated in a recent study [NCHRP 12] for more recent vehicle technology and varying pavement condition. Life-cycle cost analysis is also recognized for bridge management system development [Hudson 87].

Considering the global trend of rural population migration to cities and demand on urban mobility, there is a strong need to implement sustainable land use and transportation management policies for reducing fossil fuel use, emissions, congestion, and crashes. Strategies include mass transit to replace single-occupancy vehicles from roads. The use of LCA of agency costs and user/societal benefits can help to select innovative mass transit technologies, which may have higher initial investment costs. To illustrate life-cycle benefit and cost analysis of a personal rapid transit (PRT) system the following results, based on various assumptions, compare the benefits and costs with the base alternative of "no change in multimodal transport policy" [Uddin 12]:

- Base alternative implies the same modal mix and traffic pattern over the 10-year analysis period, given the following:

  - The construction/maintenance cost of 10-km road network over 10 years is 30 million dollars. This includes milling and repaving asphalt roads every 5 years.

  - The added user cost attributed to road maintenance is 2 million dollars over 10 years.

  - User delay cost caused by congestion averages 25 hours of delay per year per person during peak-hour traffic volume is 73 million dollars over 10 years. The cost of wasted fuel due to congestion is ignored in this calculation.

  - The total life-cycle cost over 10 years is (30 + 2 + 73) or 105 million dollars assuming zero percent discount rate. There is zero benefit or a NPV of −105 in 10 years. In 5 years the total cost is 52.5 million dollars with NPV of −52.5.

- PRT alternative: It is assumed that the 30 to 40 mi/h speed PRT system on elevated alignment will cost 4 million dollars per km with low annual operating cost because of energy-efficient, lightweight, zero-emission vehicle technology. Assume 20 percent of car owners in the daily traffic volume will use PRT instead of driving single-occupancy cars.

  - The total PRT cost of constructing 10-km stretch with 0.5 million dollar annual operating and maintenance cost is 45 million dollars cost over 10 years (at 0 percent discount rate).

  - Assume that each person using the PRT system saves annually 25 hours of delay at $16 per hour and avoids wastage of 15 gallons of fuel and oil at $3 per gallon.

  - The life-cycle benefit is 90 million over 10 years (at 0 percent discount rate), based on total user saving of 9 million dollars annually.

  - The NPV is equal to (90 − 45) or 45 million dollars over 10 years. In 5 years the NPV is zero.

- The PRT alternative is favored after 5 or more years of full load rate operation and LCA. Passenger-fare revenue will be an added benefit for the owner agency or operator, not considered in this simple analysis.

- Additionally, the societal benefit of the innovative PRT vehicle technology will be large in terms of reduction in CO_2 at about one million tons annually [Uddin 12]. Other indirect benefits include less air pollution and new PRT-related manufacturing and service jobs.

14.10.2 Built Infrastructure Assets

A NIST (National Institute of Standards and Technology) report [CDA 96] outlines the application of LCA using new technology materials, such as the use of fiber-reinforced plastics (FRP) to retrofit and repair bridges in seismic areas. According to this report, the minimum NPV of a life-cycle cost method can be used to satisfy the Intermodal Surface Transportation Efficiency Act's requirement that life-cycle costs be considered in the design of transportation-related structures, and Executive Order 12893, which requires that the costs of federal infrastructure investment be accounted for over the life span of each project [CDA 96].

The noncorrosive lightweight FRP sheet-pile technology, discussed in value engineering analysis in Sec. 9.5 for protection of coastal roads, has been studied for levee reconstruction and shown to be a preferred alternative [Durmus 12]. The FRP sheet-pile alternative is 25.8 percent cheaper in total life-cycle construction and maintenance costs over a 50-year analysis period at a 5 percent discount rate, compared to traditional concrete sheet piles. Study the reference for details.

Similarly, an LCA of cost and benefits may be used to evaluate alternative technologies in other infrastructure assets, such as renewable-energy technologies (wind turbine, solar panels, etc.), "green" buildings, and eco-friendly land development.

14.11 Application to Water and Sewer Mains

Leakage of potable water pipeline network or treated water and risk of contamination with sewage pose serious health problems in most developing countries. "In developing countries, as much as 80 percent of illness is linked to poor water quality and sanitation conditions" (Twitter tweet, August 16, 2012, via urb.im@urb_im48m RT @nonprofitcoffee). Industrialized nations are not immune to the water leakage problem caused by aging mains and accidental breaks during excavation for other construction projects.

Economic analysis to compare repair versus replacement M,R&R alternatives for water and sewer mains has been used by agencies in many cities, such as New York City, Dallas, Texas, Washington, DC, and Seattle, Washington [O'Day 84, Steinthal 84, Thomasson 82, Urban 84]. In general, replacement alternatives are more favorable when mains have deteriorated structurally and shown a consistent record of frequent

	Year Break										
Main Size	**71**	**72**	**73**	**74**	**75**	**76**	**77**	**78**	**79**	**80**	**81**
15 cm	/	/						/	///	////	///
(6 in)										/////	

Breaks per year in the last   Length of main = 250 ft.   Breaks per year per 1,000 ft =
four years = 16/4.   (16/4)/(250/1,000) = 16.

Cumulative repair cost = 360.34   ×   16/4   ×   75.477 = 108,789.53 ($)
                                         ($/break) (Breaks/year) $(CAF)^1$

Cumulative replacement = 20.5   ×   250   ×   0.1095 ×   20 = 11,223.75 ($)
cost   ($/ft)   (ft) $(CRF)^2$ (Years)

Analysis conducted for 20-year (n) cost of both repairing and replacing a given water main whenever the water-main break rate exceeds 1.8 breaks per year per 1,000 ft of main (over the last four years). The above information is given at 12.4 percent interest (i) for CAF and 9 percent (i) bonds for CRF.

$(CAF)^1$ = compound amount factor (for uniform annual series) = $[(1 + i)^n - 1]/i$

$(CRF)^2$ = capital recovery factor = $i(1 + i)^n/[(1 + i)^n - 1]$

TABLE 14-1   Economic Analysis for Dallas Water Main Alternatives [Steinthal 84]

breaks. Table 14-1 shows the results of Dallas water-main economic analysis [after Steinthal 84] to calculate the least life-cycle cost alternative based on the previous breaks record and a 20-year analysis period.

Risk analysis, including lost revenues, cost of lost time, and inconvenience for each alternative, should also be included in any analysis that considers the reduction in the user costs as "benefits." A substantial leakage-control program can provide considerably more benefits than the costs involved in the identification and repair of leaking segments, as noted by O'Day. The following example is from a major U.S. water-utility study conducted in 1981, which resulted in the location and repair of 596 leaks, and the elimination of 12.2 million gallons of wasted water per day in the 200 mi of the water-distribution system [O'Day 84]:

The leak detection costs for the utility staff, consultant fees, and equipment costs were approximately $204,000, equivalent to slightly more than $1,000 per mile for the 200 miles in the survey area. The total costs for repair were $243,000 for a total detection and repair cost of $447,000.

The value of the future water losses saved by this leak control program is equivalent to the present value of future water and waste-water treatment and pumping costs. The net present worth for these savings is estimated to be nearly $6,600,000 over a 30-year period. That is, the $447,000 leak detection and repair program will save an estimated $6,000,000 in treatment and pumping costs over a 20-year period.

The foregoing case example shows that the benefits in terms of savings are 14.42 times the costs. For a cost of $2,000 per mile, significant water loss was saved, future causes of main breaks were reduced, and only less than 1 mi out of 200 mi of water mains were replaced [O'Day 84]. An example of life-cycle cost analysis for sewer-main material-type selection is presented in Table 14-2 [AISI 90].

Data and Analysis Method	Alternative A	Alternative B	Alternative C
Description	Galvanized corrugated steel pipe (CSP)*	Asphalt-coated CSP	Reinforced concrete pipe (RCP)
Initial costs	$195,000	$214,500	$230,000
Other costs	$48,750*	—	—
Design life	50 years	50 years	50 years
Discount rate	9%	9%	9%
Current cost	$195,000 + 48,750 Total = $243,750	$214,500 + 0 Total = $214,500	$230,000 + 0 Total = $230,000
Present-worth method [present value (at 9%)]	$195,000 + (48,750 × 0.0318 = $1,550) = $196,550 (lowest)	$214,500	$230,000
Equivalent uniform annual cost (50 years at 9%)	$17,930 (lowest)	$19,560	$20,980

Differential Cash-Flow Method	Compare Alternative A and C		
	Alternative A	Alternative C	Difference (C–A)
Year 0	$195,000	$230,000	$35,000
Year 40	$48,750	—	$(48,750)
Total	$243,750	$230,000	$(13,750)
Rate of return, %	To avoid future expenditure of $48,750 in Alternative A, the added initial cost of $35,000 in Alternative C results in 0.83%, or less than 1% internal rate of return. This is a poor rate of return on the initial investment.		
Effect of salvage	Not significant; at even 30% of the original cost, present worth of salvage represents less than 5% of the initial cost.		

*Alternative A requires invert maintenance at 25% of initial cost due to a projected life of 40 years for the galvanized CSP ($48,750).

Table 14-2 Life-Cycle Cost Analysis for Sewer-Main Material Selection [AISI 90]

14.12   Application to Buildings and Land-Use Changes

Prediction of effective service life is more complex for buildings, because the structural integrity of a building depends upon many factors aside from the materials of construction and the performance of the various functional subsystems. These complexities generally have discouraged the use of routine life-cycle cost analysis for design and M,R&R programming of buildings in the past. In *Pay Now or Pay Later* [BRB 91], the Building Research Board provides general guidelines for life-cycle cost analysis and gives an example in the Department of Veterans Affairs (VA), which owns and operates the largest system of hospitals (172 medical centers, some 2,000 buildings) in the United States. The VA has used a 40-year analysis period and a 5 percent discount rate for their life-cycle cost analysis [BRB 91]. The study highlighted seven factors that could reduce the cost of ownership:

1. Reduced tie to design and construction

2. Standardization of building components and utility-system concentration

3. Improved space utilization and layout of mechanical system

4. More cost-effective maintenance and minor alternations, using materials with better performance and access for service

5. Optimization of heating, ventilation, and air conditioning

6. Better design features to improve housekeeping

7. More cost-effective rehabilitation alternatives

The total life-cycle cost of a hospital was 8.7 times the initial cost; however, the present worth of savings was estimated to be approximately 10 percent of the life-cycle cost.

The National Institute of Standards and Technology (NIST) has also developed a computer program, *Building Life-Cycle Cost* (BLCC), and is currently involved in an international effort to standardize service LCA procedures [AIJ 93, BS 92, CSA 94, Frohnsdorff 96, NIST 95]. The Building Maintenance, Repair, and Replacement Database (BMDB) computer program offered by the American Society for Testing and Materials (ASTM) can be used to estimate data for building life-cycle cost analysis [ASTM 90]. The BMDB program generates maintenance costs by applying labor and equipment rates, material-location adjustment factors, cost-escalation factors, and the discount rate. The material costs (in 1986 dollars) are based on Washington, DC, and are adjusted to other locations through the application of a geographical location adjustment factor. Table 14-3 shows an example of data generated from the BMDB program for life-cycle cost analysis [ASTM 90].

Resource and Unit Cost—Present Value of All 25 Years of Maintenance and Repair Costs ($ per Unit Measure)					
		Annual Maintenance and Repair			
(a) Component Description	**Unit Measure**	**Labor**	**Material**	**Equipment**	**Unit Cost ($)**
Architecture					
Roofing					
Roof covering					
Built-up roofing	SF	0.00488	0.03171	0.00244	0.14
Place new membrane over existing					
Modified Bitumen/ thermoplastic	SF	0.00248	0.03218	0.00119	0.09
Membrane replacement or repair					
Thermosetting	SF	0.00174	0.02202	0.00088	0.06
Membrane replacement— thermosetting					
Slate	SF	0.00259	0.01458	0.00124	0.07
Cement asbestos	SF	0.00254	0.03402	0.00122	0.09
Tile	SF	0.00217	0.02934	0.00103	0.08
Roll roofing	SF	0.00754	0.01556	0.00386	0.18
Total roof replacement-roll roof					
Shingles	SF	0.00259	0.02384	0.00140	0.08
Replace new over existing shingle					
Metal	SF	0.00204	0.01546	0.00103	0.06
Fiberglass rigid stp. roof	SF	0.00232	0.06266	0.00113	0.11
Concrete, sealed panel roof	SF	0.00601	0.01643	0.00297	0.15
Concrete, sealed panel RF4	SF	0.00544	0.01175	0.00282	0.13
Concrete sealed poured	SF	0.01374	0.08807	0.00692	0.39
Fiberglass, rigid roof	SF	0.00463	0.06266	0.00236	0.16

Table 14-3 An Example of Data Generated from the BMDB Program for Life-Cycle Cost Analysis [after ASTM 90]

Open green wooded area in and around cities being converted into concrete and asphalt surfaces and buildings lead to "heat-island" effects. These are purported to increase energy demand, smog conditions, GHG, and other harmful emissions [Uddin 09, 12]. Life-cycle

Resource and Unit Cost—Present Value of All 25 Years of Maintenance and Repair Costs ($ per Unit Measure)						
		Replacement and High-Cost Tasks				
(b) Component Description	**Unit Measure**	**Year**	**Labor**	**Material**	**Equipment**	**Unit Cost($)**
Architecture						
Roofing						
Roof covering						
Built-up roofing	SF	28	0.04938	0.7049	0.02469	1.79
Place new membrane over existing	SF	14	0.02414	0.6996	0.01207	1.23
Mod. bit./thermoplastic	SF	20	0.05659	0.8586	0.02829	2.11
Membrane replacement or repair	SF	20	0.05659	0.8586	0.02829	2.11
Thermosetting	SF	20	0.03683	0.6996	0.01841	1.51
Membrane replacement— thermosetting	SF	20	0.03683	0.6996	0.01841	1.51
Slate	SF	70	0.06885	6.042	0.03442	7.56
Cement asbestos	SF	70	0.05437	0.7519	0.02718	1.95
Tile	SF	70	0.10169	3.074	0.05084	5.32
Roll roofing	SF	10	0.04141	0.74963	0.0207	1.66
Total roof replacement— roll roof	SF	10	0.04141	0.74963	0.0207	1.66
Shingles	SF	40	0.04118	0.74497	0.02059	1.65
Replace new over existing shingle	SF	20	0.02996	0.4346	0.01498	1.1
Metal	SF	30	0.36265	2.173	0.18132	10.17
Fiberglass rigid stp. roof	SF	20	0.04543	6.0155	0.02272	7.02
Concrete, sealed panel roof	SF	60	0.06123	24.07419	0.03061	25.42
Concrete, sealed panel RF4	SF	300	0.04342	24.07419	0.02171	25.03
Concrete sealed poured	SF	500	3.81056	18.03219	1.90528	102
Fiberglass, rigid roof	SF	20	0.04133	6.0155	0.02066	6.93
Total roof replacement— fiberglass	SF	20	0.04133	6.0155	0.02066	6.93

Location: Washington, DC. Study starts three years before beneficial use. Discount rate = 10%.

TABLE 14-3 An Example of Data Generated from the BMDB Program for Life-Cycle Cost Analysis [after ASTM 90] (*Concluded*)

societal public health costs were calculated using satellite imagery–based estimates of the increased built-surface area and increased traffic volume measured in the rural town of Oxford, Mississippi. The following key results are associated with that study to evaluate the adverse impacts of the construction of a large shopping mall and surrounding commercial development during 2001 to 2004 [Uddin 05, 06]:

- The areas of constructed surfaces (asphalt, concrete, and buildings) in the study site increased 14.4 percent, replacing the areas of natural surfaces (grass, tree, and soil).

- The building area after the construction increased almost 300 percent.

- The weighted average surface temperature of the developed site area increased by 1.9°C (3.4°F), or 3.9 percent increase compared to the pre-construction condition.

- The daily maximum 8-hour average ground-level ozone (O_3) concentrations, predicted using a comprehensive multifactor O_0 forecasting model [Boriboonsomsin 04], increased by 16.7 percent, as a result of the coupled effect of land development, built infrastructure, and traffic growth.

- The public health cost of respiratory diseases and lost productivity associated with vehicle emissions and air quality degradation is 13 cents per vehicle-mile, which is about 19 percent of the total societal cost estimate for this study site.

The foregoing example should not be taken as anti-development. Rather, it illustrates that the impacts of any development go beyond only direct economics of facility costs and revenues.

14.13   Summary

Economic analysis is an essential part of infrastructure asset management. The methods and applications may vary, as shown here and in the references; they may not, however, be omitted, if good results are expected from IAMS. Many agencies have economic analysis capabilities in house, but they need to use them effectively. The methods are relatively simple and straightforward, and should be coded directly into any IAMS software adopted.

It is recommended that public agencies and professional service providers strive to adapt the new cost models and methodologies developed in recent decades to include life-cycle public health and societal costs. This can help to encourage new infrastructure technologies that go beyond considering only initial construction costs.

14.14   References

[AIJ 93] Architectural Institute of Japan, *The English Edition of Principal Guide for Service Life Planning of Buildings*, Tokyo, Japan, 1993.

[AISI 90] *Modern Sewer Design*, American Iron and Steel Institute (AISI), Washington, DC, 1990.

[Army 86] "Economic Studies for Military Construction Design-Applications," *TM 5802-1*, Department of the Army, Washington, DC, 1986.

[ASTM 90] ASTM, "Building Maintenance, Repair, and Replacement Database (BMDB) for Life-Cycle Cost Analysis," *A User's Guide to the Computer Program*, American Society for Testing and Materials, Philadelphia, Pa., 1990.

[Boriboonsomsin 04] K. Boriboonsomsin, "Transportation-Related Air Quality Modeling and Analysis Based on Remote Sensing and Geospatial Data," Ph.D. Dissertation, Department of Civil Engineering, The University of Mississippi, May 2004.

[BRB 91] Building Research Board, *Pay Now or Pay Later*, National Research Council, Washington, DC, 1991.

[BS 92] *Guide to Durability of Buildings and Building Elements, Products and Components*, BS 7543:1992, British Standards Institution, London, UK, 1992.

[CDA 96] "NIST Develops Methodology to Evaluate Life-Cycle Costs of Composites in Bridge Applications," *Composites Design and Applications*, March 1995, p. 6.

[CSA 94] *Guideline on Durability in Buildings*, CSA S478, Draft 9, Canadian Standards Association, September 1994.

[DOI 83] "Economic and Environmental Principles and Guidelines for Water and Related Land Resources Implementation Studies," *Report PB 84-199405*, U.S. Water Resources Council, Department of the Interior, Washington, DC, 1983.

[Durmus 12] A. Durmus, "Geospatial Assessment of Sustainable Built Infrastructure Assets and Flood Disaster Protection," M.S. Thesis, Department of Civil Engineering, The University of Mississippi, August 2012.

[Frohnsdorff 96] G. Frohnsdorff, "Predicting the Service Lives of Materials of Construction," *Proceedings, Fourth Materials Engineering Congress*, Washington, DC, Vol. 1, November 1996, pp. 38–53.

[Haas 11] R. Haas, "Technical and Economic Base Requirements for Effective Asset Management," *Proceedings, 14th International Flexible Pavement Conference*, Sydney, Australia, September, 2011.

[Haas 94] R. Haas, W. R. Hudson, and J. P. Zaniewski, *Modern Pavement Management*, Krieger Publishing Company, Malabar, Fla., 1994.

[Hudson 87] S. W. Hudson, R. F. Carmichael III, L. O. Moser, W. R. Hudson, and W. J. Wilkes, "Bridge Management Systems," *NCHRP Report 300*, National Cooperative Highway Research Program, Transportation Research Board, National Research Council, Washington, DC, 1987.

[Hudson 97] W. R. Hudson, R. Haas, and W. Uddin, *Infrastructure Management*, McGraw-Hill, New York, 1997.

[Murphy 98] J. J. Murphy, and M. A. Delucchi, "A Review of the Literature on the Social Cost of Motor Vehicle Use in the United States," *Journal of Transportation and Statistics*, 1(1), 1998, pp. 15–42.

[NCHRP 12] National Cooperative Highway Research Program, "Estimating the Effects of Pavement Condition on Vehicle Operating Costs," *Research Report 720*, Transportation Research Board, Washington, DC, 2012.

[NIST 95] Personal communications of W. Uddin with G. Frohnsdorff, Chief of Building Materials Division, National Institute of Standards and Technology, 1995.

[O'Day 84] D. K. O'Day, "Aging Water Supply Systems: Repair or Replace," *Infrastructure—Maintenance and Repair of Public Works, Annals of the New York Academy of Science*, Vol. 431, December 1984, pp. 241–258.

[Paterson 92] W. D. O. Paterson, and B. Attoh-Okine, "Simplified Models of Paved Road Deterioration based on HDM-III," in *Transportation Research Record 1344*, Transportation Research Board, Washington, DC, 1992.

[Small 95] K. A. Small, and C. Kazimi, "On the Costs of Air Pollution from Motor Vehicles," *Journal of Transport Economics and Policy*, January 1995, pp. 7–32.

[Steinthal 84] B. G. Steinthal, "Infrastructure Maintenance Strategies— Governmental Choices and Decision Methodologies," *Infrastructure— Maintenance and Repair of Public Works, Annals of the New York Academy of Sciences*, Vol. 431, December 1984, pp. 139–154.

[Thomasson 82] R. O. Thomasson, "In-Place Sewer Reconstruction Proves Cost-Effective," *American City and County*, February 1982.

[TRB 85] Transportation Research Board, "Life-Cycle Cost Analysis of Pavements," *Synthesis of Highway Practice No. 122*, National Cooperative Highway Research Program, National Research Council, Washington, DC, 1985.

[Uddin 87] W. Uddin, R. F. Carmichael III, and W. R. Hudson, "A Methodology for Life-Cycle Cost Analysis of Pavements Using Microcomputer," *Proceedings, 6th International Conference, Structural Design of Asphalt Pavements*, Ann Arbor, Mich., Vol. 1, 1987.

[Uddin 93] W. Uddin, "Application of User Cost and Benefit Analysis for Pavement Management and Transportation Planning," *Proceedings, 4R Conference and Road Show*, Philadelphia, Pa., December 1993, pp. 24–27.

[Uddin 05] W. Uddin, K. Boriboonsomsin, and S. Garza, "Transportation Related Environmental Impacts and Societal Costs for Life-Cycle Analysis of Costs and Benefits," *IJP–International Journal of Pavements*, ISSN 1676-2797, Vol. 4, No. 1–2, January–May 2005, pp. 92–104.

[Uddin 06] W. Uddin, "Air Quality Management Using Modern Remote Sensing and Spatial Technologies and Associated Societal Costs," *International Journal of Environmental Research and Public Health*, ISSN 1661-7827, MDPI, Vol. 3, No. 3, September 2006, pp. 235–243.

[Uddin 09] W. Uddin, C. Brown, E. S. Dooley, and B. Wodajo, "Geospatial Analysis of Remote Sensing Data to Assess Built Environment Impacts on Heat-Island Effects, Air Quality and Global Warming," *Paper No. 09-3146, CD Proceedings, 85th Annual Meeting of The Transportation Research Board*, Washington, DC, January 11–15, 2009.

[Uddin 11] W. Uddin, "Life Cycle Assessment of Sustainable Pavement Systems," *Proceedings, 5th ICONFBMP, International Conference Bituminous Mixtures and Pavement*, Thessaloniki, Greece, June 1–3, 2011.

[Uddin 12] W. Uddin, "Mobile and Area Sources of Greenhouse Gases and Abatement Strategies," Chapter 23, *Handbook of Climate Change Mitigation*, (Editors: W.-Y. Chen, J. M. Seiner, T. Suzuki, and M. Lackner), Springer, 2012, pp. 775–840.

[Urban 84] H. P. Hatry, and B. G. Steinthal, *Guides to Managing Urban Capital Series*, Vol. 4: *Guide to Selecting Maintenance Strategies for Capital Facilities*, The Urban Institute, Washington, DC, 1984.

[Watanatada 87] T. Watanatada, C. G. Harral, W. D. O. Paterson, A. M. Dhareshwar, A. Bhandari, and K. Tsunokawa, *The Highway Design and Maintenance Standards Model*, Vols. 1 and 2, Johns Hopkins University Press, Baltimore, Md., 1987.

[Winfrey 69] R. Winfrey, *Economic Analysis for Highways*, International Textbook Company, Scranton, Pen., 1969.

[Zaniewski 82] J. P. Zaniewski, B. C. Butler, G. Cunningham, G. E. Elkins, M. S. Paggi, and R. Machemehl, "Vehicle Operating Costs, Field Consumption and Pavement Type and Condition Factors," *Final Report*, Federal Highway Administration, Washington, DC, 1982.

CHAPTER 15

Prioritization, Optimization, and Work Programs

15.1 Introduction

Chapter 8 demonstrated that when the serviceability or quality of an infrastructure component reaches an unacceptable or intervention level, some action is needed. This level could, for example, be in terms of the maximum allowable number of annual service interruptions shown in Figure 14-1, the maximum allowable user-delay time on a facility, minimum flow capacity in a pipe network, and so forth. In other words, the component or facility reached a condition where action is needed. Such action might consist of M,R&R, or simply increased routine maintenance to keep the facility from exceeding the intervention level.

If a facility is currently at the intervention level, it is a "now need." Future needs would be determined on the basis of the performance or deterioration prediction model (i.e., the time or year that the facility or component is estimated to reach the intervention level, which was, for example, seven years for alternative X_1, in Figure 14-1).

If sufficient funds are available (i.e., an unconstrained budget), all needs can be addressed when they occur. The usual situation, however, for most public infrastructure is a limited or constrained budget. In such cases, priorities have to be set on what action or work will be undertaken, where, and when. This chapter is directed to the procedures that can be used to establish priorities, the role of optimization in priority analysis, budgeting issues, and the development of work programs under constrained budgets.

Optimization is particularly important in priority analysis as it serves decision-makers in allocating scarce resources. In practice though, there are several approaches to allocating resources, ranging from true optimization to varying degrees of suboptimal allocations.

There are subsequently discussed, as well as incorporating multifactor effects in optimization.

15.2   Framework for Prioritization: From Simple Ranking to Multifactor Optimization

Prioritization as a whole involves four central steps: (1) information acquisition; (2) processing and interpretation; (3) determination of current needs and future needs; and (4) priority analysis and results, as shown in Figure 15-1. Additionally, there are key associated elements of deterioration or performance models, intervention levels, budget constraints, program period selected, and available alternative strategies.

The actual models, methods, and procedures within the framework of Figure 15-1 can range from very simple to very sophisticated. For example, the prioritization might simply involve some basic inventory data, an annual (year-by-year) program period, a determination of needs, a priority analysis, and a recommended program of work selected by engineering judgment. At the other extreme it may involve a comprehensive database, a multiyear program period, multifactor deterioration models, decision analysis procedures to identify feasible alternative strategies, a mathematical programming method for priority analysis, and the establishment of work programs based on optimization.

On a hierarchical basis, the approaches can be categorized as follows:

- Optimization models based on operations research/mathematical programming (e.g., linear, dynamic, integer, etc. programming) typically used in such applications as multiproduct lines from a factory, or genetic algorithms, neural networks, etc.

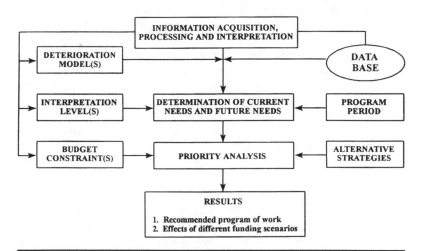

Figure 15-1   Framework for prioritization.

- Heuristic or near optimization based on marginal effectiveness (where effectiveness is a weighted measure and acts as a surrogate for difficult-to-quantity benefits).

- Business case approaches prepared by managers of individual assets or groups of assets typically focused on preservation, security, safety, etc., with the degree of optimality unknown.

- Arbitrary, best-judgment allocation may be best applicable to small agencies.

Whatever approach is used, the final result, in terms of the recommended progress of work, should answer the following questions:

1. Which projects or infrastructure components should receive action?

2. What action (M,R&R) should be applied?

3. When should the work be done?

A methodology, within the framework of Figure 15-1, that can deal with these questions concurrently will not be simple. However, if a truly optimal expenditure of available funds or budget is to be achieved, all possible combinations of which, what, and when should be considered and evaluated.

15.2.1   Selecting a Program Period

The length of the program period for M,R&R improvements is not necessarily the same as that for life-cycle economic evaluation. In fact, it would usually be less, because the latter commonly covers 20 or more years. Certainly, the length of the program period for maintenance would generally only be 1 to 5 years.

The question then becomes one of what is both a reasonable and useful program period for rehabilitation or renovation. It can range from single-year-by-single-year to multiyear programs. The former is convenient but cannot consider alternative action years for rehabilitation. Thus, timing is excluded from the optimization.

The following guidelines for rehabilitation program periods are considered reasonable and have found substantial acceptance in practice:

1. Five years for infrastructure networks or portions of networks (roads, pipe distribution systems, and so forth), where there is considerable uncertainty for future funding and/or no useful purpose is seen in developing even tentative priority programs beyond that time.

2. Ten years for other network situations, particularly when it is desired to evaluate the long-term, strategic implications of greater or lesser levels of funding.

A very practical approach is to develop a 5- or 10-year program. However, considering that uncertainties increase with time, it is more practical to fix only the first two or three years of the program, and carry out annual or biannual updates. In this way, the flexibility exists for some projects to be advanced one or two years if costs or bid prices come in below estimates. Alternatively, if costs are higher than expected, then some first-year program projects may be deferred to the second or later years.

15.3   Priority Analysis Methods

Priorities can be determined by many methods, ranging from simple subjective ranking to a true optimization in which all possible combinations of "what project, when, and where" are evaluated in terms of an output parameter or objective function, as subsequently discussed. Table 15-1 summarizes the various classes of methods, plus their advantages and disadvantages.

Simple subjective ranking is easy to apply in most situations, but the results may not be the best expenditure of funds or very defensible. Moreover, its use in other than very straightforward cases (options are few and priorities are relatively simple to determine), obviously defeats the principles of an infrastructure management system.

Class of Method	Advantages/Disadvantages
Simple subjective ranking of projects based on judgment	Quick, simple; subject to bias and inconsistency; may be far from optimal
Ranking based on parameters, such as level of service and condition	Simple and easy to use; may be far from optimal
Ranking based on parameters with economic analysis and/or risk exposure	Reasonably simple; should be closer to optimal
Optimization by mathematical programming model for year-by-year basis	Less simple; may be close to optimal; effects of timing not considered
Near optimization using a marginal cost-effectiveness approach	Reasonably simple; close to optimal results
Comprehensive optimization by mathematical programming model, taking into account the effects of "which, what, and when"	Most complex; can give optimal program (maximization of benefits, minimization of costs)
Genetic algorithm	Widely used as a search heuristic to solve an optimization problem
Neural network	Nonlinear mathematical model which can give close to optimal results

TABLE 15-1   Classes of Methods That Can Be Used in Priority Analysis

However, rankings can be weighted, for example, by using traffic, volume of flows, and amounts of usage, depending on the infrastructure involved, and a much more rational and defensible result can be achieved. Such results may still be less than optimal, especially if a significant number of possible alternative treatment and timing combinations exist for each project.

If the ranking also includes an economic analysis, a closer-to-optimal result is more likely to be ensured. Optimization on a year-by-year basis may or may not achieve an overall closer-to-optimal result than the foregoing method, but has a higher likelihood of doing so.

15.3.1 Priority Setting Based on Ranking

Capital budgeting decisions at the network level are commonly based on ranking or rating criteria. The ratings are combined in an overall priority score using appropriate weights to adjust for importance of a project and safety. For example, the city of Dallas categorizes stormwater improvement projects based on: (1) possible loss of life; and (2) floodwater in residential or commercial properties, or in surrounding areas to a depth that disrupts emergency-vehicle access [Steinthal 84].

Initial prioritization of water-main projects in Dallas and Milwaukee used ratings based on "breaks per year per 333 m (1,000 ft) of pipe" and "breaks per 33 m (100 ft)," respectively. Large water-main projects in Seattle are selected based on a composite overall rating considering the following factors [Steinthal 84]:

- Visual inspection of lining condition (0–10 points)
- Effect of the lining on water quality (1–10 points)
- Degree of pitting in the pipe (0–10 points)
- Leak listing over 10 years (0–5 points)
- Necessity of the water main for area supply (0–5 points)

Hatry and Steinthal [Urban 84] reviewed the priority-setting procedures of many public works agencies, summarized rating criteria in the following categories: Environmental Factors and Resources.

A ranking approach based on risk exposure has been described in [Antelman 08]. It is termed as Mission Dependency Index (MDI) and was developed as a joint venture between the National Aeronautical and Space Administration (NASA), the Naval Facilities Engineering Command (NAVFAC), the U.S. Coast Guard Office of Civil Engineering. The MDI uses an intra-dependency risk assessment matrix on functionality of an asset element or component and an inter-dependency risk assessment matrix between asset elements (e.g., a multifactor representation) to arrive at combined MDI scores for each asset element ranging from low to critical. Prioritization of projects using the MDI approach, in conjunction with environmental assessments and sustainability, is applicable to a wide range of infrastructure from utilities to buildings to transportation.

15.3.2   Near-Optimization Methods

Near optimization, based on a marginal cost-effectiveness method, has found considerable application in the roads-and-pavements area [Haas 94], and could also be applied in many other infrastructure areas. It can provide essentially the same results as a true optimization, and may be summarized in terms of the following steps:

1. Consider each combination of section (component), treatment alternative, and year in the program period (i.e., the "which, what, and when" combination).

2. Calculate the effectiveness (E) of each combination (i.e., the area under the performance curve, weighted by usage, length, and area; see Sec. 14.7.5).

3. Calculate the cost (C) in present worth terms, of each treatment alternative in each combination.

4. Calculate the cost-effectiveness (CE) of each combination as the ratio of effectiveness divided by costs.

5. Select the combination of treatment alternative and year for each section that has the best CE, until the budget is exhausted.

6. Calculate the marginal cost-effectiveness (MCE) of all other strategies (i.e., "which, what, and when") for each section, as follows:

$$MCE = (E_s - E_r)/(C_s - C_r) \qquad (15.1)$$

where E_s = effectiveness of the combination selected in Step 5
E_r = effectiveness of the combination for comparison
C_s = cost of the combination selected in Step 5
C_r = cost of the combination for comparison

7. If MCE is negative, or if $E_r < E_s$, then the comparative strategy is eliminated from further consideration; if not, it replaces the combination selected in Step 5.

8. The process is repeated until no further selections can be made in any year of the program period (i.e., when the budget is exhausted).

15.3.3   Prioritization Based on Optimization

Optimization based on mathematical programming models can be used for either single-year or multiyear prioritization. The results do not indicate priorities per se; rather, alternatives are selected to satisfy a specific objective function (i.e., a value is maximized, such as effectiveness, or minimized, such as costs). Formulations used in these models include several variations of linear programming and dynamic programming. Linear programming (LP) is a useful formulation for multiyear prioritization because it can be set up to model the trade-offs between project timing and benefit losses. Each possible implementation, or action, year can be treated as an independent alternative, along with the within-project alternatives and the projects or sections in the network. Thus, all possible combinations are considered and compared.

The formulation of an LP model for maximizing the total present value of m projects, each with k within-project alternatives, for a T-year programming period can be done in the following way:

$$\text{Maximize} \sum_{i=1}^{m} \sum_{j=1}^{k} \sum_{t=1}^{T} X_{ijt} \times B_{ijt} \tag{15.2}$$

$$\text{Subject to} \sum_{t=1}^{T} \sum_{j=1}^{k} X_{ijt} \leq 1 \quad \text{for} \quad i = 1, 2, \ldots, m \tag{15.3}$$

$$\sum_{t=1}^{T} \sum_{j=1}^{k} X_{ijt} \times D_{ijtt''} \leq B_t \quad \text{for} \quad i = 1, 2, \ldots, T \tag{15.4}$$

and

$$X_{ijt} \leq 0 \tag{15.5}$$

where X_{ijt} = section i (of m total sections) with alternative j (of k total treatment alternatives) in year t (of the T years in the program period)

B_{ijt} = present value of annual benefits (including salvage value) of section i, with alternative j, built in year t, all discounted to base year at a discount rate of R

$D_{ijtt''}$ = the actual construction and/or maintenance cost of section i, with alternative j, built in year t, incurred in years t''

B_t = budget for year t

X_{ijt} is a discrete variable equal to 1 if a project is selected for section i in year t, and equal to zero otherwise.

Equation (15.2) is the objective function for maximization of benefits. Maximization of net benefits (benefits minus costs) is not necessary, because construction and maintenance costs are specifically dealt within the annual budget constraints. Thus, there is no need to discount costs back to a base year to calculate the net present value of the projects.

Equation (15.3) states that an X_{ijt} combination is unique; that is, a section cannot be selected twice during the analysis period. Either it is, or is not, built. If a project is committed (by judgment, political intervention, and so forth), then the constraint becomes one of equality (i.e., $X_{ijt} = 1$).

Equation (15.4) represents the budget constraint each year in the programming period, which cannot be exceeded. It states that the summation over m total sections and k total alternative treatments of all the X_{ijt} costs must be less than or equal to the budget, B_t.

Equation (15.5), a common constraint of all linear programming problems, states that it is not possible to recapture construction costs by not constructing the project.

It should be noted that costs are considered directly in the LP model up to the end of the programming period. After that, costs cannot be taken into account because of the absence of budget constraints. It is for this reason that costs incurred in each year after the programming period are considered in the analysis indirectly by subtracting them from the respective annual benefits.

15.3.4   Multifactors in Optimization

The situation of multifactors or components can occur in an allocation scenario using optimization. For example, this could be what is sometimes referred to as a cross-optimization, where, for example, bridges compete with pavements, or water treatment and distribution compete with wastewater treatment. Multi-attribute utility (MAU) theory has been applied by [Gharaibeh 06] to transportation assets competing for limited funds, using a sample state highway network in Illinois. This procedure consists of developing single-attribute utility (SAU) functions for individual assets and then combining these SAUs into an overall MAU function. It is a relatively complex practice, which relies on decision criteria defined through experienced input, but the procedure is capable of developing objectively based funding allocations.

15.3.5   Genetic Algorithms Are Neural Networks in Optimization

Genetic algorithms are often used to solve optimization problems [Bauer 94]. They use a search heuristic on candidate solutions to create a population of strings. For each population, the "fitness" of an "individual" is evaluated, and then multiple individuals from the current population are stochastically selected to form a new population, and so on until a satisfactory fitness level is reached as the optimal solution.

Neural networks represent another widely used tool in the area of optimization or near optimization [Reed 99]. They consist of a nonlinear mathematical model of which the principles are based on how the nervous system works. Simple units in the model are linked together to generate comprehensive systems.

15.4   Budgeting and Financial Planning Issues

Several possible approaches exist for setting infrastructure expenditure budgets. They include the following:

1. Extrapolation from previous years and adjusting for additions or growth to the system (i.e., the water-distribution network has grown by 5 percent and is adjusted for inflation)

2. Demand based on condition-responsive budgeting (see Sec. 13.4), which programs work to respond to expected performance indicators or conditions

3. Budgeting based on a desired level of performance or level of service (which then becomes a financial-planning tool)

4. Budgeting based on achieving an optimum level of service for the network or system of assets as a whole

5. Budgeting based on affordability, as determined by agency planners or administrators, or by elected bodies

A key issue for all of the above approaches is whether the budgets are split or separated into different categories for expenditures of: (1) maintenance and rehabilitation; (2) new construction; and (3) reconstruction. Traditionally many larger agencies, and in particular state transportation departments, have operated by creating separate expenditure categories for maintenance and capital spending. This often seems less a matter of rational policy than of organizational structure (i.e., maintenance division and planning division might each be responsible for its own programs). In the case of the privatized road network described in Sec. 13.3, this is not an issue, since a single entity (the contractor) is responsible for all the work and is able to optimize the budget.

The first approach, setting a budget by extrapolation and minor adjustments, has been common in road, water, sewer, parks, buildings, and various other infrastructure components maintenance. It is relatively simple, and uses the experience from previous years. But it also represents business as usual, may well involve a significantly less than optimal use of funds, and may tend to discourage innovation.

The second approach, demand-based or condition-responsive budgeting, particularly in the case of maintenance, is quite rational because it should involve only necessary expenditures. However, it can also result in actual expenditures that vary greatly from the budget. For example, severe, unexpected circumstances, such as an extreme winter or flooding, could result in a high number of breaks or interruptions of service requiring extra maintenance.

Budgeting based on a desired level of performance, the third approach, can also be quite rational, and can result in a high degree of system stability. For example, consider a city's network of recreational paths, which are assessed in terms of a composite-condition index

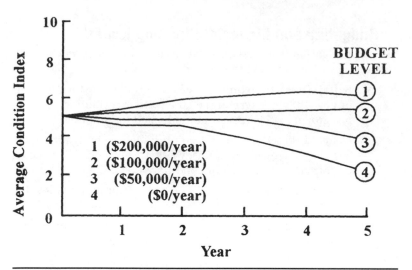

FIGURE 15-2 Variation of the average condition index of a city's recreational path network with different annual budget levels.

on a scale of 0 to 10. Figure 15-2 shows the variation of the average condition index of the network for four budget levels ranging from $0 per year to $200,000 per year. The current average condition index is approximately 5, and would stay at that level for budget level 2 ($100,000 per year). If that average condition index represents the minimum desired level of performance, then at least $100,000 per year needs to be spent on M,R&R.

The fourth approach, budgets set to optimize some factor, such as benefits, is perhaps the most rational, but may run up against the issue of affordability. In this approach, expenditures would be at a level where the marginal benefits achieved approach zero; in other words, increasing the budget would not bring any increased benefits and would represent "gold-plating." A good example would be a road network, where VOC savings have increased thanks to increased expenditures for smoother pavements or greater capacity. At some point, however, the VOCs level out, and additional expenditures would represent gold-plating and would not be warranted.

A budgeting approach based on affordability seems to have become quite popular in the case of roads, hospitals, and various other public assets. However, such an approach can result in serious problems and a decline in asset value if the affordability is set at too low a level. Such an approach only works at all in the case of reasonably affluent agencies.

15.5 Budget Allocation Issues

Budget allocation can become a particularly difficult issue when different functional classes and/or different degrees of traffic/ usage are involved in the same analysis. For example, on a purely economic basis (i.e., to maximize benefits) a state transport authority would expend its budget on freeways in the major urban centers, and allocate little, if anything, to low-volume rural roads because of the large traffic differential. As another example, hospitals in small towns or remote locations with relatively few patients would receive minimal funding based on total service alone. However, in most cases this would be quite politically and socially unacceptable, since everyone deserves some level of hospital availability.

Consequently, policy or political decisions usually specify that at least a minimum level of service and accessibility be maintained for such low-volume facilities, and some budget is allocated to meet these minimum targets. This illustrates that policy decisions are interwoven with levels of economic optimization.

15.6   Financing Models

The counterpart of budgeting is of course financial. Obviously any budgeting approach has to be based on having the necessary financing or revenues available. In the public sector, this usually includes one or more of the following, depending on whether the agency is federal, provincial, or state or local:

- User charges for the service involved (e.g., monthly use of water, toll charges on an expressway, etc.)
- Property taxes (a major source of revenues for local agencies)
- General taxation (e.g., on income, consumer goods, fuel, etc.)
- Transfers from senior governments (e.g., from a state or province to a local agency)
- Other, such as sales of public property, fines for traffic violations, etc.
- Transfer of the asset to a PPP, which can range from finance to design and build to operations

The latter is becoming a choice for many areas of infrastructure in many areas around the world. Current initiatives and status in this area can be found in various publications, including those periodically generated by the World Bank, in proceedings (PPPs) of the First International Conference on Public-Private-Partnerships in 2013 in Dalian, China (ppp-china.dlut.edu.cn) and in other publications [Haas 13, Queiroz 13].

The financing of infrastructure, particularly "green" urban infrastructure is addressed in a comprehensive OECD publication [Merk 12]. It suggests the following instruments as applicable:

- Public-Private-Partnerships
- Tax-investment financing (attracting private finance through future tax revenues)

- Development changes from developers
- Loans, bonds, and carbon finance/charges

Many international examples exist of financing models for PPPs, including Canada's E-407 Express Toll Route bypassing Toronto, the 2012 Infrastructure Bank concepts of New York City and Chicago, announcements of loans from international banks for infrastructure construction in African countries, and others. While numerous references can be cited, readers can find web-based information in such sources as Google and Wikipedia.

15.7   Work Programs

The final result of a prioritization, or optimization, process, regardless of type, should be a recommended work program. This is usually in terms of a listing of sections, components, or units; their location; the type of work or action to be undertaken on each (i.e., pipe relining and road-surface treatment); and the estimated cost. Usually, this is done on a yearly basis. The next operational step would be to establish a work schedule, allocation of resources (people, materials, and equipment), and quality control/assurance planning in accordance with the work program.

It should be noted that any work program is only as good as the data, models, and assumptions used in developing it. Frequently, data is collected well in advance of the program development, and conditions may have changed between the time the data was collected and the program is formulated. In addition, the priority, or optimization, model may not capture all of the factors that need to be considered for each individual section in a network. In large organizations, the programming is often performed by the central office. Hence, there may be a need for adjustments by the local engineer to ensure the applicability of the program. The infrastructure asset management system and the prioritization activity provide decision support; it does not make the decisions. These are always reserved for the decision-maker or governing body.

15.8   Institutional Issues

The success of any asset management system is highly dependent on addressing and resolving institutional issues. While these vary with agencies, location, types of assets involved, political jurisdiction, and many other factors, following is a list common to most public agencies:

- Organizational structure: centralized versus local decisions; simple versus large/comprehensive organization; in-house application of performance indicators; capability of adapting to change; impact of funding on the structure; retention of institutional knowledge

- Organizational decision making: adequacy of information for policy, strategic, network, and project levels; incorporation of risk exposure; transparency of the decision process; incorporated of long-term sustainability requirements

- Location of component systems within the overall AMS: separate or combined offices; communication channels; degree of integration/"silo" effects; retention of distinctiveness of component systems

- Employment of technology: maintaining in-house expertise; investments required or appropriate for new or improved technologies; assessing new technologies

- Skills: existing experience and training base; periodic upgrades; skill sets needed by the leaders of tomorrow; in-house skills/knowledge requirements vis-a-vis outsourced/purchased skills; replacing losses through retirements and resignations/succession planning

- PPPs in appropriate asset areas; political and social acceptability; benefits and costs for stakeholders; ensuring appropriate business rules and practices

- Valuation of assets; method(s) most appropriate to the infrastructure assets (e.g., market value, replacement value, written-down replacement cost, etc.); tracking change of asset value over time; communication with stakeholders

- Communication of asset condition for stakeholders; appropriate measures; frequency; method (e.g., web site, period publications, etc.); understandability; responding to queries

15.9 References

[Antelman 08] Albert Antelman, James J. Demprey, and Bill Brodt, "Mission Dependency Index—A Metric for Determining Infrastructure Criticality," *ASCE Monograph on Infrastructure Reporting and Asset Management*, American Society of Civil Engineers, 2008.

[Bauer 94] R. J. Bauer, *Genetic Algorithm and Investment Strategies*," John Wiley & Sons, New York, 1994.

[Gharaibeh 06] N. G. Gharaibeh, Y. C. Chia, and P. L. Gurian, "Decision Methodology for Allocating Funds Across Transportation Infrastructure Assets," *ASCE Journal of Infrastructure Systems*, American Society of Civil Engineers, Vol. 12, No. 1, 2006, pp. 1–9.

[Haas 13] R. Haas, and L. C. Falls, "Realistic Long Term Warranty Sustainability Requirements for Network Level PPPs," *Proceedings 1st International Conference on Public-Private-Partnerships*, Dalian, China, August, 2013.

[Haas 94] R. Haas, W. R. Hudson, and J. Zaniewski, *Modern Pavement Management*, Krieger Publishing Company, Malabar, Fla., 1994.

[Merk 12] O. Merk, S. Saussier, C. Starapoli, E. Slack, and J.-H. Kim, "Financing Green Urban Infrastructure," *OECD Working Papers*, Organization for Economic Cooperation and Development, Paris, 2012.

[Queiroz 13] Cesar Queiroz and A. L. Martinez, "Legal Frameworks for Successful Public-Private Partnerships," In The Routledge Companion to Public-Private Partnerships, edited by Piet de Vries and Etienne B. Yehoue, Routledge, New York, pp. 75–94. http://www.routledge.com/books/details/9780415781992/ accessed April 18, 2013.

[Reed 99] R. D. Reed, and R. J. Marks, *Neural Smithing: Supervised Learning in Feedforward Artificial Neural Networks*, MIT Press, Massachusetts Institute of Technology, Cambridge, Ma., 1999.

[Steinthal 84] B. G. Steinthal, "Infrastructure Maintenance Strategies—Governmental Choices and Decision Methodologies," *Infrastructure—Maintenance and Repair of Public Works, Annals of the New York Academy of Sciences*, Vol. 431, December 1984, pp. 139–154.

[Urban 84] H. P. Hatry and B. G. Steinthal, *Guides to Managing Urban Capital Series*, Vol. 4: *Guide to Selecting Maintenance Strategies for Capital Facilities*, The Urban Institute, Washington, DC, 1984.

PART 5

IAMS Development and Implementation, Examples

CHAPTER 16

Concept of Integrated Infrastructure Asset Management Systems

16.1 Background

Since many good management systems have been developed for a particular type of facility, such as pavements and bridges, it is reasonable to assume that an integrated management system can be developed to cover two or more related facilities. The purpose of this chapter is to prove this assumption and to show how it has been fulfilled.

The objective of implementing an integrated IAMS in an agency, public or private, is to provide decision-makers with processed quantitative data they can use to examine the impact of alternative funding scenarios. IAMS is an organized approach to helping them manage a set of infrastructure more effectively and efficiently.

Various individual site management systems are already in place in many agencies, but they are not always coordinated in a comprehensive way. Coordinated management subsystems can establish formal procedures for recommending candidate projects and evaluating different strategies for correcting deficiencies and assessing trends of future needs. Specifically, a total budget can properly be divided among various asset classes such as pavements, bridges, and safety. To tie these factors together, an integrated infrastructure asset management system (AMS) must incorporate forecasting models to predict trends of asset condition, assess needs, and analyze future funding or budget scenarios. An AMS can help decision-makers in the development of both short- and long-term programs.

For several decades, pavement management systems (PMSs) have provided important and valuable guidance for AMS development in three general areas. These include:

1. State transportation systems (as prescribed in U.S. federal guidelines)

   • Pavements
   • Bridges
   • Safety
   • Congestion
   • Public transportation
   • Intermodal facilities

2. City infrastructure; same as for state transportation systems, plus

   • Water
   • Sewer
   • Traffic signals, signs, and markings
   • Emergency services
   • Electricity
   • Garbage collection
   • Recycling
   • Drainage
   • Park facilities

3. Major unitized facilities—public and private

   • Airports
   • Nuclear power plants
   • Refineries
   • Parks and recreation areas
   • Other

After describing the framework for appropriate integrated systems, these three general application areas are discussed in the following sections.

16.2 Framework of Integrated Systems

As individual decision-support systems are developed and mature with use, this is important to consider ways in which they can and should be coordinated. There is also a need to determine how the results should be integrated to produce information for strategic planning as a part of the executive management process using an executive information system (EIS). In general, individual systems (PMS, BMS, ITS, traffic monitoring, and so forth) have been developed with only informal coordination among the various system developers. Correcting such a major deficiency by integrating systems will provide policymakers with a tool for investing transportation and related infrastructure funds most economically and efficiently. Transportation policy, for example, is thus no longer developed in a vacuum with AMS. Other state and national priorities, such as economic development, are directly influenced by transportation decisions and vice versa.

Multiple-criteria decisions are best analyzed interactively, allowing the judgment of the decision-makers to be used directly. Such interaction permits decisions based on an assessment of relative allowable trade-offs among feasible criteria. Multivariate analysis techniques have been developed to assist in this type of analysis for industry [Hoskins 88, Hsiung 91]. Sufficient structure and details need to be incorporated into the AMS so the system represents the physical environment reasonably well, but detail can be limited to reduce the burden placed on the decision-makers.

A general framework for infrastructure management was presented in Chapter 1 (see Figure 1-7). An IAMS using minimal assumptions will permit decision making via any desired technique using the response information from the total management system, or on any portions thereof. Screening or ranking of acceptably good alternatives can be based on a combination of qualitative and quantitative responses, depending on the type of information available, the degree of sophistication of the decision-makers, and the degree of detail considered essential in the management system.

16.3   Common Aspects of Management Systems

Good management systems have many things in common, including analysis modules, a central database that is easily updated, compatible economic-analysis models, optimization models, and a graphical interface with a GIS (geographical information system). Also, any infrastructure or facilities management system must allocate resources between capital improvements and maintenance, as discussed in Chapter 15. These common aspects of management systems are discussed in the following sections.

16.3.1   Central Database

A good database and data-processing system is the heart of any management system. A database is not a management system itself but is required for one. A management system can only operate on the data made available to it. Systems that are set up independently with separate databases often duplicate many data items, but, more importantly, they are likely to have inconsistent location and identification information. An integrated management system operating on a coordinated central database, which contains all data needed for any of the management systems or subsystems, eliminates duplication and inconsistencies. The database must be flexible so that when the modular analysis components are upgraded and/or new components are added, it can also be updated or modified to accommodate the data needed for the new subsystems. The database should also be designed to accommodate expected additions of new infrastructure types. See Chapters 4 and 5 for details on database design and development.

16.3.2   Modular Analysis Tools

All management systems need analysis tools that can manipulate the pertinent data from the central database to produce useful results and recommendations for management. Such analysis tools include the generation of summary statistics and graphs, economic analyses, benefit-cost analyses, decision methodologies, optimization routines, statistical analyses, and deterioration-rate analyses. An integrated system should have the flexibility to use modular analysis tools that can be easily updated as new technologies are developed. These tools should be consistent across the various infrastructure components that may be managed under the integrated AMS. Such consistency would include common or compatible optimization methodologies and life-cycle cost analyses.

Modularity of a system requires that the analysis tools be independent of the database. Unit cost information, deterioration rates, infrastructure types, functional classes, and so forth are all kept updated in the central database. The analysis tools are defined relative to the data needs, the models or analyses are performed, and the outputs are returned to the management system. As new modules are developed, each of these three main aspects must be included considering the rest of the system. The inputs are extracted from the central database, the analyses are performed in modular subroutines, and the control, interaction, and output review is performed at the user-interface level. Consistency in data needs and interface and output types is important to maintaining a seamless, integrated system. In joint analyses, such as corridor analyses involving multiple assets, say pavements, bridges, signs, and safety, common data from all systems must be used appropriately.

16.3.3   Geographically Based Location/Identification

Geographical information systems (GISs) have come of age and are seeing widespread adoption around the world. Today any integrated management system should be developed with a GIS. A GIS provides a coherent, common geographical database location-referencing system, spatial visualization of data, and a platform for relational databases that minimizes errors and confusion in collecting, processing, and storing data.

16.3.4 User-Friendly Graphical Interface

The outputs from any management system are critical to its acceptance. Detailed printouts or tabular data are not easily used by managers to make decisions. Current systems have capabilities that include on-screen push-button controls, graphical views of the work environment, visual outputs and reports, on-screen photographs and videos, sound, and other utility-improvement features. Such graphical interfaces not only enhance the use of the AMS output, they also make the system more powerful and enjoyable to use.

16.3.5 Compatible Benefit-Cost Economic Analyses

Any infrastructure management system requires a valid life-cycle cost and/or benefit analysis. All such analyses should be consistent across all infrastructure subsystems within an agency, such as state, city, private entity, or airport. Using an AMS will help ensure economic consistency, including unit cost, definition of benefits and effectiveness criteria, discount rates, analysis period, definition of user and agency costs, level of service criteria, and goals.

16.3.6 Global Consideration and Allocation of Resources

If an agency has separate individual management systems, then they must allocate resources within the agency separately for individual subsystems. Thus separate budgets are set up for water, sewer, pavements, traffic control, bridges, and so forth. This will likely result in suboptimization of the total budget, even though expenditures within a specific facility category may be optimal. An AMS should provide global consideration and allocation of resources. Certain constraints exist, however, for funding "set-asides," such as bridge, safety, and other improvement programs. These can be handled as constraints within the individual subsystems.

16.3.7 Evaluation of Maintenance versus Capital Improvements

All facilities and infrastructure eventually deteriorate and ultimately become functionally or structurally unserviceable. Structural deterioration can produce major safety and operational problems. Facilities that are in poor physical condition present aesthetic, operational, and safety concerns. Management systems must be able to recommend budget division into timely maintenance or needed capital improvements. This requires accurate modeling and analysis of the costs and effects of maintenance activities that extend life or preserve facilities so they can be compared against capital-improvement replacement costs.

16.4 Common Aspects of PMS and BMS

The team that originally developed much of the pavement management philosophy, particularly the original, innovative thinking, was also selected by the U.S. National Cooperative Highway Research Program (NCHRP) to undertake the first definition of a BMS [Hudson 81, 87, NCHRP 73]. The same concepts and application of system technology apply, even though the specific details vary. Both systems involve a change in the way an agency does business. Each requires a clear understanding of inputs, performance models, and outputs of the system. Both systems require the use of accurate models to estimate the behavior of the facility and models to evaluate cost. Each requires a system output function that can be optimized in relation to costs and benefits.

In essence, the development of BMS made extensive and efficient use of the earlier PMS development and experience. This lesson should be applicable to other individual management systems and to integrated infrastructure management systems.

16.5 State-Level Management Systems

State-level infrastructure asset management systems are generally fairly well developed in 40 percent of U.S. states. As described previously, PMS development and implementation experience has provided valuable guidance in terms of both concepts and application. Other states have been slower in implementation but are moving forward slowly. An integrated system framework should provide an umbrella for thorough and consistent development. Each of the six management systems prescribed by U.S. federal guidelines, as listed in Sec. 16.1, can and should be a subsystem in an overall AMS framework. They should be able to operate from a common database that contains all data for each individual management system.

Sinha and Fwa [Sinha 87] outlined a concept of a total highway management system in which they envision a comprehensive highway management system as a three-dimensional matrix, as depicted in Figure 16-1.

The three dimensions are highway facility, operational function, and system objective. Table 16-1 lists the possible elements of each of the three dimensions. The suggested three-dimensional matrix structure indicates that a highway agency has a number of facilities to be managed in the highway system. The overall objectives of the agency could be well defined, and the effort for accomplishing these objectives divides the management task into a group of functions. Each facility in the system requires all of the management functions, and through these management functions the overall system objectives are accomplished. The Figure 16-1, framework views the highway management process as multiple-objective.

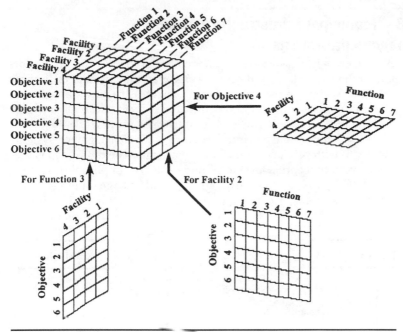

FIGURE 16-1 Three-dimensional matrix structure of a highway management system [Sinha 87].

An initial integrated IAMS for states in the United States, could be comprised of at least the following individual management systems (or subsystems in the AMS): (1) pavement; (2) bridge; (3) safety; (4) congestion; (5) public transportation; and (6) intermodal—plus others that are now being added. The structure should be flexible enough to accommodate changes in the form of the subsystems and in the model inputs and outputs, as well as dividing a subsystem into smaller elements for ease of simulation.

Elements of the Highway Facility Dimension	Elements of the Operational Function Dimension	Elements of the System Objectives Dimension
Pavement	Planning	Services
Bridge	Design	Condition
Roadside	Construction	Safety
Traffic-control devices	Condition evaluation	Cost
—	Maintenance	Socioecononomic factors
—	Improvements	Energy
—	—	Data management

TABLE 16-1 Elements of Highway System Dimensions [after Sinha 87]

16.6 Municipal Infrastructure Asset Management Systems

An integrated infrastructure management system is equally important at the city or local authority level. Given the diversity of services provided within a typical municipality, an infrastructure management system can assist administrators and engineers to effectively manage and maintain services. This section illustrates frameworks that have been used for municipal infrastructure asset management systems (MIAMS) with primary municipal subsystems. The discussion is presented on a conceptual level to provide a widely applicable outline.

One way to visualize the integration and linkages among the different subsystems, according to the Texas Department of Transportation (TxDOT), is shown in Figure 16-2. A representation of model

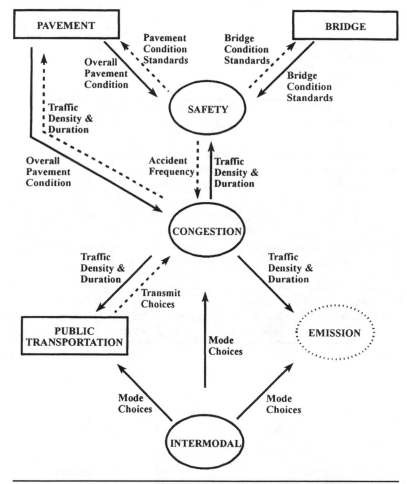

FIGURE 16-2   Major model integration and links (according to TxDOT).

interactions for the various subsystems is shown in Table 16-1. There is a strong link between public transportation and bridge and pavement management. A good illustration is provided by larger cities, where many arterial streets are being damaged by heavily loaded buses. In the case of metropolitan planning organization (MPO) interactions, there will be particularly strong links among bridges, pavement, congestion, safety, and public transportation.

16.6.1 Systems Definition for MIAMS

The definition of municipal infrastructure, in terms of public works, given by the American Public Works Association (APWA) [Stone 1974] is still valid.

> Public works are the physical structures and facilities that are developed or acquired by public agencies to house governmental functions and provide water, power, waste disposal, transportation, and similar services to facilitate the achievement of common social and economic objectives.

The APWA goes on to list 18 categories of public works and environmental facilities that are included in this definition. Six categories of municipal infrastructure that are clustered together by industry and professional interest group are listed as follows:

- Roads group (roads, streets, and bridges)
- Transportation services group (transit, rail, ports, and airports)
- Water group (water, waste water, all water systems including water ways)
- Waste-management group (solid-waste management systems)
- Buildings and outdoor sports group
- Energy production and distribution group (electric and gas)

The potential comprehensive MIAMS system and the identified subsystems linkages are illustrated in Figure 16-3.

Almost all cities deal with this set of subsystems, most or all of which normally are owned or controlled by the municipality. They are commonly large enough to require network-level oversight. It follows that the five most important systems, as shown in Figure 16-3, should be selected for initial development. With a comprehensive MIAMS, all subsystems would be developed and incorporated into the system; however, it is necessary to start with two or three and add parts incrementally.

Once the system elements and boundaries have been defined, the next step is definition of the goals and objectives. Goals can be defined as the idealized "end states" toward which a plan is expected to move. Objectives are operational statements of goals. They must be measurable and attainable, where attainability is defined without

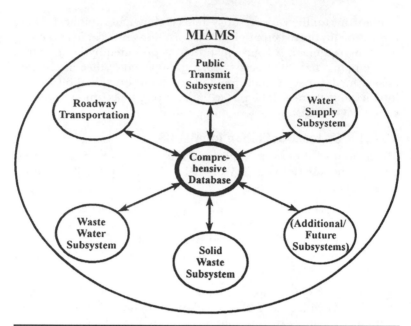

FIGURE 16-3 MIAMS system and subsystems identified by APWA.

reference to availability of resources or to budgetary restrictions. By definition, goals are broad and general in nature, while objectives are precise statements of what is to be done to reach a goal. One perspective on the goals and objectives of a MIAMS can be obtained through the definition of the goals and objectives of each of the subsystems identified, the overall goal being the achievement of "common social and economic objectives."

The next step in the systems-development process is to create subsystem models. This involves formulating, calibrating, and validating basic models for each system or subsystem. They should be designed to predict of the consequences of alternative plans, without the necessity of direct experimentation on the real-world system. At this point, the basic requirements for predicting the consequences of various alternative plans should have been developed. However, before generating alternatives, a set of criteria is required to assess goal achievement of the alternative plans.

To facilitate understanding of the requirements, models, processes, and so forth involved in each subsystem, some kind of basic systems analysis procedure, or analysis flow chart, should be developed for each of the subsystems shown in Figure 16-3.

Efficiently collecting and disseminating useful and timely information is critically important to managing any enterprise. The first important issue for developing an overall infrastructure management system is the creation of a centralized database that allows a flow of information to and from each subsystem, and also from each activity as needed. This results in a more comprehensive, efficient, and timely flow of information (data) for the MIAMS as a whole, and for each subsystem in particular.

Although the specific information requirements of each subsystem will vary, certain broad information concepts apply to all subsystems. Each of the subsystems requires section identification information, preferably a common system that allows cross-referencing among subsystems. Each also requires historical data on system performance and system demand. Each requires information on subsystem-specific particulars, such as design and construction parameters, geometric data, and maintenance records. Finally, each subsystem requires tracking of economic effectiveness and both agency and user costs of service.

Ease of integration into one common municipal database is a major consideration in the design of these data-requirements. For future development, it is suggested that data should also be classified as "required" or "desired."

16.6.2   Geographical Information Systems in MIAMS

In recent years, GISs have been widely used in various management fields, such as natural-resource management, wildlife protection, and environment protection. GISs have also been applied to several areas of infrastructure, including transportation systems and components, such as pavements and bridges. GIS applications for the Urban Roadway Management System (GIS-URMS) have been investigated at the University of Texas [Zhang 93, 96]. This work demonstrates the strong potential of GISs as a platform for developing an integrated overall infrastructure management system.

A GIS is a computerized database management system designed to manage and visualize spatially defined data, such as shown in Figure 4-5. The database contains information on spatially distributed entities that occur as points, lines, or polygons; and tools for capturing, storing, retrieving, displaying, interrelating, and analyzing locational as well as nonlocational data. Details are provided in Chapter 4.

There are three elements in a GIS: the geographic database, the attribute database, and the georelational data structure.

The *geographic database* processes three inherent characteristics: position, description, and variation with time. It can be structured in either grid (raster), vector, or triangulated irregular network (TIN) format.

The *attribute database* deals with all descriptive nongeographic information (variables, names, and characteristics) that identify a geographic feature. For instance, a lake can be identified by such attributes as its name, water level, water quality, chemical composition, salinity, and water temperature.

The *georelational data structure* deals with one of the main function of a GIS, which is to establish the relationship between the location of features in the geographic database and their corresponding descriptions in the attribute database. GIS performs the linkage between locational and attribute data by means of a georelational data structure.

16.6.3   MIAMS Central Database

One of the important issues in developing an overall MIAMS is data exchange and communication both among the subsystems and with other external systems. There are three well-known approaches to data management:

- Relational approach
- Hierarchical approach
- Network approach

The relational database management systems (DBMs) available for microcomputers at the time of writing this book included dBase, FoxPro, Oracle, and Rbase. A relational DBMS can effectively perform data exchanges and communications among the proposed subsystems via common data fields, such as location ID, segment ID, or other items in the database. Figure 4-3 illustrates a possible integrated database for MIAMS.

16.7   Unitized Facilities Management Systems

The third important area of application of overall asset management is for what can be called "unitized facilities." These include, for example, airports, nuclear power plants, and major refineries. Each fulfills a major function, but requires a number of individual, independent subsystems to be operational.

For example, an airport requires roads, parking facilities, terminal facilities, aircraft parking aprons, taxiways, runways, and aircraft-control facilities. Each type of other such facility has a number of subsystems that must be combined to produce the required results. In this regard, the main function of an IAMS is to integrate the various subsystems and their relationships to each other. The capacity of an airport, for example, can be controlled by any one of a number of subsystems. If access roads are inadequate and/or if parking is inadequate, then the overall airport capacity is reduced. The same can be said for terminal buildings, check-in facilities, baggage-handling facilities, number of aircraft gates, aircraft parking facilities, take-off and landing facilities, and air-traffic-control equipment. Poor management in any of these subsystems can limit airport capacity and functionality.

The concepts of asset management can be applied to integrate airport subsystems effectively. This includes consideration of passengers, so that common database items include scheduling of take-offs and landings, related arrival of passengers, traffic delay, weather delays, weather information, passenger demand, and other factors.

16.7.1 Differences and Common Factors with Other Infrastructure Facilities

Unitized facilities are in some ways easier—and in some ways harder—to handle than other management-system applications. Ease in handling arises from the fact that most facilities are located at a common location, such as an airport site. This is compounded when an agency, such as the Port Authority of New York and New Jersey, must operate three airports within one metropolitan area. Nevertheless, the process of handling cohesive facilities in close proximity to each other is a good application of asset management.

Difficult aspects of the problem, which vary from, say, pavement or bridge management, include the broad diversity of functions that must be dealt with. In the case of airports, this ranges all the way from ground access to the airport through baggage handling to the complexity of air-traffic control. The scope of this chapter does not allow for detailed examination of these common and diverse factors. However, the benefits of unitized asset management systems (UAMS) can be fully realized with a reasonable amount of research and development support for these areas. Agencies should, however, look carefully at commercially available systems from companies with broad experience in developing and implementing Commercial Off-The-Shelf (COTS).

Items related to this concept include the potential use of GIS within the facility itself to address the flow of passengers and the proximity (location) of facilities. A relational database tying all the factors together is also critically important to a unitized asset management system.

A major factor in UAMS is the need for models that describe the operation of each of the individual subsystems. In fact, such models are already required for operating the facilities without a UAMS, although they are often not explicitly available. In other words, attempts to create a UAMS do not increase the difficulty of developing individual models. It does make explicit the need for such models, and it requires the management team to carefully consider the input and output of the individual models and their interactions. However, as previously discovered in bridge management and pavement management, while this explicit requirement may tend to make the current operating team uncomfortable, and seemingly apply a certain amount of pressure, it is very beneficial because it requires an explicit consideration of major factors by managers. However, this is often a road block to change in the agency.

In summary, UAMSs are currently the least developed areas of asset management systems. Thus, they now offer the most potential. Private companies, such as Walmart, have developed, to some degree, a UAMS for handling their inventory of goods and services.

They also have developed a common store configuration and common training methods so that uniformity and consistency exists throughout their company stores over large geographic areas. It is recommended that consideration be given to UAMS development, particularly for public facilities like airports and parks, and facilities that require strong regulations, such as nuclear power plants. A UAMS can, for example, evaluate the overall aspects of the disposal of waste better than studying the parts of the process separately. The steps involved include the following: (1) apply principles for general management systems as presented herein; (2) define all aspects of the facility; (3) properly model the system being managed; (4) develop an integrated database for the system; (5) evaluate the problems that occur; (6) build in monitoring procedures to define and evaluate problem areas; and (7) devise coordinated methods to correct problems in the most cost-effective way in terms of action and timing.

16.8   Implementation and Institutional Issues

Considerable effort has been invested in the development and implementation of pavement and bridge management systems. Nevertheless, as pointed out by many authors [Smith 91, Haas 94], the greatest problems associated with the successful implementation of PMSs are institutional issues. There are agencies that started development of a PMS in the mid-1970s that are still ineffective in its use. It is necessary to address the problem of institutional issues and the reluctance of agency personnel to change, to help ensure that the integrated systems developed in the future actually get implemented.

Great progress has been made in this area for state DOTs, but less progress has been made in airports and other areas. An excellent example of a long-running IAMS that has overcome complex institutional issues is StreetSaver® developed for the Metropolitan Transportation Commission (MTC) in 1984. Special thanks are due to Roger Smith, Texas A&M University, Sui Tan, MTC, and other authors cited in references below for providing this detailed summary.

16.8.1   Success with PMS, San Francisco Bay Area, 1984 to Date

The MTC is the transportation planning, coordinating, and financing agency for the nine-county San Francisco (SF) Bay Area on the U.S. west coast that also includes 100 cities. It functions as both the state-level regional transportation planning agency and the federal-level MPO for about 7.3 million residents, about 42,600 lane miles of local roads and streets, 23 transit agencies, and a network of 7 toll bridges. It is headquartered in Oakland, California, [MTC 12a] and works collaboratively with numerous local agency transportation partners in the Bay Area and Caltrans, which is responsible for the construction, operation, and maintenance of 6,500 lane miles of state highways in the Bay Area. As the regional MPO, MTC is a steward of federal funds and is accountable for how regional funds, including those made available for local street and road maintenance, are spent [Romell 11].

In 1981, at the urging of several SF Bay Area public works directors, MTC conducted a study that identified a shortfall of funding for local roads and street maintenance and rehabilitation (M&R) of about $100 million a year for the Bay Area's 17,000 mi of streets and roads. It estimated that the amount of deferred maintenance fund needs was in the range of $300 million to $500 million at that time. After it was determined that the local agencies needed a decision-support system to help manage their road and street M&R needs, a study was conducted that resulted in development of one of the first pavement management software programs developed to assist local agencies with network-level M&R questions. The MTC pavement management program was started with six Bay Area community users as a pilot program in 1984. After modifications and adjustments further tailoring it to the specific needs of cities and counties, the full program got under way in 1986 [MTC 12b]. The pavement management program and software was later named StreetSaver® and is currently used by all Bay Area agencies and over 250 other agencies across the United States, most located in California, Oregon, and Washington. A user group continues to guide improvements and additions to the software and other components.

The success of MTC's PMS program, now the longest-running municipal PMS in the world, is in part because of the support MTC provides the users including training, on-call assistance, assistance in addressing budgeting issues, assistance with keeping the system current, and continuous feedback [MTC 12b]. The robustness, relative simplicity of the user interface, continuous improvement, and support for the software also make it easier for users to effectively use the software in their pavement management activities. MTC instituted the Pavement Technical Assistance Program (PTAP) about 15 years ago to help local agencies in the SF Bay Area implement and maintain their pavement management programs. While implementation was important early in the process, now most of the more recent effort has directed at keeping their data current and helping agencies prepare reports and presentations that public works personnel can use to demonstrate road and street M&R needs to city councils, county boards, the MPO, and the state. The actual assistance is provided by a group of consulting firms who have pavement management experience and have developed expertise in pavement inspection and use of the StreetSaver® software. The program is funded with federal grant funds and supports assistance to about

50 Bay Area agencies a year [MTC 12b]. It includes a comprehensive data quality management plan [Smith 11] and has helped develop a cadre of several consulting firms who provide pavement management assistance and training to many of the local agencies using StreetSaver® outside the Bay Area.

MTC conducted a number of needs studies using the data from StreetSaver® and analysis tools within the software that were presented in publications such as the "Bay Area Transportation State of the System" in 2006 [MTC 06]. A local Streets and Roads Working Group was formed of public works personnel in the Bay Area that acts as an advocate for better funding of roads and streets M&R needs. They advise MTC on roads and streets funding needs, and with their help and the StreetSaver® pavement management software, MTC was able to develop a regional funding policy for local roads and street funding that allocates funding not just on need, but on performance. The policy rewards jurisdictions that focus their efforts on pavement preservation rather than "worst first" funding. Despite the dour outlook, there has been a sixfold increase in regional investment in local roads and streets over the last several years, due in large part to the information generated using data from and analysis using StreetSaver® [Romell 11].

The MTC pavement management program and analysis tool within the software, StreetSaver®, have been used by local agencies, the regional transportation-planning agency and the MPO, federal agencies, and private entities to demonstrate the funding needs for local roads and streets, to analyze various investment and allocation approaches, and support implementation of performance based allocation of funds. Although the use of the software has not always resulted in increased funding, it has resulted in funding increases in individual agencies, increased allocation of federal funds to local agency roads and streets in the Bay Area, better allocation of funds to more preservation treatments compared to "worst first," and a host of other benefits that are difficult to attribute directly to the use of StreetSaver® but which have developed as more agencies adopted and used StreetSaver®.

16.9   Outline of a Comprehensive IIAMS

An ambitious integrated infrastructure asset management system (IIAMS) initiative was undertaken in Kuwait [Abdel Kader 96, Al-Kulaib 97]. The main individual systems in the IIAMS, as developed for the Ministry of Public Works and the Ministry of Water and Electricity, include the following list (see also Figure 16-4). While there has been little published information on the implementation of this system, the list below can serve the reader as a guide of factors to consider in their own systems.

This was a very ambitious project and there is no mention of phasing training or education. This opens the possibility that the

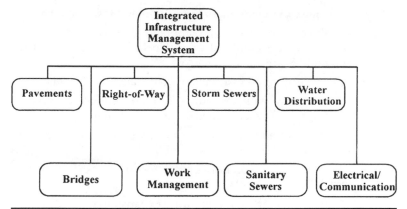

FIGURE 16-4 Scope of Kuwait's IIAMS.

project was stalled by complexity, lack of long-term funding and low acceptance by staff and users due to lack of training.

I. Roadway systems
- Pavement management system (PMS)
- Bridge management system (BMS)
- Right-of-way features management system (RFMS)

II. Utility systems
Sanitary-sewer network management system (SANS)
- Storm-sewer network management system (STMS)
- Water main network management system (WNS)
- Electrical/communication management system (ECMS)

III. System integration
- Spatial data management (SDM)
- Work management system (WMS)

Key features of Kuwait's IIAMS include the following [Abdel Kader 96]:

- Interpretation of the various system modules at both the data and user-interface levels
- Use of fourth-generation language (4GL) database managers that are standard query language (SQL) compliant
- Implementation of the IIAMS in a client/server environment, web based
- Database independence of the system modules, so that client applications can access data in various 4GL databases
- Ability of client to use a graphical user interface (GUI)
- A comprehensive IIAMS database containing data fields for all infrastructure components

- Capability of managing data on an integrated global basis
- Capability of producing a comprehensive set of understandable tabular and graphical reports, including color-coded maps and digitized images
- System security, to restrict user access at several levels
- Compatibility with existing GIS, management information system, and maintenance management system
- Capability of network analysis on any individual component system and on a multiple-system basis
- Ability to operate on both a local area network (LAN) and a wide area network (WAN)
- Open architecture to allow future expansion
- Capability of operating with minimum response times
- Capability of assessing the condition/performance of each system component individually and collectively
- Ability to produce optimal multiyear priority programs of rehabilitation and maintenance for each IIAMS component system and for the IIAMS as a whole
- Ability to process service requests, workload analysis, performance tracking, and work-order historical information through the work-management system
- Ability to automatically adjust maintenance and rehabilitation strategies to varying service levels set by the user
- Ability to automatically identify and assess combined projects (e.g., roads and sanitary)
- Interactive with the budget process
- Ability to manage historical information
- Ability to provide a fully integrated, network-level electronic map of any individual types of infrastructure to be interfaced with any future GIS vendors
- Capability of receiving input data and producing reports in both Arabic and English

It is possible that large companies such as Walmart and major shopping centers have developed integrated management systems. If so, they have not been publicly described or are likely proprietary. The list of vendors, Table 18-1 offers a list of possible sources of such systems.

16.10 Smaller Agency Case Studies

The previous section outlines a comprehensive integrated AMS that was never implemented due to a change of leadership in the agency or perhaps because it was more comprehensive than the agency was

Geographics—Infrastructure—Agency			
Population	1 million	Road miles	1,100
Surface area	607 sq mi	Number of bridges	130
Public works staff	440	Operation budget	>$50 million

TABLE 16-2 Agency's Infrastructure Infographics

prepared for. This is itself a lesson in implementation. Remember, it is important to match the breadth and complexity of a proposed AMS to the agency's ability to absorb and accept it.

To follow up, three implemented case studies are described below for one county and two smaller cities in the United States. They are in smaller scale and provide a more focused opportunity.

16.10.1 Case Study 1: Pinellas County, Florida

Pinellas County is the most densely populated county in Florida and is also a very popular vacation destination. The Pinellas County Public Works Department is responsible for all county transportation infrastructures (Table 16-2). In order to meet the demands of a growing population and tourism industry while facing budget reductions, the Public Works Department has implemented an asset management system.

The following benefits of implementing an AMS were identified in a management audit:

- Able to justify needed resources, identify needed work, and to develop optimized work plans to maximize the resources.

- Create an automated system to inform management of work planned and completed.

- Establish an efficient approach to planning, budgeting, and performing work while institutionalizing best management practices in the agency.

16.10.1.1 Implementation

The county identified maintenance management as the first area to concentrate on. After a rigorous software-selection process, the county selected the AgileAssets, Inc. Maintenance Manager software program. In parallel, the county also worked on developing new business practices and policies. The county selected, purchased, configured, and installed a web-based maintenance management system (MMS), which went live in October 2006. The software configuration took three months. Management systems for traffic, stormwater, drainage, mosquito control, and vegetation were integrated and configured into MMS, as were GIS and vector control, performance and program plans, and other systems. Other activities included data

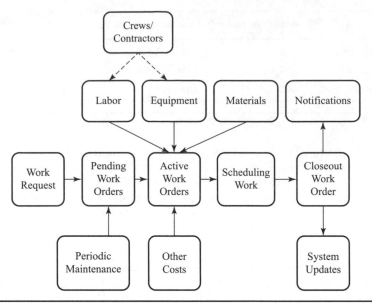

FIGURE 16-5   User-defined work-order flow process.

conversion and integration with other county systems, and training 75 systems users in 14 locations.

Initially, the implementation focused on 23 work activities that were identified as consuming 80 percent of the resources. By 2010, more than 140 work activities were totally managed with the MMS. This solution replaced more than 40 old software systems and databases.

Public Works now fully tracks all maintenance work with resources, cost, location, and accomplishment. This includes tracking 100 percent of all service requests; in addition, out-of-yard hours for each piece of equipment are recorded, and productivity analyses are performed to compare actual accomplishments to planned work. The typical work-order flow process defined in the system is shown in Figure 16-5.

An example of the benefits that are derived from properly managing resources is the county's mowing operations. Productivity increased while needed resources and complaints decreased as shown in Figure 16-6. The results were attributed to more efficient use of equipment and staff, staging staff closer to the work, developing routine schedules, establishing quality and quantity standards, and peer review to add motivation and sharing lessons learned, both positive and negative experiences.

After successfully implementing the MMS, the county expanded into broader asset management. In May 2009, the web-based AgileAssets pavement management system and bridge management system also went live. By expanding with a common system foundation, the county can now integrate MMS, BMS, and PMS to more efficiently plan and coordinate all work activities within the

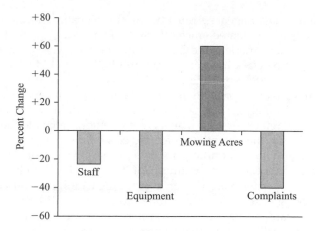

FIGURE 16-6   Mowing improvements attributed to efficient use of resources.

county right-of-ways. In 2010, Pinellas County Public Works listed the following benefits [Bartlett 11]:

- The commitment of senior management, supervisors, and all staff members to participate, discuss, and jointly decide on the best solutions produced in a common goal and improved the team spirit in the organization. The new systems also make it easier to cooperate.

- Efficiency, decision making, organizational development, accountability, planning, reporting, speed of information gathering, and transparency. All improved in the organization.

- The AgileAssets systems eliminated the need to acquire two other computer systems budgeted at >$500,000.

- The mowing department alone saved more than $1.7 million in 2010 by a better match between quantity, quality, inventory, and methods.

- Since 2003 the labor pool was reduced by 81,657 hours and 70 pieces of equipment. Yet more work is done.

- A comparison of the productivity in units of work per hour from 2007 to 2008 showed a saving of 45 percent.

- In 2004 it was anticipated that the annual savings from new systems would be $2 million to $3 million, but the documented budget reduction is >$6 million with another $6 million in process improvements·

16.10.2   Case Study 2: City of Newark, Delaware

Even smaller agencies can benefit from asset management. Typically with very limited resources available to effectively manage infrastructure, asset management is crucial for smaller agencies.

That said, smaller agencies must be careful to ensure that asset management is implemented efficiently and effectively.

One such example is the City of Newark, Delaware [Kercher 11]. The City of Newark has been using an older basic PMS for several years to manage 80 centerline miles of streets. Each year, the city would perform a pavement condition survey, enter the data, and perform a repair needs analysis. The software program would select the most appropriate repair for each management section, calculate the repair cost, and rank the potential repair projects by simple priority using condition and level of traffic as the two primary weighting factors.

Over the years, the condition of the pavement network has continued to deteriorate (lower average PCI rating score) and the total backlog of repair costs continued to grow. In 2010, the new city manager wanted a PMS that could project the future condition of the network as well as optimize the use of limited resources and set and analyze performance goals. The city selected the AgileAssets Pavement Analyst system to provide the desired functionality.

Data in the original software program included inventory attributes, pavement condition history, pavement type, distress type, treatment type and costs, and basic decision-tree framework. This data was imported directly into the new system. In order to complete the configuration of the new system, individual distress indices and a composite PCI were developed along with a family of performance models for four repair categories: preventive maintenance, functional rehabilitation, structural rehabilitation, and reconstruction.

The annual budgeting process starts with the city's anticipated work plan containing current and future planned projects. Then optimized multiconstraint analyses are performed to determine the PCI of the overall network for different funding levels and constraints. After review and discussion, the city determines best funding levels. Once the funding levels are selected, the cost-effective work plan for the next five years is generated. Listing all projects including type, cost, and year of repair needed to meet the constraints of the analysis.

Results of the initial analysis performed by the city's consultant clearly showed the importance of using a software program capable of projecting future conditions but most importantly, the benefits of being able to perform optimization analyses. Performing a 10-year analysis with the then current $1 million per year level of funding, the worst-first analysis indicated that the network PCI would decrease from 72 to 44. When the multiconstraint analysis allowed the software to freely optimize the budget in a "maximize network benefit" scenario, the PCI only decreased to 56. More importantly, the backlog by repair cost analysis clearly illustrated the monetary benefit of optimization. For the worst-first analysis, the backlog of repair costs grew from $14 million to $38 million while the backlog for the optimized analysis resulted in a backlog of $29 million. Whereas the optimized analysis reduced the deterioration by $9 million over a 10-year period (Figure 16-7).

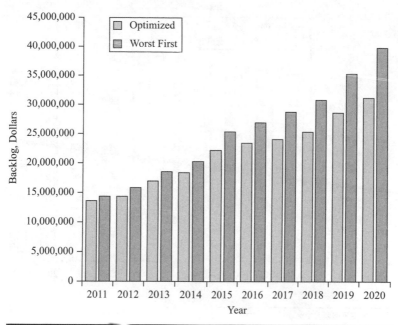

FIGURE 16-7 Funding level to achievo a PCI equal to 77 in Year 10.

In 2010, after presentation of the multiconstraint results, the council decided that it was not acceptable to have the network deteriorate so much over the next 10 years. The city decided that the network PCI should be increase by five points to a PCI of 77. An analysis was run to determine the level of annual funding over the next 10 years to achieve the desired goal (Figure 16-8). The city council approved the funding increase to meet the desired goal of PCI

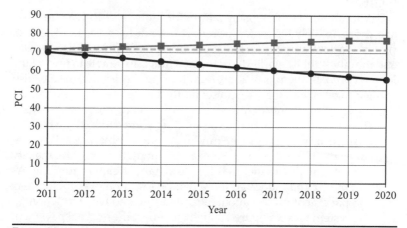

FIGURE 16-8 Funding level to achieve a PCI equal to 77 in year 10.

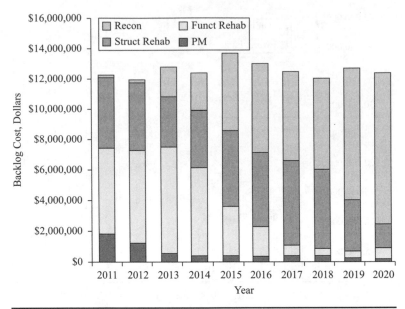

FIGURE 16-9   Backlog of repair costs for the new funding level [Kercher 11].

equal to 77 in 10 years. Also, the analysis showed that in 10 years the overall backlog would be reduced to $12 million, which is less than in 2010 (Figure 16-9). The city also agreed to pursue a pavement preservation program to support the optimized work plan, which includes a significant amount of surface coats and thin pavement overlays.

16.10.3   Case Study 3: City of Wilmington, Delaware

This example of the City of Wilmington, Delaware, shows how an agency is planning to expand from PMS into broader AMS beyond transportation.

Wilmington has been using a PMS for more than 10 years to manage approximately 165 centerline miles of streets [Kercher 12]. The PMS provides a summary report of pavement repairs with associated costs, and the city's public works department wanted to be able to predict the long-term consequences of short-term funding decisions. The city also wanted to be able to analyze and set performance goal and selected the AgileAssets Pavement Analyst software to meet its functional needs.

The city had no prior dedicated funding source for paving repairs. Streets were only repaved when funding was available, typically going several years without paving. Since implementation of PMS, the city has been able to justify the need for a constant stream of funding. As a result, the city has allocated more than $11 million over the past six years to pavement preservation, rehabilitation, and reconstruction.

The system was configured initially to reflect the original pavement management system setup. However, it was decided after the initial implementation that the city wanted several PMS modifications included a project-level patching analysis. Since a significant amount of utility patching was performed in the city each year as a result of repairs to deteriorating sewer and water pipes, some more than 100 years old, a "significant patching" analysis decision framework was created.

The intent of the "significant patching" program was to identify potential pavement segments that could be significantly improved by patching one or more areas that are highly distressed rather than rehabilitating the entire segment. In order to justify funding for the "significant patching" program, software modifications included project-level decision trees, performance models, improvements rules, etc. necessary to provide project-level scenario analysis capabilities in order to capture the benefit derived from a dedicated "significant patching" program.

16.10.3.1   System Expansion

Paving projects in Wilmington involve other improvements to sidewalks including ADA noncompliance issues, drainage and both city-owned and privately-owned utilities. In an effort to improve both short-term and long-term budgeting, communications and project coordination, the city desired to expand beyond pavement management into broader asset management.

The city's consultant has started to expand the system to include sidewalks/ADA improvements, utilities (in-house sewer and water utilities first—then external utilities such as gas and electric later) and drainage. This will help greatly to improve communications and project coordination internally across departments, as well as externally with the state DOT, state transit, the local school district, and private utility companies.

The pedestrian facilities portion of the management system will include an inventory of sidewalks, ADA noncompliance issues and missing pedestrian facilities. This system will include the capability of developing a dynamic ADA Compliance Transition Plan, including ramps, sidewalks, obstructions, crosswalks (pavers) and transit stops. Entering completed repair projects as well as newly reported ADA noncompliance issues into the software will update the transition plan thereby providing engineers and planners with the latest information to prioritize selected projects by several factors including land use (density, zoning, etc.) safe routes to schools and proximity to transit stops and schools. Benefits of the pedestrian facilities management system include:

- Ability to effectively and efficiently develop ADA compliance transition plans individually or in conjunction with connectivity and/or safe routes to school plans.
- Automatic updates to the transition plan as improvements are accomplished. This can be achieved through entering

completed capital projects into construction histories as well as completed work orders.

- Integration with other assets allows work plans to be coordinated with interrelated assets such as pavements, drainage, utilities, signs, signals, etc. Right-of-way management systems can be very helpful by providing information relating to locations of right-of-ways, easements, utilities, etc.

- As connectivity work is completed, it can be added to the active pedestrian facilities inventory.

Utility asset management initially addressed two main concerns, coordination of planned in-ground utility repairs with paving and drainage improvement projects and the improved management of the city's utility assets. The utility assets will include horizontal (linear) assets such as sewer collection and water distribution piping systems and associated appurtenances such as manholes, valves, fire hydrants, pressure-relief valves, etc. Vertical (point) assets will include treatment plants, pump stations, wells, and elevated storage tanks.

Setup of the utility asset management system follows the pavement management systems model. There will be pipe and material types, decision trees, repair types and performance models. One main difference between utility asset management and most pavement management systems is that minimizing risk instead of maximizing benefit will be the primary focus of the optimized scenario analyses for funding.

16.11   Summary

Many other examples could be cited, but these will give you the reader a clear picture of what you can accomplish. We offer our thanks to Mr. Alan Kercher of Kercher Engineering, Newark, Delaware, for preparing these descriptions of actual cases in which he has been involved and which are described in the references.

16.12   References

[Abdel Kader 96] M. O. Abdel Kader, and A. A. Al-Kulaib, "Kuwait Integrated Infrastructure Management System," *Proceedings, International Road Federation Asia-Pacific Regional Conference*, Taipei, November 1996.

[Al-Kulaib 97] A. A. Al-Kulaib, M. O. Abdel Kader, M. A. Karan, J. Powell, and G. Johnston, "Integrating Road Management with Other Infrastructure Assets Management in Kuwait," prepared for *the International Road Federation XIII World Meeting*, Toronto, Ont., June, 1997.

[Bartlett 11] S. Bartlett, and J. Thurman, "Getting the Most from Public Works Dollars Using Technology," *Presentation at the AAI User Group Conference*, Austin, Tex. 2011.

[Haas 94] R. Haas, W. R. Hudson, and J. P. Zaniewski, *Modern Pavement Management*, Kreiger Publishing Company, Malabar, Fla., 1994.

[Hoskins 88] J. C. Hoskins, and D. M. Himmelblau, "Automatic Process Control Using Reinforcement Learning in Artificial Neural Networks," *Proceedings, 1st Annual Meeting, International Neural Network Society,* Boston, Mass., 1988.

[Hsiung 91] J. T. Hsiung, and D. M. Himmelblau, "Development of Control Strategies via Artificial Neural Networks and Reinforcement Learning," *Proceedings 1991 ACC,* Boston, Mass., 1991.

[Hudson 81] W. R. Hudson, and R. Haas, "Development, Issues, and Process of Pavement Management," *Pavement Management: Proceedings of National Workshops, FHWA-TS-82-203,* Federal Highway Administration, Washington, DC, June 1981, pp. 25–35.

[Hudson 87] S. W. Hudson, R. F. Carmichael III, L. O. Moser, W. R. Hudson, and W. J. Wilkes, "Bridge Management Systems," *NCHRP Report 300,* National Cooperative Research Program, Transportation Research Board, Washington, DC, 1987.

[Kercher 11] A. Kercher, "Pavement Management Report," Kercher Engineering Inc., Newark, Delaware, June 2011.

[Kercher 12] A. Kercher, "Expansion of Asset Management for the City of Wilmington," Presentation by Kercher Engineering, Inc., Wilmington, Delaware City Public Works Department, October 2012.

[MTC 06] Metropolitan Transportation Commission (MTC), "Bay Area Transportation State of the System," Oakland, Ca., 2006.

[MTC 12a] MTC Homepage, "About MTC," Oakland, CA. http://www.mtc.ca.gov /about_mtc/about.htm, accessed December 18, 2012.

[MTC 12b] MTC, "Pavement Management Program (PMP) Homepage," Oakland, CA. http://www mtc.ca.gov/services/pmp/, accessed December 18, 2012.

[NCHRP 73] W. R. Hudson, and B. F. McCullough, "Flexible Pavement Design and Management—Systems Formulation," *NCHRP Report 139,* Highway Research Board, National Research Council, Washington, DC, 1973.

[Romell 11] T. Romell and S. Tan, "Performance-Based Accountability Using a Pavement Management System," *8th International Conference on Managing Pavement Assets,* Santiago, Chile, 2011.

[Sinha 87] K. C. Sinha, and T. F. Fwa, "On the Concept of Total Highway Management,"*Transportation Research Record 1229,* Transportation Research Board, National Research Council, Washington, DC, 1987.

[Smith 91] R. E. Smith, "Addressing Institutional Barriers to Implementing a PMS," *Pavement Management Implementation,* American Society for Testing and Materials, Philadelphia, Pa., 1991.

[Smith 11] Roger Smith, Sui Tan, and Carlos Chang-Albitres, "Distress Data Collection Quality Management in a Regional Agency," *Eighth International Conference on Managing Pavement Assets,* Santiago, Chile, 2011.

[Stone 1974] D. C. Stone, *Professional Education in Public Works/Environmental Engineering and Administration,* American Public Works Association, Chicago, Ill., 1974.

[Zhang 93] Z. Zhang, W. R. Hudson, T. Dossey, and J. Weissman, "GIS Applications in Urban Roadway Management Systems," Center for Transportation Research, The University of Texas at Austin, Tex., April 1993.

[Zhang 96] Z. Zhang, "A GIS Based and Multimedia Integrated Infrastructure Management System," *Ph.D. Thesis,* The University of Texas at Austin, Tex., August 1996.

CHAPTER 17

Visual IMS: An Illustrative Infrastructure Management System and Applications

17.1 Introduction

In previous chapters, various concepts and components of IAMSs were discussed. The most important step required to gain the real benefits of asset management systems, however, is to apply these concepts and integrate the components into an active working system. Although no comprehensive working asset management system will be perfect in its initial stage of development, it is the essential starting point from which subsequent improvements can be made.

Ideally this chapter would explain the working details of current, as of 2012, asset management systems or infrastructure asset management systems as they have sometimes been called. It is not possible to do this since the best, most widely used systems are proprietary, and details of operations are not available. Therefore, we will inform the reader in three phases: (1) first, we will describe the details of a working system that is public information, Visual IMS [Zhang 96], which illustrates to the reader how the parts of an AMS/IAMS work; (2) we will provide the reader with a discussion and list of known AMS/IAMS venders circa 2012; and (3) we will detail three of the best-known and most successful vendors, including input from known clients and vendor sources. These last two tasks are accomplished in Chapter 18.

Thus, this chapter introduces Visual IMS, a working GIS-based and multimedia-integrated infrastructure management system that was developed at the University of Texas at Austin [Zhang 96]. The general description, main structure, and major operational functions of Visual IMS are presented, along with some data-flow diagrams and computer-screen examples.

17.2   General Description of Visual IMS

Visual IMS is a GIS-based and multimedia-integrated infrastructure management system developed primarily for application by public works agencies. Currently, three subsystems are included: roads, water supply, and wastewater. The main purpose of Visual IMS is to provide a series of rational, well-ordered analyses of input data so that all infrastructures involved can be effectively maintained above the specified minimum level of service at a minimum overall cost. Visual IMS is programmed in Avenue, an object-oriented GIS programming language capable of handling the operations of both numerical data and graphical objects. Visual IMS runs under the Windows environment and is facilitated with user-friendly GUIs. It can be operated as a stand-alone IMS, or used to enhance the user-friendliness and visual capability of any other IMS. The key difference between the Visual IMS and other decision-support systems is that Visual IMS is able to handle a wide variety of data and information visually, analyze them spatially, and present the results graphically, taking advantage of both GIS and multimedia technologies.

The operation of Visual IMS is divided into two parts: (1) analytical operations based on decision models; and (2) visualizing/mapping operations based on GIS and its integration with multimedia.

17.3   Analytical Operation of Visual IMS

The analytical functions of Visual IMS are divided into six categories, which are implemented as six function modules: (1) database; (2) model setup; (3) performance evaluation; (4) planning; (5) budget allocation; and (6) proximity analysis. Figure 17-1 shows the six categories and the specific functions under each category. Each subsystem (road, water, or wastewater) has its own separate modules for the first four categories. The budget allocation and proximity analysis modules are for the entire system, and include all three subsystems in each, respectively.

17.3.1   Database and Data Input

A database is an essential component of any decision-support system, such as Visual IMS. Five groups of input data are required by the Visual IMS program: (1) geographic data that define the coordinate location of particular infrastructure features; (2) inventory data that provide an accounting of the existing physical infrastructure features; (3) condition data that are essential for conducting various performance-related evaluations and analyses; (4) policy and cost data (in terms of weighting factor and unit costs); and (5) multimedia data, such as videos and images, for visualizing infrastructure conditions.

There are primarily two ways to prepare the required database. One is to convert an existing database into a Visual IMS compatible

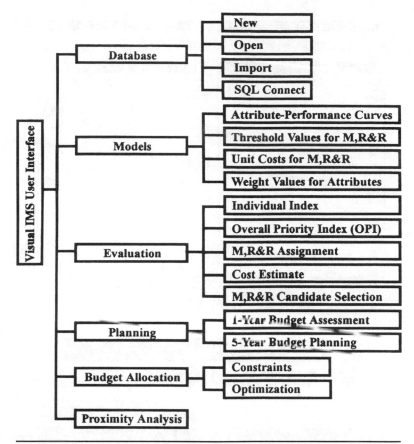

FIGURE 17-1   Six categories of analytical functions in Visual IMS [after Zhang 96].

database; the other is to create a new database and fill it with data collected in the field. Figure 17-2 is a typical data-input window in Visual IMS. For economic reasons, users should always examine the existing database for use with Visual IMS before carrying out field data collection. With its open interface for data import and export, Visual IMS is capable of accommodating a wide variety of existing data. Data stored in any other SQL-compliant database can also be accessed by Visual IMS through its SQL connection, as illustrated in Figure 17-3.

17.3.2   Model Setup

As explained earlier, Visual IMS was designed and developed as a generic system rather than an application-specific system. The models implemented in Visual IMS are basically generic models. Users should exercise caution when these models are employed with their default parameters. The purpose of model setup is to calibrate and/or replace the default parameters and curves using local

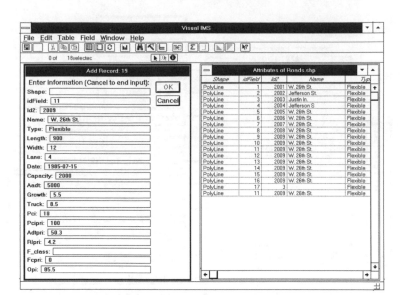

Figure 17-2 A typical data-input screen [after Zhang 96].

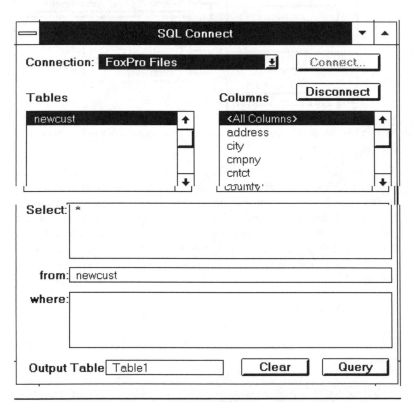

Figure 17-3 An example of establishing SQL connection to retrieve data [after Zhang 96].

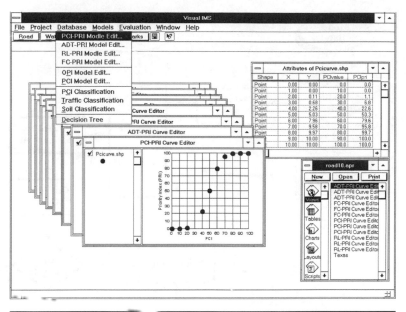

FIGURE 17-4   Graphical model editor in Visual IMS [after Zhang 96].

experience and practice. Curves and parameters that usually require modification or update during model setup include: (1) membership curves between infrastructure attribute value and normalized performance value; (2) performance threshold values for M,R&R; (3) unit costs for M,R&R; and (4) weight values for infrastructure attributes. To ease the process of model setup, Visual IMS is facilitated with user-friendly graphical model editors, as illustrated in Figure 17-4. These graphical model editors allow users to interactively modify model curves on the computer screen.

17.3.3   Performance Evaluation

One of the main uses of performance evaluation is selecting and ranking candidate projects for M,R&R [NCHRP 79, Haas 94]. The inputs for the evaluation module include: (1) attribute measurements of infrastructure [such as Pavement Condition Index (PCI) for the road subsystem]; (2) unit costs for M,R&R alternatives; and (3) available budget. Figure 17-5 illustrates the process of selecting M,R&R candidates according to the overall priority index (OPI) model. An OPI model is developed by considering a weighted combination of all of the attributes that are important to the M,R&R decision-making process. The M,R&R candidate selection starts with an OPI evaluation, which uses the OPI model and the condition data stored in the attribute database. A maintenance-and-rehabilitation action class is then assigned to each M,R&R section based on its infrastructure condition index value. The cost of the assigned M,R&R is calculated for each section.

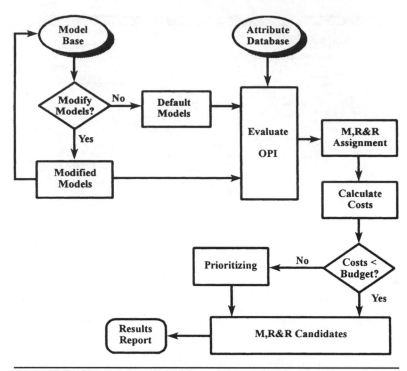

Figure 17-5 The process of M,R&R candidate selection [after Zhang 96].

If the total estimated cost for all such sections is within the available budget limit, M,R&R can be done on all sections as they are needed. Otherwise, the M,R&R sections are prioritized by their OPI values. A final list of M,R&R candidate projects is generated by adding the section with the next higher priority to the list until the budget limit is reached.

17.3.4 Budget Planning

The budget-planning module is based on the infrastructure performance index (PI) and the performance-prediction model presented in detail by Zhang [Zhang 96]. Figure 17-6 shows the analysis process for budget planning. The algorithm starts by calculating the length-weighted average of performance for each infrastructure system. The calculated average PI is then compared with the user-desired PI. If the calculated PI is below the desired PI, the algorithm proceeds to assign M,R&R actions, calculate costs, prioritize project sections, and re-evaluate the PI values. The total cost is then written to the output report. If the calculated PI is higher than the desired PI, the algorithm proceeds directly to predict the PI for the next year. This process is repeated until it reaches the last year of the five-year planning period. As an example, the typical input and

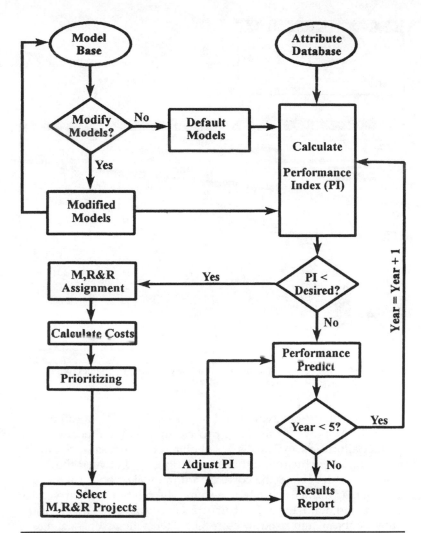

Figure 17-6   Analysis process of budget planning [after Zhang 96].

output values for a five-year budget-planning analysis are shown in Figure 17-7.

The inputs are the average performance levels expected by users for the planning period; the outputs include the M,R&R candidate projects and the budgets needed for achieving the specified performance. The output of yearly budgets is presented in both tabular and graphical formats.

17.3.5   Budget Allocation

The objective of the budget-allocation module is to maximize the overall performance of the infrastructure system with the

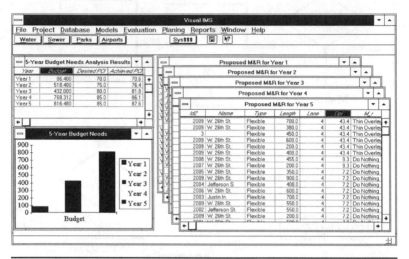

FIGURE 17-7 Typical input and output of a five-year budget-planning analysis [after Zhang 96].

constraint that the available budget is limited [Aguilar 73]. Figure 17-8 shows the analysis process for budget allocation. The budget-allocation analysis is based on a set of curves stored in the model base. Each curve defines a relationship between the available budget level and the corresponding level of performance for a particular infrastructure subsystem. The relationship curves are constructed with the infrastructure condition data in the attribute database, and need to be recalibrated whenever a change is made to the condition data. To ensure that a particular subsystem receives at least a minimum amount of money, users can specify a minimum budget level for each subsystem. Of course, the sum of the minimum budgets must not exceed the total available budget. The optimization algorithm then performs the analysis and reports the results. Figure 17-9 shows the typical input and output of an optimizing budget-allocation analysis among different infrastructure subsystems.

17.3.6 Proximity Analysis

One of the most important spatial-analysis functions in Visual IMS, proximity analysis, is aimed at improving the overall efficiency of M,R&R. The most effective approach to achieve this

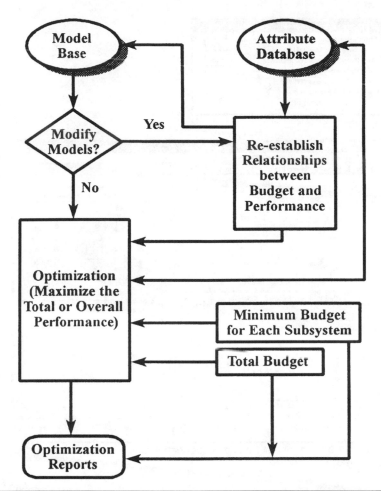

FIGURE 17-8   Analysis process of the optimization of budget allocation [after Zhang 96].

purpose is to schedule all M,R&R activities for the related infra-structure systems in a systematically coordinated manner. In this way, all M,R&R works that share the same geographical location can be planned and carried out together in the most efficient and economical sequence. For example, any problem associated with the wastewater pipelines should be identified and fixed before applying an overlay in that GIS section. The inputs for proximity analysis include: (1) geographical data that defines the geometric location of the infrastructure involved; and (2) M,R&R candidate projects obtained from the evaluation module. The output from proximity analysis is the M,R&R status of those infrastructure features that are connected with a specific infrastructure feature scheduled for M,R&R.

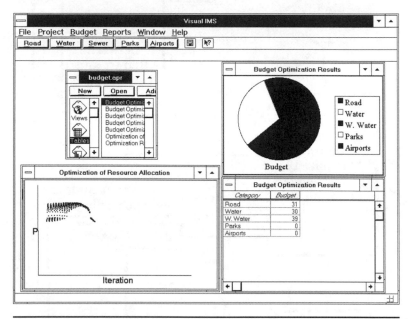

FIGURE 17-9   Typical input and output of optimizing budget allocation [after Zhang 96].

17.4   Visualizing/Mapping Operation of Visual IMS

Another important part of the Visual IMS program is the visualizing/ mapping operation. It enhances the process of analytical operation and supports the presentation of analysis results. As shown in Figure 17-10, visualizing/mapping functions that are useful for infrastructure management can be classified into six categories: (1) map display; (2) feature selection; (3) spatial query; (4) spatial analysis; (5) multi-media link; and (6) map production.

Map display. Map display includes a wide variety of operations for map manipulation, such as zooming into a specific area of the map for a closer look at certain infrastructure features and editing the map legends to produce the best visual effect. New infrastructure systems can be added to the system through the map-adding/ deleting function. Furthermore, infrastructure attribute data can

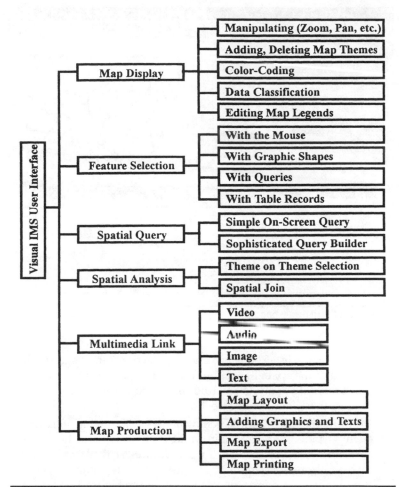

Figure 17-10 Six categories of visualizing/mapping functions in Visual IMS [after Zhang 96].

be classified and color-coded on the map for better presentation, as illustrated in Figure 17-11.

Feature selection. Feature selection allows some of the infrastructure features to be selected interactively for analysis. The selection can be done with the mouse, graphic shapes drawn on the theme, records in a table associated with the map, or even a query operation.

Spatial query. Spatial query is useful in examining information related to a specific infrastructure feature or a set of features that meet certain criteria. For example, the sophisticated query builder can help to find all the pavement sections with a PCI less than 55 and a traffic volume (AADT) greater than 35,000 and highlight them.

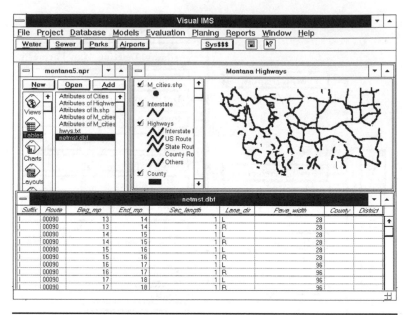

Figure 17-11 Highway base map and related pavement information [after Zhang 96].

Spatial analysis. "Theme-on-theme selection" and "spatial join" are two of the spatial-analysis functions available in Visual IMS. The proximity-analysis module discussed earlier is built on the basis of theme-on-theme selection.

Multimedia link. Multimedia link allows multimedia information, such as video, audio, image, and text, to be attached to the corresponding infrastructure map features. The attached multi-media information can then be accessed and used simultaneously through an on-screen query, as illustrated in Figure 17-12.

Map production. The effective presentation of analysis results is an important part of IMS. With Visual IMS, analysis results and related information can be produced as tables, charts, and maps, or as a combination of forms with a customized layout. The final document can be exported as a graphic file and/or printed as a hard copy.

17.5 Hardware, Operating System, and Memory

For successful installation and operation of Visual IMS on a personal computer (PC), the computer must meet the required minimum configuration outlined below, operate under the required operating system (OS), and have adequate hard-disk space, random access memory (RAM), and virtual memory. These requirements are discussed as follows.

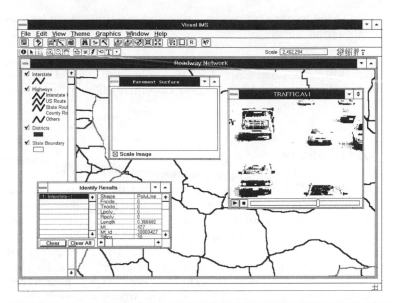

FIGURE 17-12   Multimedia information capability in Visual IMS [after Zhang 96].

Hardware requirements. The hardware requirements include computer CPU (central processing unit), hard disk, monitor, pointing device, sound card, and speakers.

- *Computer CPU.* The computer should be IBM-compatible and have at least an 80386 or higher microprocessor. It also should have a hard disk and a media/disk driver.

- *Hard-disk space.* The installation of Visual IMS requires at least 20 MB of hard-disk space available on the computer.

- *Monitor.* A video graphic adapter (VGA) or higher-resolution monitor is required.

- *Pointing device.* A Microsoft mouse or compatible pointing device is required.

- *Multimedia device.* Sound card, CD-ROM, and speakers must be properly installed and configured if the user wants to employ all the multimedia functions in Visual IMS.

Operating system requirement. To run Visual IMS on a PC, MS DOS 5.0 or later and one of the following Windows OSs are required: Microsoft Windows 3.1, Windows for Workgroups 3.11, Windows NT 3.1, or their later versions.

Memory requirement. A minimum of 8 MB RAM is required to run Visual IMS. However, 16 MB RAM or higher is recommended for more efficient operation, especially if the user wants to run other applications simultaneously with Visual IMS. A minimum of 12 MB file-swap space also is required as permanent or temporary virtual memory.

Note that the requirements presented so far are for running Visual IMS on a stand-alone PC. To implement Visual IMS in a distributed network, such as a LAN environment, different hardware and software are required.

17.6 User Training

The GIS-oriented nature and the multimedia-integrated features of Visual IMS are somewhat more complicated than simple PC programs, and it is expected that a medium level of user training will be required. Some basic knowledge of GIS will be helpful to personnel in understanding the concepts and principles of Visual IMS.

It is estimated that two days of training will be required for an engineer who has a normal level of computer knowledge (i.e., a person who understands a Windows OS and knows how to run Windows-application software) to perform analyses with the basic functions in Visual IMS. More sophisticated applications with Visual IMS will require additional training, depending on the level of prior knowledge.

The primary purpose of this early version of Visual IMS is to demonstrate the concepts, approaches, and models developed under this study. All of the major function modules have been included in this version of the program, and when future work is invested, further improvements can be made to the Visual IMS program.

17.7 Application of Working Infrastructure Management Systems

Unless it is implemented or applied to solve practical engineering problems, an IMS has little value. This section presents several case studies involving the application of Visual IMS. These case studies demonstrate how an IMS can be used not only as a comprehensive management system, but also as an enhancement to existing management systems. The application of the IMS concept to other management problems is also illustrated with a case study.

Three case studies were developed as part of the effort to test the Visual IMS program and to manifest the concepts presented in previous chapters. Case Study 1 shows how an IMS can be implemented to work with an existing management system. Case Study 2 illustrates how the analytical capabilities of the IMS can be used to answer engineering questions in infrastructure management. Case Study 3 serves as an example of how the concepts of IMS can be used for other similar management applications.

17.7.1   Case Study 1: Montana PMS

An important implementation of IMS was carried out by the Texas Research and Development Foundation (TRDF) to support a PMS it had developed for the state of Montana. As indicated earlier, the purpose of this case study was to demonstrate how an IMS could be used with an existing management system to improve the user-friendliness and visual capability of the system.

The TRDF PMS for the state of Montana is a Windows-based system that was developed with PowerBuilder, an SQL-compliant database management system [TRDI 96]. Although the PMS is capable of conducting various sophisticated analyses for pavement management, it does not have a GIS interface, which is increasingly demanded by users who want to present analysis results geographically. Major tasks for this case study included: (1) developing the interface between the IMS and the TRDF PMS; (2) preparing a base map of the Montana highway system; (3) relating pavement-attribute data records to their corresponding map features; and (4) incorporating multimedia data. Both desktop PC and laptop PC computers were used for this case study.

17.7.1.1   Interface between Visual IMS and TRDF PMS

The main purpose of such an interface is to facilitate data communications between the two computer programs. Two methods were employed for this purpose. The first method was to export the analysis results from the TRDF PMS as a dBase file to be read directly by the IMS. The results were used as input for the IMS. The second method was to establish an SQL connection between Visual IMS and the TRDF PMSs database so that the result tables of PMS analyses could be directly retrieved by the IMS.

17.7.1.2   Base-Map Preparation

The "National Transportation Atlas Data Bases: 1995" (NTAD-95) was used to prepare the Montana base map. NTAD-95 was developed and distributed by the Bureau of Transportation Statistics of the U.S. DOT free of charge [USDOT 95]. The original data are stored as ASCII text files in fixed digit format—a format that cannot be directly used by any GIS software. To solve this problem, a computer program was written to subtract the longitude/latitude coordinates that define the Montana highway network from the ASCII file and to rewrite the subtracted coordinate data into another GIS-readable text file, as illustrated in Figure 17-13.

17.7.1.3   Relating Pavement Attribute to Base-Map Features

Pavement sections are represented by the milepost method in the analysis results of the TRDF PMS for Montana. The relationship between the analysis results and their corresponding map features was established by creating a common ID for the pavement-data table based on

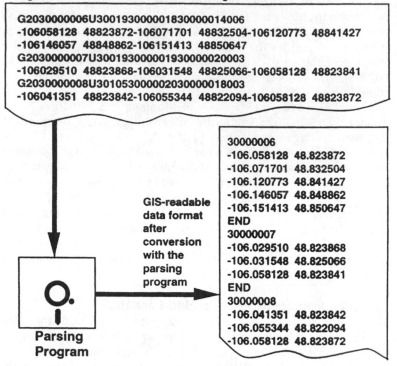

Original NTAD-95 data stored in fixed-digit format

```
G2030000006U300193000001830000014006
-106058128 48823872-106071701 48832504-106120773 48841427
-106146057 48848862-106151413 48850647
G2030000007U300193000001930000020003
-106029510 48823868-106031548 48825066-106058128 48823841
G2030000008U301053000002030000018003
-106041351 48823842-106055344 48822094-106058128 48823872
```

GIS-readable data format after conversion with the parsing program

```
30000006
-106.058128 48.823872
-106.071701 48.832504
-106.120773 48.841427
-106.146057 48.848862
-106.151413 48.850647
END
30000007
-106.029510 48.823868
-106.031548 48.825066
-106.058128 48.823841
END
30000008
-106.041351 48.823842
-106.055344 48.822094
-106.058128 48.823872
```

Parsing Program

FIGURE 17-13   Conversion of NATD-95 into a GIS-readable format.

its mileposts. The common ID was then used to either "join" or "link" the pavement data to their corresponding sections on the base map.

17.7.1.4   Multimedia Data

Some multimedia data were also prepared for the case study, including: (1) traffic video in *.AVI format; (2) pavement surface image in *.BMP format; (3) audio comment in *.WAV format; and (4) information summary in *.TXT format.

17.7.2   Case Study 2: Analytical Capability

In Case Study 1, the analytical part of the problem was largely handled by the TRDF PMS. The intention of Case Study 2, therefore, is to demonstrate the analytical capabilities of Visual IMS with some common problems. Due to the limitation of time and resources, the data used for Case Study 2 approximates data but is not real data. The infrastructure features on the base map were created through interactive on-screen digitizing, using an aerial photograph as a background. The database tables of infrastructure condition were developed by using the Visual IMS database module. All six categories of analytical functions have been tested with the developed infrastructure condition database. An example of the typical input and output for a five-year budget-planning analysis are presented in Figure 17-7. The input for the analysis was the average performance levels expected by the user for the planning period; the output included the M,R&R candidate projects (highlighted records) and the budgets needed for achieving the specified performance. The output of yearly budgets was presented in both tabular and graphical formats. According to the outcome of this case study, all function modules have produced reasonable results.

17.7.3   Case Study 3: Conceptual TMS for Cobb County DOT

Case Study 3 was carried out by the TRDF as a direct application of the concepts of Visual IMS. In December 1995, the Cobb County Department of Transportation (Cobb DOT) requested proposals for the development of a transportation management system (TMS) for their surface-transportation network. The Cobb DOT required six management modules to be included in the TMS:

1. Road inventory module
2. Traffic congestion module
3. Bridge maintenance module
4. Pavement management module
5. Traffic signal module
6. Work order scheduling module

Since these six modules must be integrated and operated under one master system, it was decided to use GIS as the integration platform. The objective of Case Study 3 was to conduct a conceptual design for the TMS. After analyzing the requirements of the Cobb County TMS, the concepts and frameworks of Visual IMS were found readily applicable to the development of a TMS. As a result, a Visual TMS was conceptually developed for the Cobb DOT. Figure 17-14 depicts the major function modules and their relationships in the Visual TMS designed for the Cobb DOT.

Many other AMSs have been developed since Visual IMS, but all are proprietary and thus details are not available. It is safe to say, however, that they in general follow the concepts of Visual IMS, thus it remains a model for study. It should be recognized that any application software developed in the 1990s or earlier needs to be updated to stay compatible with later OS versions such as Windows 2000, XP, Vista, and Windows 7 and 8.

In the next chapter, we will look at the information that is available on the proprietary commercial off-the-shelf (COTS) systems and their developers and vendors. The knowledge of Visual IMS concepts in the chapter will aid the reader in evaluating these COTS systems for possible use.

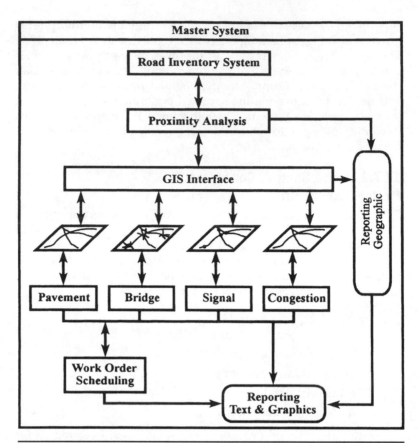

FIGURE 17-14 Visual TMS designed conceptually for Cobb County DOT.

17.8 References

[Aguilar 73] R. J. Aguilar, *Systems Analysis and Design in Engineering, Architecture, Construction, and Planning*, Prentice-Hall, Englewood Cliffs, N.J., 1973.

[Haas 94] R. Haas, W. R. Hudson, and J. Zaniewski, *Modern Pavement Management*, Krieger Publishing Company, Malabar, Fla., 1994.

[NCHRP 79] "Pavement Management System Development," *NCHRP Report 215*, National Cooperative Highway Research Program, Transportation Research Board, Washington, DC, November 1979.

[TRDI 96] *TRDI Visual PMS User Guide*, Texas Research and Development Inc., Austin, Tex., 1996.

[USDOT 95] *National Transportation Atlas Data Bases: 1995*, BTS-CD-06, Bureau of Transportation Statistics, U.S. Department of Transportation, Washington, DC, 1995.

[Zhang 96] Z. Zhang, *A GIS Based and Multimedia Integrated Infrastructure Management System, Ph.D. thesis*, the University of Texas at Austin, Austin, Tex., August 1996.

CHAPTER 18

Available Asset Management System and Commercial Off-the-Shelf Providers

18.1   Introduction

Since the inception of management systems beginning with pavement management system [Hudson 70, Haas 78, 94] and bridge management system in 1987 [NCHRP 87], software for asset management system (AMS) has generally been provided in one of four ways: (1) in-house development, (2) individual agency development by a consultant, (3) purchase and customization of existing software, or (4) licensing or subscription of commercial off-the-shelf (COTS) software from qualified venders.

As expected, the development approach has generally transitioned from 1 through 4 over several decades. Since there was no commercial AMS in the 1970s, agencies undertook in-house development of PMS (Texas, Washington, Maryland, etc.). Some agencies recognized they did not have the needed capabilities in-house and hired consultants to develop AMS (Arizona, Kansas, etc.). By 1985, commercial systems were available, and some agencies purchased simple PMS packages from Deighton and ARE Inc. (now AgileAssets Inc.). From 1995 onward most agencies and software providers realized that software obtained in these first three ways was static and quickly became dated, like using Windows 1 in 2005 after Windows 5 was available. From 2000 onward most AMS software has been obtained by subscription and licensing of COTS systems where improved and modular updates can be regularly made. Agencies that did not use COTS have struggled. We will deal no further with development types 1 through 3 but will discuss COTS only since they make up most new systems since 2000. Even efforts by agencies to develop hybrid systems in-house have bogged down [Hatcher 08, 09]. There is an important line between agencies working with a vendor to adapt a system during implementation and agencies having to develop and

429

input their own models and major components into a standard system. While the latter is a good sales gimmick, the former is quicker, cheaper, better, and more productive in the long run.

18.2   Review of COTS Asset Management Software

18.2.1   Available COTS Vendors across Asset Types

Two sources were found who have searched out COTS vendors for a variety of asset types [Brunquet 07, Mizusawa 09] and summarized them for convenience and information. We have extracted their lists and prepared a third list from current sources. Many such vendors are small, and some have been subsumed into other companies while others simply vanished. These sources give the reader a representative list of resources and information that individual potential users can investigate.

18.2.2   Dominant COTS in Infrastructure Asset Management

It is beyond the scope of this book for us to evaluate vendors. The prior reviews [Brunquet 07, Mizusawa 09] show that three vendors dominate the field of state agency COTS in North America—Deighton, Stantec, and AgileAssets. Available sources are summarized for the reader. Appropriate references from which the reader can search for details are provided.

18.2.3   Canadian Evaluation

In 2007, Juliette Brunquet from the Ecole Centrale de Lille, an exchange student at the University of Waterloo with Professor Ralph Haas, compiled a report [Brunquet 07] entitled "Comparative Assessment of Road Asset Management Systems Software Packages." In her report, she outlines details of COTS and summarizes various properties used in her evaluation. For more information the reader is directed to that reference.

Brunquet condenses the list of software packages to a select four. Atlas-EXOR (www.exorcorp.com), dTIMS Deighton (www.deighton .com), EMS-WASP-EMSSolutions (www.ems-solutions.com), and AgileAssets Management Suite-AgileAssets (www.agileassets.com). Atlas-EXOR has since been purchased by Bentley (www.bentley.com). The Atlas-EXOR team is a partnership based on EXOR's network manager [Brunquet 07]. Since EXOR has subsequently been bought by Bentley, it is hard to know exactly how applicable this source of software is now, but it is worth the reader's time to examine and look up more recent details.

EMS-WASP-EMSSolutions is primarily an Australian-New Zealand implementation company that applies its asset management software in the utilities industry. It has done little work in the road asset management area as of 2007. However, as Brunquet indicated they may have the potential to expand, and readers can update themselves on that possibility.

The remaining two companies, AgileAssets of Austin, Texas, and Deighton of Ontario, Canada, remain active players in the asset management arena. They both started with PMSs; AgileAssets, as early as 1970, under the name of Austin Research Engineers Inc. and Deighton, beginning in 1983, under the leadership of Rick Deighton. After an early start in pavement management they have developed in different directions. ARE Inc. morphed into AgileAssets, which now produces 17 different management packages for assets that are important to state, county, and city agencies. They also combine these modules into asset management systems of various sizes depending on the desires of the particular customer. AgileAssets expanded past pavement management with bridge management in 1987 [TRDI 96] and has been expanding its AMS ever since.

Deighton started with small implementations in various DOTs in the United States and Canada about 1983, generally, initiating a simple PMS. They currently report having software used by 19 state DOTs in the United States and profess implementations in up to 100 other locations worldwide. However, this is an ever-changing kaleidoscope, The state of Utah, for example, has recently implemented the AgileAssets PMS to supplement the Deighton PMS because they felt they needed better analyses and the ability to develop performance models for their pavements. Indiana has adopted AgileAssets maintenance management in tandem with its early Deighton work. AgileAssets now has PMS in 14 U.S. states plus many other locations as shown later. We have added Stantec, Inc. (www.stantec.com) to our later detailed discussions.

18.2.4   Road Management COTS System's Catalogue Version 2

In 2009 Daisuke Mizusawa completed a report at the University of Delaware with Dr. Sue McNeil for the World Bank funded by the Japanese Consulting Trust [Mizusawa 09]. Table 18-1 lists the systems which Mizusawa deemed to be appropriate for possible use by the World Bank et al. As seen from the listing, the author cast a wide net to include partial systems of various kinds, which may not fulfill the complete requirements for AMSs. Nevertheless, this independent reference is worth study in case the reader wishes to deal with a subpart of pavement management for example.

18.2.5   Further References to Available Systems

We have also assembled a detailed list of providers of various types of asset management software available as of July 2012. That list is given in Table 18-2. No attempt is made to evaluate the details of these various companies. Some are not very active in areas like transportation. That does not mean that they do not provide good

Category	System	Provider/Developer	Available in This Catalog?
Pavement management systems	• Agile Assets Pavement Analyst V5	AgileAssets	No
	• AIM for Roads	Axiom Decision Systems	No
	• Desyroute	Laboratoire Central des Ponts et Chaussees	No
	• HDM-4 Version 2	HDM Global	Yes (but not complete)
	• HERS-ST	FHWA	Yes
	• MARCH PMS	Data Collection Ltd.	Yes (but not complete)
	• MicroPAVER	US Army-ERDC	Yes
	• PAVEMENTtview®	CarteGraph	Yes (but not complete)
	• RealCost	FHWA	Yes
	• RONET	The World Bank	Yes
	• SMEC PMS	SMEC	Yes (but not complete)
	• Stantec PMS	Stantec	Yes
	• WDM PMS	WDM Ltd.	No
Bridge management systems	• AgileAssets	AgileAssets	No
	• AIM for Bridges	Axiom Decision System, Inc.	No
	• Exor Structures Manager	Exor Corporation Ltd.	Yes
	• NBIAS	Cambridge Systematics, Inc.	Yes
	• PONTIS		Yes
	• RoSy® BMS	AASHTO	No
	• Stantec BMS	Grontmij | Carl Bro Stantec	Yes
Integrated management systems	• AssetManager NT	AASHTO	Yes
	• AssetManager PT	AASHTO	Yes
	• Confirm	MapInfo	No
	• dTIMS CT	Deighton Associates Ltd.	Yes
	• Exor Highways	Exor Corporation Ltd.	Yes
	• HIMS	HIMS Ltd.	Yes
	• ICON	GoodPointe Technology	Yes (but not complete)
	• INSIGHT	Symology Ltd.	Yes
	• ROMAPS	Roughton International	No
	• RoSy® PMS	Grontmij | Carl Bro	Yes
	• SMART	Ramboll	Yes

TABLE 18-1 List of COTS Systems [Mizusawa 09]

Name	Description	Web Address
Accela Inc.	Focused in government permitting, but also pursues municipal asset management projects.	http://www.accela.com/
Advitam	French company specializing in asset management; office in VA and NYC.	http://www.advitam-group.com/
Aegis ITS Inc.	Aegis ITS provides Intelligent Transportation Systems (ITS) services.	http://www.aegisits.com
Aleier Inc.	Aleier provides a comprehensive asset management solution that incorporate its proprietary computerized maintenance management system (CMMS).	http://www.aleier.com
AllMax Software Inc.	No information	
Applied Data Systems	No information	http://www.adsi-fm.com
Aquitas Solutions Inc.	No information	
ARA	Founded in 1979, ARA is dedicated to produce innovative solutions that tackle critical national problems in National Security and Infrastructure among other areas.	http://www.ara.com/
Asset Mgmt Engineering Inc.	No information	
Assetworks Inc.	30 years of fleet management services; expanding in other asset management asset classes.	http://www.assetworks.com
Axiom Decision Systems Inc.	Axiom provides services and products for the management of infrastructure assets for cities, counties, states.	http://www.axiomds.com/
Azteca Systems	Cityworks empowers GIS to manage both physical infrastructure and land-focused asset management, validated by ESRI.	http://www.cityworks.com/
Bentley Systems Inc. (Exor)	See Exor.	http://www.bentley.com/

TABLE 18-2   List of Providers

Name	Description	Web Address
Cambridge Systematics	As transportation specialists the company provides innovative policy and planning solutions in transportation areas.	http://www.camsys.com
CarteGraph	CarteGraph is a cloud-based Operations Management System.	http://www.cartegraph.com/
CCG Faster	No information	http://www.ccgsystems.com
CFAWinn	No information	http://www.cfasoftware.com
CGI (AMS Advantage ERP)	CGI provides governments our full-service and focused capabilities built upon proven frameworks.	http://www.cgi.com
CHEC Pavement Management Systems	CHEC Management Systems, Inc., formally known as CMSI, specializes in development and implementation of infrastructure management systems.	http://www.cmsicorp.com/
Chevin Fleet Solutions	No information	http://www.chevinfleet.com
Ciber	CIBER® is a leading international system integration consultancy with superior value-priced services for both private and government sector clients.	http://www.ciber.com/
CIP Planner	No information	
CitiTech Systems Inc.	CitiTech Management Software is designed to help cities and counties manage infrastructure assets, department resources and work requirements more efficiently.	http://www.cititech.com/
Collective Fleet	No information	http://www.collectivedata.com
COWI	COWI, a leading northern European consulting group, provides state-of-the-art services within the fields of engineering, environmental science and economics.	http://www.cowi.com
Data Transfer Solutions LLC	DTS is an approved vendor for the State of Florida IT Consulting Services contract. The categories covered include Analysis and Design, Development and Integration.	http://www.edats.com/

TABLE 18-2   List of Providers (*Continued*)

Name	Description	Web Address
Deighton Associates Limited	Deighton develops and markets software solutions for asset managers to locate, categorize, and manage the lifecycle of all assets, business operations, and initiatives.	http://www .deighton.com
Delcan Corporation	Delcan provides a broad range of integrated systems and infrastructure solutions with focus on transportation and water sectors.	http://www .delcan.com/
Design Information Technology	No information	
DiExSys LLC	No information	http://diexsys .com/
DP Solutions Inc.	No information	http://www.dpsi .com
E H Wachs	No information	
EDI	EDI has good record of customer service.	http://www.edatai .com
EMA Inc.	For over 30 years EMA has focused on the technology and business management needs of utility, government, and manufacturing organizations.	http://www.ema- inc.com
Engineering Management Applications, Inc., USA	Established in 1991 in Maryland it provides specialized infrastructure management software services and strives to bridge gaps between technology innovations and industry applications.	http://emainc-usa .com/
Engineering Mapping Solutions	No information	
EPAC Software Technologies	No information	http://www .epacst.com
ESRI	No information	http://www.esri .com
Exor/Bentley	Exor software system is a part of Bentley's AssetWise platform, which provides information for asset management. Exor provides a total infrastructure asset management solution.	http://www .exorcorp.com

TABLE 18-2   List of Providers (*Continued*)

Name	Description	Web Address
Fugro	No information	http://www.fugro.com/
GBA Master Series	GBA Master Series, Inc., (gbaMS) focuses exclusively on the design, development and implementation of commercial off-the-shelf (COTS) software applications for public works and utilities.	http://www.gbams.com/
GeoDecisions /Gannett Fleming	Gannett Fleming has been providing quality, full-service, multi-disciplinary engineering services and technical innovation in the United States and abroad since 1915.	http://www.geodecisions.com/
Golder Associates	A global company specializing in ground engineering and environmental services through technical excellence, innovative solutions and client service.	http://www.golder.ca
GoodPointe Technology	Its management system software helps owners of public and private infrastructure to more effectively meet their management and maintenance needs.	http://www.goodpointe.com/
IBM	No information	http://www.us.ibm.com
InCircuit Development Corporation	Provider of Enterprise Asset Management solutions. For over a decade, we have worked with managers in government.	http://www.incircuit.com
Indus/MDSI	Indus International Inc., a Service Delivery Management (SDM™) solution provider and MDSI Mobile Data Solutions Incorporated have merged.	
Infor	Formerly, Hansen Information Technologies, it is a supplier of application software that helps manage the operations of over 450 governments.	http://infor.com/
InspectTech Systems Inc.	No information	
Intergraph	No information	http://www.intergraph.com

TABLE 18-2 List of Providers (*Continued*)

Name	Description	Web Address
ITIS Corporation	ITIS Corporation is an IT consulting firm specializing in "Integrated Transportation Information Systems," with a focus upon Transportation Safety and ROW Management.	http://itis-corp1 .com/
JACOBS Engineering	No information	http://www .jacobs.com/
Lawson Software	No information	http://www .lawson.com
Ledge Light Technologies, Inc.	No information	
Lucity	Formerly GBA Master Series.	http://www.lucity .com/
MaintStar Inc.	No information	http://www .mainstar.com
Maximus	Maximus offers government and industry a range of unique services, products, and solutions,	http://maximus .com/
MDS Technologies, Inc.	No information	
MicroMain	MicroMain Corporation provides solutions for facility operations professionals and leads in the development and implementation of CMMS/EAM and CAFM software.	http://micromain .com/
MMM Group	MMM is recognized for providing quality, cost-effective and technically excellent multidisciplinary engineering solutions for a diverse range of clients.	http://www.mmm .ca/
Mott MacDonald	Mott MacDonald is a £1 billion business working in all sectors from transport, energy, buildings, water and the environment to health and education, industry and communications.	http://www .mottmac.com/
MRO	MRO Software is the leading provider of strategic asset and service management solutions. Maximo Enterprise Suite, the Company's flagship solution, is delivered on web-architecture.	http://mro.com/

TABLE 18-2   List of Providers (*Continued*)

Name	Description	Web Address
Opus International	International multidisciplinary consultancy.	http://www.opus.co.nz/
Rauros	Rauros is a consulting company founded in 2002, linked to the development of new technologies in the civil engineering areas.	http://www.rauroszm.com/quienes_somos.aspx
Riva Modeling	Riva Modeling Systems Inc. was formed to sell and support strategic long range physical asset planning software, originally named Riva Online, a toolkit of software.	http://www.rivamodeling.com/
RKV Technologies, Inc.	RKV Technologies is a provider of information technology solutions for businesses and governments.	http://www.rkvtechnologies.com/
Ron Turley Associates	No information	http://www.rtafleet.com
SAP	Governments—federal/central, state/provincial and local—deploy SAP's enterprise solutions in a shared services delivery mode.	http://www.sap.com
Sierra Systems	Sierra Systems serves clients by offering a full range of management consulting, project implementation and application support services.	http://www.sierrasystems.com
Stantec	A well-known Canadian provider of PMS and other consultancy services.	www.stantec.com
Starboard Consulting LLC	No information	
Sungard Public Sector Inc.	No information	http://www.hteinc.com
Tata Consulting	International consulting firm from India.	
Technology Associates International Corp.	No information	
Technology Consortium LLC.	Technology Consortium LLC is an information technology consulting firm serving clients in the United States.	http://www.tech-consortium.com/

TABLE 18-2 List of Providers (*Continued*)

Name	Description	Web Address
Timmons Group	No information	http://www.timmons.com/
TISCOR	No information	
TMA Systems, LLC	No information	http://www.tmasystems.com
Total Resource Management	No information	http://www.trmnet.com
UR International, Inc.	No information	http://www.urinternational.com
WilburSmith Assoc.	Wilbur Smith Associates is a full-service infrastructure consulting firm providing a unique blend of planning, design, toll, economic and construction-related services.	http://www.wilbursmith.com/index.htm
WilsonMiller	No information	
Woolpert Inc	Woolpert provides consultancy services to plan, budget, implement, and manage projects to get the job done. Successful projects involve collaboration.	http://www.woolpert.com
WSP IMC	WSP IMC is a management and engineering consulting company offering professional services.	http://www.wspgroup.com
WTH Technology, Inc.	No information	

TABLE 18-2   List of Providers (*Concluded*)

software. It is up the reader to search details out through websites, personal contacts, and evaluations.

Based on the foregoing discussions and the authors' knowledge of the asset management field we will limit detailed discussion to the following three vendors: Deighton Ltd, Ontario, Canada; Stantec, Canada; AgileAssets, Inc., United States. These sources clearly have the largest penetration into asset management for the transportation field especially in North America. They have an even stronger penetration in the PMS field since that was the type of AMS first initiated in the late 1960s. It is important to note also that maintenance management systems (MMS) form a strong basis for developing an asset management system because maintenance deals with most types of assets. AgileAssets has an early history of development of MMS.

Limiting our examples to three vendors should not imply a slight to any other COTS vendors. The ones described best illustrate the evaluation process agencies should go through to procure COTS software.

18.2.6   Guidelines for Evaluating Vendors

This book does not evaluate or compare the various vendors. However, readers can use the following guidelines in their evaluations. The first step is to contact the vendor and access their website to obtain published information. The following steps may help readers evaluate potential software vendors.

1. Start with the company website and brochures, evaluate their claims critically.

2. Search for published technical references from actual users who have evaluated and described implementation and use of the software.

3. Contact existing users, at least two or three, because one user may be satisfied with the status quo. Any system not updated in the last three or four years is suspect.

4. Request the vendor to present a detailed software demonstration at an appropriate site. Prepare questions in advance and be prepared to deal with overview generalities that vendors often present. These questions and related concerns might include, for example,

   a. Specifics about the guts of the system including performance models used. Some COTS use default values not applicable to your agency.
   b. Does the system use real optimization as opposed to prioritization or incremental cost benefit? The word "optimization" is often used to mask a number of approaches.
   c. What individual subsystems and models does the AMS actually use, not what they advertise, nor what they *will be* developing, but what they have actually implemented.
   d. Ask about individual subsystems, how they operate, and integrate.

5. Ask about vendor user group meetings if held on an annual basis and attend, where you can meet and discuss software with several current users of the software.

6. Get price quotes and specifics what modules are included. Don't buy on price. Some providers will offer a skeleton AMS at low cost and then charge extra for important details. In general you get what you pay for.

7. Find out system outputs to compare with your agency's specific needs. Watch out for incremental creep or "bait and switch." Some vendors sell in low and encourage you to add on extras, much like a car salesman. This can be costly and delay your system development tremendously.

Some COTS providers claim open architecture, adaptable for any type of asset ranging from light poles to vehicle fleets to pavement sections. Others provide complete but customizable software for individual assets. Proven performance is the key so investigate before buying. Description of three leading providers is provided in the following sections.

18.2.7   Deighton Associates Limited

18.2.7.1   dTIMS CT—Deighton Associates Limited

The following information is extracted from [Deighton 10] and two users of their AMS software [Hatcher 08, 09].

Deighton has been in business for over 25 years and is a leader in transportation infrastructure asset management. Deighton has evolved from a small engineering firm providing client-specific applications for PMS into an international software development organization. They have now PMS implementations in the following states: CO, CT, LA, IN, ME, MO, NH, ND, OH, RI, SD, UT, and VT. According to the company their clients want to apply the advanced capabilities within dTIMS CT to their other assets, such as roadway assets, bridges, and subsurface utilities. Deighton reports partnerships with companies with other expertise.

They state that their more progressive clients are using dTIMS CT to manage not only pavements and bridges but safety and traffic data and then incorporate it into an entire network analysis. It is not clear who and how many there are of these progressive clients. As far as it can be verified most of their clients still use Deighton PMS modules only. They also report that today dTIMS CT is in use around the world to manage many different assets for over 400 agencies worldwide. There is no indication of who these 400 clients are.

Clearly, Deighton provides a useful PMS for its clients as evidenced by the 18 to 19 agencies that currently use their system in the United States. Deighton is in the process of developing a larger asset management system. What is not clear is how fast the change is developing and how many agencies are using multiple subsystems. Some information is available from two published reports and one or two individual communications related to the Deighton systems, as described below.

18.2.7.2   Asset Management Implementation Framework for Colorado Department of Transportation

In the September 2010 Deighton report submitted to the Colorado Department of Transportation [Deighton 10], they describe an implementation plan for an asset management system. No subsequent publications that indicate the progress of that implementation were found.

While the Deighton literature implies a complete asset management system, various portions of [Deighton 10], suggest otherwise; for example the report states "this would enable CDOT to measure and analyze safety and mobility as assets in the *future*." And "As the AMS matures, CDOT will be able to concurrently analyze assets such as roads, bridges, structures, culverts, traffic safety, etc. and optimize budgets across these same assets."

Other clarification about the status of dTIMS CT can be gained from the report, which states, "As the PMS will form the basis of the AMS and is implemented in dTIMS CT already, very little initial work must be done and yearly maintenance will be minor." PMS clearly forms the basis for dTIMS CT. "Bridge management at CDOT is completed through the use of manual inspections and semi-automated processes using PONTIS and other software packages" as also stated.

18.2.7.3   Summary

Thus, while Deighton reports that dTIMS CT is a full-fledged AMS, this seems inadequately supported by their own report. While dTIMS CT has the prospects of becoming an AMS if diligently pursued, the reader should be aware that dTIMS CT may not yet be fully developed.

18.2.7.4   dTIMS CT Implementation in New Zealand

Other information on the details and concepts of dTIMS CT is provided in [Hatcher 09]. According to the report, the Highways Agency [New Zealand Transport Agency (NCTA) formerly Transit New Zealand], with the assistance of international consultants and others have undertaken the implementation of an AMS in New Zealand. After reviewing several options, the Agency undertook a contract for Deighton dTIMS CT software.

Hatcher points out that the software configuration for dTIMS CT "can be undertaken by the owner but Deighton will provide consulting services 'at an additional cost.'" Unfortunately, Hatcher is in error. When he implies that one vendor, AgileAssets has hard-coded business logic while dTIMS CT can be configured by the owner of the software "consulting services provided by the vendor and certified consultants." These errors fail to recognize that AgileAssets has provided management services for maintenance of all transportation assets and individual management systems for fleet, safety, bridges, etc. for many years. He suggests the dTIMS CT is "capable of modeling all assets in the road corridor. However, the majority of implementations they have made so far are for pavements and surfacing."

Hatcher [08, 09] describes a number of missteps and delays that have occurred where a consortium of local agencies and representatives in the transportation agency have tried to do their own implementation. This points out the value of good COTS software properly implemented by people that understand its details and its use in lieu of "in-house" implementation. Hatcher suggests "avoid providing definitive answers in a PMS." We do not agree with this philosophy.

18.2.8 AgileAssets Inc. Asset Management Capabilities

AgileAssets is another widely used provider of COTS asset management systems. They began in 1970 originally as ARE, Inc., then TRDI, Inc., morphing to AgileAssets Inc. in 1994. In addition to early PMS implementation, they developed the initial basic concepts of bridge management [NCHRP 87].

As is the case with others, AgileAssets does not usually share details of its software but much information can be gained from their website (www.agileassets.com) and published literature. Their product overview [Agile 12] is shown in Table 18-3, which outlines 17 separate COTS systems that are available. AgileAssets lists 34 agencies using their software. Eleven of these use an AMS involving three or four Agile systems.

18.2.8.1 Summary of AgileAssets System Capabilities

As with all COTS providers, AgileAssets provides no specific details on the internal operations of their system beyond that given in Table 18-3. Some insights however, can be obtained from published literature [Hudson 11, Bhargava 11, 13, Azam 12] and users of the system. Idaho Transportation Department and AgileAssets staff describe integration of PMS and MMS [Hudson 11], as shown in Figure 18-1.

Pavement Analyst™ (PMS)	Safety Analyst™
Bridge Analyst™ (BMS)	Sign Manager™
Mobile Apps (Field Data Collection)	Trade-Off Analyst™ (Funding Allocation Across Assets)
Bridge Inspector™	Signal and ITS Manager™
Mobility Analyst™ (Traffic and Congestion)	Utilities Manager™
Facilities Manager™ (Building Communications, etc.)	System Foundation
Network Manager™	Utilities Analyst™
Fleet & Equipment Manager™	Telecom Manager™
Maintenance Manager™ (MMS)	

TABLE 18-3 AgileAssets Inc., Products Overview

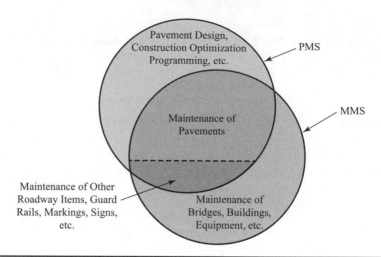

FIGURE 18-1   Interface of pavement management and maintenance management [after Hudson 11].

18.2.8.2   Idaho's Integrated AMS

Component review of the Idaho PMS system is described in [Perrone 10, Pilson 10, Scheinberg 10] and summarized in Figure 18-2. The output to other systems shown include MMS specifically.

The PMS functions in Figure 18-2 can be divided into the following categories: (1) configuration controls the way the software uses data contained in the PMS database, (2) a database is the repository of the PMS data collected by various organizations of units within the agency, (3) analysis provides the functions used to analyze pavement performance and optimize work programs, and finally, and (4) the reporting defines the ways that the system user can present the pavement data and the results of the analyses.

Idaho uses an integer optimization methodology with multiconstraint and multiyear analysis [Scheinberg 10]. As of 2012, 14 U.S. states were using AgileAssets' integer optimization methodology, while other states are using simpler methodologies, such as incremental benefit-cost analysis, general linear optimization, and simple ranking methods.

The MMS implemented in Idaho is a web-based system, which produces integrated asset maintenance management from strategic planning to daily activities [Pilson 10]. In its key concepts the MMS handles planning, scheduling, execution, and management of maintenance work on agency infrastructure assets. It also integrates with other available modules and external client subsystems to obtain aggregated data and provide transaction data.

Figure 18-3 shows how MMS and PMS provide key elements to a total asset management system and how from the initiation of PMS and MMS, integration in the IDT framework for AMS has rapidly developed.

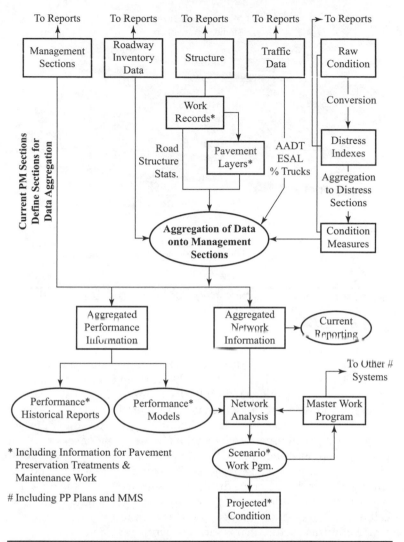

FIGURE 18-2   PMS system flow framework, as implemented in Idaho [after Perone 10, Pilson 10, Scheinberg 10].

18.2.8.3 Other Examples

North Carolina DOT has implemented an integrated AgileAssets system comprised of PMS, MMS, BMS, and asset trade-off analyst (ATOA) [Bhargava 13]. This reference presents the framework and applications of the integration as well as case studies which reveal the positive and stabilizing impact that maintenance has on network condition [Bhargava 11]. In [Azam 13] the same team outlines the concepts of its network safety screening, which is being implemented in the SMS (safety management system) for West Virginia DOT.

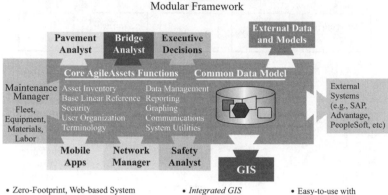

- Zero-Footprint, Web-based System
- All Transportation Assets
- *Agency-specific Models*
- Integrated Asset & Maintenance Management

- *Integrated GIS Mapping* Capabilities
- Secure and Scalable to Thousands of Users

- Easy-to-use with *Sophisticated Analysis*
- Powerful Reporting Tools

FIGURE 18-3   Asset management integration [after Pilson 10].

18.2.9   Stantec

18.2.9.1   Introduction

According to company brochures and their website (www.stantec .com), Stantec, formed in 1954 by Dr. Don Stanley, is a global consulting firm of 11,000 employees with 190 locations in North America, plus other international offices, which is "One Team Providing Integrated Solutions." Headquartered in Edmonton, Alberta, they provide public- and private-sector clients with a wide range of services from planning, engineering and architecture to environmental sciences to facilities and infrastructure projects. The world of Stantec as reported, simply put, "is the water we drink, the routes we travel, the buildings we visit, the industries in which we work and the neighborhoods we call home." In essence, Stantec says it is committed to a sustainable world of infrastructure that is safe, efficient, and stable.

18.2.9.2   Technology

Stantec employs a large suite of engineering, architecture, science, and other technologies with an open enterprise-based architecture and specialized packages for management of individual infrastructure assets involving, for example, the following basic functions:

- Data collection tailored to the asset
- Condition assessments
- Data processing
- Geographic information systems
- Global Positioning System (GPS)

- Infrastructure asset performance modeling and assessment
- Life-cycle analysis
- Software development and implementation
- Management systems for pavements, bridges, traffic, maintenance, water treatment and distribution, waste management, and others

In employing these technologies, the firm deals with large amounts of data. This requires extensive in-house computing facilities and expertise, plus the use of outsourced data storage and more recently the use of cloud computing. For example, stormwater management projects involve detailed GIS base maps, ESRI's Arcview software and GPS for locations, all integrated with an overall infrastructure management plan for system wide operations, system improvements, etc.

18.2.9.3 Infrastructure Sectors
Stantec services are provided in the following infrastructure sectors:

- Buildings
- Industrial
- Environment
- Urban land
- Transportation (airports, border crossings, bridges, intelligent transportation systems, parking, rail, roads and highways, toll facilities, and transit)

Each sector has a number of sub-areas, as shown in the list; for example, for transportation. These range from nine up to twenty areas for the industrial sector.

Accessing their website (www.stantec.com) enables anyone interested to click on a sector, to scan what services are offered in the various areas of the sector, and to obtain information on example projects by clicking on the individual photographs at the bottom of the screen.

18.2.9.4 Examples of Major Projects
Stantec has implemented pavement management in the state DOTs of Arizona, Minnesota, New Jersey, South Carolina, Tennessee, and Virginia. Virginia replaced Stantec in 2010. Stantec has also implemented systems in several Canadian provinces and municipalities involving over 60,000 km of roads [Haas 12].

The following is a representative selection to illustrate the range of infrastructure management projects the firm has carried out:

- City of Phoenix Water Services Inventory
- Barbados Solid Waste Management
- Buffalo Wild Wings Architectural Prototype, Arizona

- National Institute for Nanotechnology, Edmonton (Project Management)
- Nashik Airport Design, India
- Kuwait Integrated Infrastructure Maintenance Management System
- City of Phoenix Water Services Inventory
- Marvin Creek Neighborhood Development, North Carolina
- City of Claremont Parking Structure, California
- Macdonald's Development Program, Multiple Locations (since 1994)
- US 68 Double Crossover Interchange, Lexington, Kentucky

18.3 References

[Agile 12] AgileAssets website, www.agileassets.com, accessed December 10, 2012.

[Azam 13] Md. S. Azam, U. Manepalli, and P. Laumet, "Network Safety Screening in the Context of Agency Specific Screening Criteria," prepared for the *92nd Annual Meeting of the Transportation Research Board*, Washington, DC, January 2013.

[Bhargava 11] A. Bhargava, P. Laumet, C. Pilson, and C. Clemmons, "Using an Integrated Asset Management System in North Carolina for Performance Management, Planning, and Decision Making," paper for presentation at the *91st Annual Meeting of the Transportation Research Board*, Washington, DC, January 2011.

[Bhargava 13] A. Bhargava, A. Galenko, and T. Scheinberg, "Asset Management Optimization Models: Model Size Reduction in the Context of Pavement Management System," prepared for presentation at the *92nd Annual Meeting of the Transportation Research Board* and publication in the *Transportation Research Record*, Washington, DC, January 2013.

[Brunquet 07] J. Brunquet, "Comparative Assessment of Road Asset Management Systems Software Packages," University of Waterloo, Canada, July 2007.

[Deighton 10] Deighton Associates Limited, "Asset Management Implementation Framework for Colorado Department of Transportation," Colorado DOT, Denver, Colorado, September 2010.

[Haas 78] R. Haas, and W. R. Hudson, *Pavement Management Systems*, Kreiger Press, Malabar, Florida, 1978.

[Haas 94] R. Haas, W. R. Hudson, and J. P. Zaniewski, *Modern Pavement Management*, Krieger Press, Malabar, Florida, January 1994.

[Haas 12] R. Haas, K. Helali, A. Abdelhalim, and A. Ayed, "Performance Measures for Inter-Agency Comparison of Road Networks Preservation," Proc., Transp. Assoc. of Canada Annual Conf., Fredericton, October 2012.

[Hatcher 08] W. Hatcher, and T. F. P. Henning, "Lessons Learnt: New Zealand National Pavement Performance Model Implementation," paper presented at the *7th International Conference on Managing Pavement Assets*, Calgary, Alberta, 2008.

[Hatcher 09] A. W. Hatcher, "Highway Agency Integrated Asset Management Program Decision Support Tools-Current State of Art," Summary Report, draft document, August 2009.

[Hudson 70] W. R. Hudson, F. N. Finn, B. F. McCullough, K. Nair, and B. A. Vallerga, "Systems Approach Applied to Pavement design and Research," *Res. Rept. 123-1*, published jointly by the Texas Highway Department, Center for Highway Research of the University of Texas at Austin and Texas Transportation Institute of Texas A&M University, March 1970.

[Hudson 11] S. W. Hudson, K. Strauss, et al. "Improving PMS by Simultaneous Integration of MMS," presented at the *8th International Conference on Managing Pavement Assets*, Santiago, Chile, November 15–19, 2011.

[Mizusawa 09] M. Daisuke, "Road Management Commercial Off-the-Shelf Systems Catalog," Version 2.0, University of Delaware, February 2009.

[NCHRP 87] S. W. Hudson, R. F. Carmichael, L. O. Moser, and W. R. Hudson, "Bridge Management Systems," *NCHRP Program Report 300*, Transportation Research Board, December 1987.

[Perrone 10] E. Perrone, "Bootcamp Training—Pavement Management," ITD MAPS Project, Version 1.0, conducted by AgileAssets, Inc. for Idaho DOT, June 2010.

[Pilson 10] C. Pilson, "Bootcamp Training Manual for Maintenance Management," ITD MAPS Project, Version 1.0, conducted by AgileAssets, Inc. for Idaho DOT, updated January 2010.

[Scheinberg 10] T. Scheinberg, and P. C. Anastasopoulos, "Pavement Preservation Programming: A Multi-Year Multiconstraint Optimization Methodology," prepared for presentation at the *89th Annual Meeting of the Transportation Research Board*, Washington, DC, January 2010.

[TRDI 96] Texas Research & Development Foundation (TRDI), "Peru Bridge Management System," Ministry of Transportation, Communications, Housing and Construction, Republic of Peru, 1996.

CHAPTER 19

Benefits of Implementing an AMS

19.1 Introduction

Public infrastructure represents a major asset for the citizens and for agencies to manage effectively. Some use the term "asset management system (AMS)" and some "infrastructure management system (IMS)," but they are essentially synonymous. Using AMS herein, it must first be emphasized that good AMS is not business as usual. Far too people have proclaimed in various conferences and state meetings that "we are doing asset management," when in fact they are only using concepts rather than a functional management system. AMS often requires a change in the way existing agency personnel do their thinking and day-to-day work, that is, a change in paradigm. No one is initially comfortable with change. Information may be essentially raw data, abundant, sometimes cheap, easy to acquire, and sometimes hard to sift through. It is often provided in great quantities. Intelligence, which is processed, evaluated, and meaningful data, is more difficult to acquire and more expensive but is also much more important to AMS.

Models, methodologies, and procedures for AMS provide the function of transforming raw data into information and intelligence, an essential service for administrators and decision-makers. It is aimed at reducing risk in policy and budgeting decisions with respect to particular projects. For such a process to be viable, several prerequisites must be fulfilled, namely:

- Decision-makers must appreciate the importance of this service. They should also be capable of understanding the usefulness and limitations of the process. AMSs must be available, usable, and credible. This is not always the case, as resources applied to the process have often been inadequate or mandated.

- Public policies involving risks are most likely to be deemed acceptable when based on credible technical and engineering information, and where sound trade-off and impact analyses have been performed.

Most man-made systems (public and private), including infrastructure assets, are planned, designed, constructed, operated, and modified or rehabilitated under conditions of uncertainty and risk. Thus, the assessment, quantification, evaluation, and analysis of risk and reliability in the context of systematic asset management are essential to the process of good decision making.

19.2   Sources of Concern on Asset Management

While AMS has been around in one form or another for over two decades, there is still too often skepticism at middle- and upper-management levels in public agencies. The following list summarizes some causes of these concerns:

1. Insufficient planning and resources allocated for AMS implementation

2. Operation of the AMS is delegated to existing staff who do not understand or believe in the process

3. Lack of incentives to properly implement and document AMS

4. Insufficient basic research for the models in the AMS

5. Misuse and incorrect application of models

6. Overemphasis on use of "black box" computer packages

7. Proliferation of models and methods, and lack of systematic inventory of available methods

8. Improper calibration, validation, and verification of some AMS models

9. Lack of communication among developers, users, and beneficiaries of AMS

10. Lack of recognition of an AMS as "means" not "ends"

11. Lack of an interdisciplinary team in the development process, such as a statistician on the team

12. Strengths, weaknesses, and limiting assumptions of AMS often unrecognized by decision-makers

13. Insufficient data available to properly use the methods in early implementation

14. Inadequate understanding of the true costs and benefits of AMS

How can one go about "selling" an AMS? The process of communicating ideas about AMS is as important as the content of the communication. Determining what change should occur (what to do) is easier than determining how to bring about that change (how to do it).

A change as great as the implementation of AMS requires a carefully coordinated sequence of events, and changes in policy, training, personnel, and practice executed over a period of time. The process of implementing change is as important as the changes themselves, particularly if getting the changes adopted and used is desired. For at least 30 years it has been thought that as new leaders and decision-makers emerge in agencies, they would accept and even demand modern management systems to help them make decisions. This has proven to be true in 30 to 40 percent of state DOTs. However, the remaining majority have either ignored AMS and not used the results or have actually refused to implement an AMS in their agency. Many different reasons are given for these failures within the agency. Nevertheless it remains true that AMS is not effectively used in many agencies.

A number of organizational barriers can exist that work against change in general, and against the adoption of modern systems technology in particular. A social system, such as a public agency, often tends to resent change and to keep old policies and practices stable because they are perceived as "good." Major changes, such as developing and implementing an AMS, provoke complex reactions. Unless the interrelationships between different parts of an organization are accurately identified, the organization may subvert the intended change by reaction or inertia.

A major example of this inertia occurred recently in a major U.S. state DOT. The state legislature and the transportation commission appointed a blue-ribbon commission of highway stakeholders to study the problem of deteriorating highway quality in the state and to try to provide solutions for additional funding or methodologies to improve the quality of the transportation network. The group concluded their one-year study by indicating that the state needed to develop a much better AMS, including better processing and use of highway-condition data and integration of pavements with bridge management and congestion management. The agency response was to set up in-house research and hire local universities to develop Band-Aids on their existing 30-year-old PMSs and BMSs and did not consider integrating the two nor adding congestion and safety management to the package. It is not clear why this agency seemed to have ignored such a strong mandate from stakeholders and the legislature, but one possible reason relates to their organization. This state divides its highway department into a number of individual units or divisions, where the administrator of each unit has significant autonomy in choosing projects within its geographical area to expend the lump sum fund allotment they receive from the central office. At least some and possibly a majority of these divisions/district administrators desire to retain their autonomy in selecting projects and allotting funds within their division. Therefore, they may be resisting the development of a true AMS because they do not see it as assisting them in making the right decisions but rather see it as a way for the central office to usurp their authority and dictate funding allocations to them.

Thus they may be resisting the change to an AMS because they do not yet see that it is a major tool to help them within their own subdivision rather than a tool for dictating their actions.

This misunderstanding could probably be eliminated if the division administrators were called together and educated for one or two days on the details and the output and purpose of a good AMS. It is intended to assist them, not to dictate to them and it can show them the true financial and increased service-level benefits within their divisions of the state. This is only one of many examples that could be cited, but it is an important one since the agency seemingly is circumventing mandate from its legislature and major stakeholders within the state. It is also significant because the quality of the highways within the state is deteriorating but that deterioration is being largely blamed on lack of funding to provide needed reconstruction and maintenance. Clearly funding reductions have some impact but a good AMS can minimize this impact and clearly would improve the overall quality of the state's highway system.

Specifically, to successfully implement an AMS the following must occur. In the case study just described, Task 3, achieving substantial organizational agreement, did not seem to occur.

Task 1. Clarify issues of choice.

Task 2. Fully inform the decision-makers.

Task 3. Achieve substantial, effective, organizational agreement on a course of action that is feasible, equitable, and desirable.

Considerable attention is needed on the best way to communicate the benefits of AMS. There are several ways to do this, but it is clear that good, well-documented case-study examples of AMS are useful in communicating its benefits. But there must be specific details provided in the examples. Currently these may best be obtained at user-group meetings of AMS vendors as outlined in Chapter 18 or visiting agencies with successful AMS.

Any approach to the issues raised here requires the consideration of the costs as well as the benefits of performing asset management and implementing an AMS. The issues of cost are treated in the following section. Benefits of various kinds are treated subsequently.

19.3   Costs Related to AMS

There are several types of costs associated with AMS. The two primary ones are:

1. The costs of developing and operating an AMS, along with the necessary and appropriate data for using the AMS and keeping it current—that is, the cost of the AMS process itself.

2. The actual cost of the infrastructure itself; more correctly, the cost savings resulting from good management.

In reality, it is ultimate savings in total life-cycle expenditures that reflect the effectiveness of AMS. The total investment and related costs must be considered, along with savings and benefits that are realized from the effective AMS.

Many problems occur in assessing all of the costs outlined above. Apparent costs can vary greatly, depending on accounting procedures and methodologies within a given organization. Some agencies do not consider overhead or indirect costs at all when they are carrying out activities within their agency. On the other hand, if the same activities are done by contract or the use of consultants, then indirect costs are clearly included in the bid or final contract price for the work. All cost figures must be considered on a common basis, and should include all costs both direct and indirect of doing the necessary work.

The difficulty in obtaining cost information to be used in evaluating cost-effectiveness is due partially to the fact that many agencies do not have a fully implemented AMS, and those that do often have not kept effective comparisons and records of development costs. In the case of the facilities themselves, the costs associated with construction can be documented, but they usually occur over many years, and a common basis of comparison is difficult to obtain. More substantially, the costs of maintaining infrastructure are difficult to define, and few agencies have good maintenance cost information.

Several approaches can be considered. These involve pulling together costs on a semi-hypothetical basis for general comparisons. Case studies from individual agencies also can be useful. Particularly, cost information on pavement and bridge management in which contractor-related AMS activities have been carried out can be useful.

19.4   Benefits Associated with AMS

Table 19-1 presents a variety of detailed products and benefits associated with AMS. Some of these are couched in terms of government benefits to senior management and decision makers, and technical-level benefits for operators of an AMS, including designers and maintenance personnel. Table 19-2 summarizes some of the benefits that occur at agency level (using highway examples). Table 19-3, on the other hand, lists many of the costs the agency may incur. Table 19-4, later in the chapter, presents a case-study economic comparison, including costs and benefits of AMS. Although not comprehensive, it points out the factors involved, and shows the significant benefits that can accrue [Haas 95].

A key issue is that some benefits are quantifiable and measurable, but others are subjective and general in nature. Benefits must be brought to a common basis if they are to be compared with costs

A. General Factors	
Benefits Involved	**Costs Involved**
Realization of the magnitude of the investment and better chances of making correct decisions on spending funds Improved intra-agency coordination Improved use of technology Improved communication	Developing the AMS Data collection Operating the system
B. Elected Representatives Receive	
Benefits	**Costs**
Ability to defend/justify funding maintenance and rehabilitation Having assurance that programs represent best expenditure of public funds Reduced pressure from constituents to make political program modifications Getting objective answers to the implications of • Lower levels of funding • Lower standards of level of service (LOS)	Development and implementation Assigning personnel Operating of the system
C. Senior Management Receives	
Benefits	**Costs**
Comprehensive, assessment of • Current status of the network • Expected future status (predictions) Objectively based answers to • What level of funding is required to keep the current LOS? • The implications of large or smaller budgets • The implications of deferred work • The implications of lower LOS Being able to objectively justify capital spending and maintenance programs to the elected council, legislature, or public Having the assurance that the recommended program represents the best use of available funds Being able to define the "management fee" as a percent of the capital and maintenance spending (i.e., the cost of the AMS divided by the spending)	Data collection Developing and installing the AMS Staffing On-going costs of operating the AMS Assigning personnel

TABLE 19-1 Benefits and Costs That Occur with AMS

D. Technical Personnel Receive	
Benefits	**Costs**
Explicit recognition of administrative and operating elements of the organization	Having to make changes in the ways things have been done
Increased awareness how to use available technology; improved communication with design, construction, maintenance, planning, etc.	Possible extra time and effort to upgrade skills and learn new procedures
Satisfaction of providing best value for available funds	Training and education costs
Capability of assessing the implications of less funds, lower standards, etc. (as previously mentioned) but also capability of making the case for higher standards	Modification and retraining for existing activities

TABLE 19-1   Benefits and Costs That Occur with AMS (*Concluded*)

1. Project maintenance and rehabilitation needs and budgets for maintenance and rehabilitation
2. Justification for requesting funds
3. Measure effectiveness of funds spent
4. Reduce user cost by providing a known level-of-service (LOS)
5. More effective expenditures on the system
6. Improved methods for:
   - Planning
   - Design
   - Construction
   - Maintenance and rehabilitation
   - Research models
7. Improved safety
8. Optimum programming for projects
9. Knowledge of statewide conditions and maintenance and rehabilitation needs by assets
10. Improved serviceability level of asset portfolios
11. Savings in user costs
12. Use of database to develop/improve performance models
13. Reduce rehabilitation costs because of timely action on maintenance and repairs

TABLE 19-2   Benefits of Asset Management at the Agency Level

1. Labor for task force within the department to implement the AMS
2. Consulting services or cost for buying the system based on the objectives and scope defined by an agency
3. Data collection (often already in place but better use results from AMS)
   a. Agency personnel (engineers, technicians, equipment operators)
   b. Training
   c. Condition-monitoring equipment
      - Vehicles for transportation
      - Data loggers for condition surveys
      - Automated distress survey devices
      - Nondestructive testing equipment
      - Roughness measuring equipment
      - Drilling and coring equipment
      - Construction acceptance (testing)
   d. Traffic-control services
   e. Travel and subsistence cost for data collection
   f. Traffic or use data (e.g., AADT, ESAL, and WIM)
4. Data processing (often existing)
   a. Personnel (system analyst, programmers, and technicians)
   b. Equipment (computers, terminals, storage and backup devices, and calculators)
   c. Supplies
5. Preparing data analysis and reports (actually a saving with automated AMS reports)
   a. Personnel (engineers, system analyst, and technicians)
   b. Computers and terminals
   c. Supplies
6. AMS operation and maintenance (can be fixed by purchasing a COTS AMS
   a. Personnel (engineers, system analyst, technicians, computer support staff)
   b. Computer and data collection equipment maintenance
7. Retraining of agency personnel associated with asset management and AMS database
8. Costs of administering the AMS

TABLE 19-3 Costs of Asset Management at the Agency Level

in a truly analytical way. A big problem with AMS today is that user benefits are almost totally excluded from the calculations of AMS decision making. It was considered sufficient in the past to provide "adequate facilities" to serve the public. The idea of improved benefits accruing to the user based on better service and lower operating costs has not been adequately exploited within United States, although it is widely used in World Bank calculations for developing country projects.

Although some work has been done, information of a comprehensive nature about the benefits of timely maintenance and rehabilitation compared to its costs is still inadequate. Another benefit of AMS is an improved awareness of the management process. The authors have seen this occur during the teaching of pavement and asset management graduate-level courses at the Universities of Texas, Waterloo, and Mississippi, where persons taking the course for a variety of reasons (some were assigned by their supervisors) turn into real advocates of AMS, and went back to their jobs with a renewed enthusiasm for providing good facilities for public use.

A final illustration of AMS benefits that are hard to quantify is the spin-off of AMS technology to other areas. After 10 years of PMS use, for example, attention was turned to the development of BMS (bridge management systems), which has now become standard nationwide in the United States [Hudson 87] and many other countries, some based on a software package called Pontis [AASHTO 05].

19.5   Evaluation Methodologies

Some attention has been given to the various methods for comparing costs and benefits. The authors examined numerous references (cited at the end of this chapter) treating many of the widely used methods, ranging from discriminate analysis to general decision theory. Of the many methodologies evaluated, the most likely candidates are reviewed here.

There is conflict between goals and objectives in certain groups. No amount of rational data will eliminate all such conflict. It is impossible to please all individuals and groups playing their various roles at different times and places in the management process. Decision making must, however, be improved if better facilities are to be provided long term at the lowest cost.

19.5.1   Benefit-Cost Criterion

Perhaps the best-known criterion for measuring the efficiency of an activity is the benefit-cost analysis or the benefit-cost ratio. Efficiency is discussed in terms of benefit-cost analysis because other variations, such as the rate of return, are sufficiently similar to the benefit-cost analysis to have the same strengths and weaknesses. Because of its theoretical foundation, the benefit-cost analysis provides a conceptually sound basis for effective comparisons. In practice, however, many difficulties arise that reduce its usefulness. The biggest drawback to benefit-cost analyses arises from the difficulty in breaking the factors into either the cost or benefit category, and the difficulty of actually measuring the true cost and the true benefit. There are many intangibles in benefit-cost analyses, which can be treated in three ways:

1. They may be rated subjectively and included in the analysis.

2. When subjective scaling is not possible, verbal descriptions of intangible benefits can be provided, in addition to the measured costs and benefits, and these can be used to aid the decision-maker.

3. Unfortunately, the most common treatment for intangibles is to simply ignore them.

In other words, benefit-cost analyses are preoccupied with measurable impacts and often omit intangible benefits. This omission results in a sort of inflexible narrowness that invalidates the benefit-cost approach or makes it at best misleading.

Another problem associated with benefit-cost analyses are the issues of "who pays the cost" and "who receives the benefits." It is not always easy to ascertain whether the benefits and costs occur to the same stakeholders. For example, the improved programming of maintenance funds is a benefit of AMS that accrues to the agency, and thus to the public or taxpayers. The cost, on the other hand, may involve a change in agency work assignments, or the requirement of additional training for various agency personnel. They (the personnel) must be convinced to willingly "pay this extra cost," even though the benefits accrue to someone else. Many are not willing to do this.

19.5.2 Excess Benefits

There are many variations of cost-benefit analyses. One of these is the comparison of cost and benefits by calculating the excess of benefits over costs. A simple case-study comparison of this methodology is shown in Table 19-4 [Haas 95]. It has no particular value over other

	Present Dollars		
Costs	**AMS Costs**	**Needs Study**	**Difference**
Capital costs	$8,671,961	$8,627,988	$43,974
Maintenance costs	$1,804,406	$1,798,209	$6,197
Study costs*	$75,000	$0	$75,000
Total costs of AMS	$10,551,367	$10,426,153	$125,214
User benefits of AMS	$110,013,920	$98,847,056	$11,166,864
Net benefits of AMS	$99,462,553	$88,420,903	$11,041,650

*AMS cost based on $200/mi. Note that the Needs Study has been assigned no extra cost.

TABLE 19-4 An Early Case Study: Economic Comparison of a Yearly Priority List and a Subjective Programming Approach [Haas 95]

benefit-cost ratio methods and will not be considered further, except as a part of the benefit-cost ratio concept.

19.5.3 Goal Achievement

As you attempt to use the various methods for analyzing costs and benefits, and review criticisms of benefit-cost and similar evaluation procedures, you may be dismayed with the seemingly overwhelming complexity facing the decision-maker. You also gain some appreciation for the problem of the chief executive officer (CEO) or the manager, who must react and give solutions for similar complex problems every day. The question arises: What, if anything, can be done to improve decision-making procedures? One technique for broadening the evaluation and decision-making process is known as "the goal-achievement method" [Dickey 83]. In this approach, potential alternatives are assessed in terms of impacts in comparison to the objectives proposed. Quantifiable measures are employed in this process, although some may be subjective and even probabilistic. In general, the procedure is to establish various criteria or "goals" for each alternative methodology. Quantitative measures or subjective estimates are then made for each criterion for each of the alternatives. These are then standardized and compared on the basis of a total score of 100 to see which alternatives most completely fulfill or achieve the goals of the decision-maker(s). This method is useful only when goals are scrupulously assigned.

19.5.4 Cost-Effectiveness Technique

The cost-effectiveness technique (C/E) is an alternative approach to the goals-achievement procedure. C/E is a less sophisticated procedure and works on the basic premise that better decisions will arise when clear and relevant data are supplied to the decision-maker. No specific attempt is made to put all benefits and costs in common units, such as the dollar. In his book, Dickey [Dickey 83] quotes Thomas and Schofer on the C/E approach:

> Because many of the consequences and outputs from a transportation system are intangible and otherwise difficult to evaluate in some common metric, the decisions regarding the conversion to a single dimension and hence the plan selection decisions are necessarily subjective in nature, at least at the present time.
>
> What might be more useful at this time is a technique for providing the kind of informational support for the selection among plans which recognizes the complex nature of these decisions. Such a decision-supporting framework does not attempt to *make* decisions, but instead *structures the information* required for making a subjective, but systematically *enlightened* choice. At the same time, however, the framework must be sufficiently flexible to permit the adoption of more sophisticated techniques, such as analytical

methods for realistically implementing benefit-cost analysis or ranking schemes, when such techniques are appropriate.

Dickey lists three criteria that any framework for evaluation should satisfy:

1. It should be capable of assimilating benefit-cost and similar methodological results, in addition to other information.

2. It should have a strong orientation toward a system of values, goals, and objectives.

3. It should allow for the clear comparison of trade-offs or compromises between objectives, and/or should make explicit the relative gains and losses from various alternatives.

Effectiveness is defined as "the degree to which an alternative achieves its objective." This definition, by itself, helps to overcome one of the major objections to the benefit-cost approach, in that goals are specified explicitly and are not covered by an all-encompassing "benefit" term. In benefit-cost analysis, for example, benefits are related to reductions in user cost, user-operating costs, and user time, but in a particular situation, any one of these factors may be of little concern. The objectives to be met may be associated with an entirely different set of factors specified explicitly, and not covered by an all-encompassing "benefit" term.

The value of the cost-effectiveness approach is threefold:

1. It simulates, to some extent, the process by which decisions are made.

2. It allows for clearer delegation of responsibility between analysts and decision-makers.

3. It makes it easier to provide relative information, structured in an understandable form so that the choice process is simplified.

19.5.5   Search and Choice

In Manheim's classic work on transportation systems analysis [Manheim 69], he deals with "search and choice in transport systems analysis," and outlines what he calls PSP (problem-solving process) for use in dynamic modeling of decision making. He outlines this process as shown in Figure 19-1.

According to Manheim, the focus of PSP is on actions. Because search and selection procedures concern the basic processes of generating and selecting actions, these procedures are at the heart of the PSP. However, there are other activities that occur in PSP to assist search and selection, and to revise the context in which they operate. In particular, goal formulation and revision procedures are important.

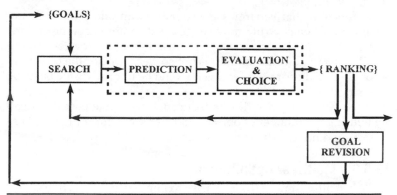

Figure 19-1   Basic cycle of PSP (problem-solving process) [after Manheim 69].

Although the PSP activity is a valid process for a decision-maker to go through, it does not appear to be an appropriate methodology for use in the next phase of AMS.

19.5.6   Statistical Decision Theory

It is easy to forget that the world is a very uncertain place and become fascinated by specific answers produced by complex models and elaborate calculations, such as benefit-cost analyses. In truth, there is always uncertainty in any analyses. Uncertainty in transportation is of at least three types: demand (such as traffic), technology, and goals. No matter how elaborate the prediction models are, or how much data are collected, there will always be uncertainty about predictions of future traffic, environment, pavement performance and life, as well as future maintenance costs and inflation. Statistical hierarchical decision processes are also outlined by Manheim [Manheim 70]. He has developed a statistical decision approach to complicated transportation planning theories. There are, however, limitations to statistical decision theory, as described in [Raiffa 97, Hammond 09] and to date this method does not appear to be applicable to AMS.

19.5.7   Discriminate Analysis

Discriminate analysis and classification are multivariate techniques concerned with separating distinct sets of objects and with allocating new objects to previously defined groups. To obtain the best results, analysts seek discriminates whose numerical values are such that the collections (groups) are separated as much, or as distinctly, as possible. The goal of classification is to sort objects into two or more labeled classes. The emphasis is on deriving a rule or rules that are used to optimally assign a new object to the labeled classes.

A function that separates may serve as an allocator, and conversely an allocatory rule may suggest a discriminatory procedure. In practice, the distinction between discrimination (or separation) and classification (or allocation) is not so clear. One of the objectives of conducting discriminate analysis would be to provide the basis for a classification rule.

The methodology of discriminate analysis, while useful in dealing with a large number of objects, does not seem appropriate for use in AMS, and no further review is recommended [Lee 85].

19.5.8   Multiyear Optimization

While benefit-cost ratio (B/C) is easy to implement, it has two severe flaws:

- The method works with one constraint only.

- As such, it can be used to solve only a single period allocation problem, whereas AMS fund allocation is multiyear and has several constraints.

For a single year, consider the following problem and three constraints:

1. Minimize total 10-year budget.
2. Provide 80 percent of all assets with a performance index of at least 80 (on 1 to 100 scale).
3. Provide 95 percent of major assets with a performance index of a least 86.
4. Provide 90 percent of all assets in district with an index of at least 90.

It is not possible to use B/C to solve this problem even for one year since there are several constraints.

Allocating funds year by year can result in poor maintenance plans. Since the model is not aware of time dimensions it tends to use many cheap fixes. In contrast, when optimizing over a 10-year horizon, a multiyear algorithm may use more expensive but high-quality treatments and get better average performance over 10 years while saving money. It is possible to save 20 percent when switching from year by year B/C methods to multiyear optimization. True optimization does take longer to calculate but new algorithms in use have reduced the time to 10 minutes or less for a big network.

19.5.9   Other Methodologies

Other methodologies are used for decision making. Among these are terms such as benefits/risk analysis and "preference and value trade-offs." Within the confines of this book, a considerable amount of effort has been assigned to examining methodologies. No additional methodologies were found that appear to be adequate to the subject of AMS.

Any evaluation used must be meaningful to the decision-maker. After all, the purpose is to present information that will be useful in convincing decision-makers to implement improved AMS methods. This concept of acceptability to the decision-maker must be given full attention in developing AMS.

19.5.10   Summary Comments on Methodologies

Three of the methodologies outlined above may be useful in clarifying the value of AMS. Certainly, benefit-cost analyses cannot be dismissed—comparing costs and benefits for a sample case-study network is an effective way of convincing decision-makers of the value of AMS. Cost-effectiveness techniques also bear additional study, and should be given strong consideration.

Finally, the general concept of goal-achievement methods should be explored. It is not clear exactly how the method might be applied, since it involves an examination of the goals of each individual decision-maker. This might be handled with hypothetical examples, if interviews with several decision-makers could be arranged for information-gathering purposes. However, in the last 15 years, no major AMS software has used goal-achievement methods.

19.5.11   Data Needs

It is important to address the data that would be needed for each of the methods considered. In all methods, costs data are needed: the costs of developing and using AMS, as well as the total costs of the components in the infrastructure site or network. Table 19-1 summarizes in four categories the types of cost data that should be included. Some of these data exist, but must be collected carefully. Other cost data will be easier to obtain, and more details on the kinds of cost information needed are given in Table 19-3.

In addition to cost data, which are common to all methods, the other data needs vary among the methods as follows.

19.5.11.1   Costs-Benefits

For the costs-benefits method, data will be needed on the benefits outlined in Tables 19-1 and 19-2. It will be necessary to locate quantitative information on as many of the benefit classes as possible. In many cases, it is difficult to locate exact financial information, and therefore estimates will have to suffice. It is also necessary to assess source and disposition of benefits; that is, "who pays the cost and who receives the benefits."

19.5.11.2   Goal Achievement

The goal-achievement method will require inputs on the goals of CEOs in regard to their infrastructure network. The goals must be assessed in quantitative terms or in terms of criteria. They will take the form of factors such as "higher levels of serviceability," "longer average life," and "reduced user delay time." It is anticipated that several interviews with agency administrators will be needed to gather data on such goals and criteria.

19.5.11.3 Cost-Effectiveness

Cost-effectiveness requires clearly structured cost data. More importantly, a set of goals must be specified explicitly; these are determined from decision-maker discussions and opinions. Final objectives must be set forth from the goals in clear, qualitative terms. Also, the rankings that CEOs assign among these objectives must be known.

19.5.11.4 Multiyear Optimization

This method will calculate the minimum budgets required to use constraints selected for study.

19.6 Summary

It appears that adequate information can be gathered to make valid comparisons, which can be presented in a unique and meaningful way to the target audience.

Costs and benefits are hard to quantify. For this reason, cost-effectiveness and goal-achievement methods need to be examined in more detail.

Much is said in this chapter about better defining the benefits of AMS, and this should be done. It is clear, however, from the developments in this field over the past 10 years that the overall benefits of good management are pervasive and large. This is clearly noted in the private sector, and must also be true in the public sector. The principles and methods provided in this book can help to fulfill this important goal.

19.7 References

[AASHTO 05] "Technical Manual of Pontis (Release 4.4)," American Association of State Highways and Transportation Officials (AASHTO), 2005.

[ASCE 82] "Multiobjective Analysis in Water Resources," *Proceedings of the Engineering Foundation Conference,* ASCE Water Resources Planning and Management Division, New York, November 1982.

[Dickey 83] J. W. Dickey, W. J. Diewald, A. G. Hobeika, C. J. Hurst, N. T. Stephens, R. C. Stuart, and R. D. Walker, *Metropolitan Transportation Planning,* 2nd ed., Hemisphere Publishing, McGraw-Hill Book Company, New York, 1983.

[Goicoechea 82] A. Goicoechea, D. R. Hansen, and L. Duckstein, *Multiobjective Decision Analysis with Engineering and Business Applications,* John Wiley and Sons, New York, 1982.

[Haas 78] R. Haas, and W. R. Hudson, *Pavement Management Systems,* McGraw-Hill, New York, 1978.

[Haas 95] R. Haas, and W. R. Hudson, "The Customers for Pavement and Maximizing Their Satisfaction," *Proceedings, Canada Transportation Conference,* Victoria, B.C., October 1995.

[Hammond 09] J. S. Hammond, R. L. Keeney, and H. Raiffa, "The Hidden Traps in Decision Making," *HBR On Point Enhanced Edition,* Digital, March 3, 2009.

[Hudson 87] S. W. Hudson, R. F. Carmichael III, L. O. Moser, and W. R. Hudson, "Bridge Management Systems," *National Cooperative Highway Research Program Report 300,* Transportation Research Board, National Research Council, Washington, DC, December 1987.

[Keeney 93] R. L. Keeney, and H. Raiffa, *Decisions with Multiple Objectives: Preferences and Value Tradeoffs,* John Wiley and Sons, New York, 1993.

[Lee 84] H. Lee, and W. R. Hudson, "Development of a Program Level Pavement Management System for Texas," *Research Report No. 307-3,* Center for Transportation Research, The University of Texas at Austin, December 1984.

[Lee 85] H. Lee, W. R. Hudson, and C. L. Saraf, "Development of Unified Ranking Systems for Rigid and Flexible Pavements in Texas," *Research Report No. 307-4F,* Center for Transportation Research, The University of Texas at Austin, November 1985.

[Luce 57] R. D. Luce, and H. Raiffa, *Games and Decisions,* John Wiley and Sons, New York, 1957, Chapter 13.

[Manheim 69] M. L. Manheim, "Search and Choice in Transport Systems Analysis," paper sponsored by Committee on Transportation System Evaluation and presented at the 48th Annual Transportation Research Board Meeting, *Transportation Research Record 293,* Washington, DC, 1969, pp. 54–82.

[Manheim 70] M. L. Manheim, "Decision Theories in Transportation Planning," *Special Report No. 108,* Highway Research Board, National Research Council, Washington, DC, 1970.

[Manheim 75] M. L. Manheim et al., "Transportation Decision Making: A Guide to Social and Environmental Considerations," *National Cooperative Highway Research Program Report No. 156,* Transportation Research Board, National Research Council, Washington, DC, 1975.

[NCHRP 87] S. W. Hudson, R. F. Carmichael, L. O. Moser, and W. R. Hudson, "Bridge Management Systems," *NCHRP Program Report 300,* Transportation Research Board, National Research Council, Washington, DC, December 1987.

[Raiffa 97] H. Raiffa, *Decision Analysis—Introductory Lectures on Choices under Uncertainty,* Addison-Wesley, Reading, Mass., 1997.

[Smeaton 81] W. K. Smeaton, M. A. Karan, C. Bauman, and G. A. Thompson, *Case Illustration of Long-Term Pavement Management and Update,* Can. Tech. Asphalt Association, November 1981.

[UCOWR 81] "Risk-Benefit Analysis in Water Resources Planning and Management," *Proceedings, UCOWR-Sponsored Engineering Foundation Conference,* Y. Y. Haimes (ed.), Center for Large Scale Systems and Policy Analysis, Plenum Press, New York, 1981.

CHAPTER **20**

Sustainability, Environmental Stewardship, and Asset Management

20.1 Introduction

Sustainability is a valid and widely accepted concept with a number of definitions and uses. For example, the U.S. DOT's Federal Highway Administration (FHWA) now calls a sustainable pavement one that can be maintained and renewed to provide continuous long life. We prefer to simply call that "good life-cycle performance." While sustainability is perhaps one of the most overused words in the English language, applied to almost everything involving environmental and ecological concerns, the concept is in fact particularly important to long-life assets.

Sustainability and environmental stewardship are often coupled in discussion. However, they are different but interrelated. Sustainable infrastructure will certainly involve environmental stewardship. On the other hand, good environmental stewardship might indicate the need for strong air pollution limits that reduce the life of the assets under consideration, in this case, reducing the sustainability. We can discuss these two separately. As far back as 1987, the Brundtland Commission [Brundtland 87] provided the following global definition for sustainability: "Sustainable development is development that meets the needs of the present without compromising the ability of future generations to meet their own needs." This definition is broad enough to still serve us today. But there is an issue of "needs" not always being the same as "wants" by individuals, groups or even whole societies.

The FHWA has a definition for sustainable transportation [NCHRP 11]. It is defined as "... providing exceptional mobility and access in a manner that meets development needs without compromising the quality of life for future generations. A sustainable transportation system is safe, healthy, affordable, renewable, operates fairly, and limits emissions and the use of new and nonrenewable resources." Sustainability and environmental stewardship, therefore, remain broad concepts and ideas. Some action is being taken but much more is yet to be done in the way of quantifiable inclusion of these concepts in asset management systems.

Such delay in action is not unusual. Pavement management was conceptualized in the mid-1960s and early implementation was simplistic, taking distress surveys and allocating funds to fix the worst pavements first. The project-level concept of considering maintenance alongside design was hard to grasp and even harder to sell to state agencies. Since that time these systems have developed into full-blown quantitative, multidimensional asset management packages in worldwide use. The same is true for bridge management. Asset management was broadly conceptualized in the 1990s to encourage decision-makers to consider resource allocation across an array of assets, pavements, bridges, safety, etc. and not allocate money to each in isolation. This conceptual approach was helpful and probably resulted in better fund distribution. But it has only been in about the last decade that quantitative asset management has evolved using bottom-up detailed data and analyzing it to make more realistic decisions across all assets.

Sustainability and environmental stewardship are now in a conceptual stage as a part of asset management. Informed decisions are being made about these issues as we will see in case studies discussed herein. But there are so many new factors with untested results that quantitative proven answers are hard to produce. As someone once said, "predictions are always difficult to make and especially about the future." Make no mistake, it is the future effects of sustainability and environmental conditions that we must predict. Further, we have to recognize that being "green" is not sufficient. Sustainability, if intended for the long term, must include continuity and stability with regard to the following [Haas 13]:

- Life-cycle management framework
- Adequate resources/preservation of investment
- New ideas and innovation
- Institutional efficiency and effectiveness
- Data and information/analysis
- Technology improvements
- Retention and renewal/knowledge management

20.2   Sustainability Related to Climate Change Adaptations

20.2.1   Consideration of Climate Change Risks

In 2010 to 2012 extreme heat and drought, superstorms, hurricanes, floods occurred in the United States and many countries worldwide perhaps caused by climate change mechanisms. Extreme weather-related natural disasters affect public mobility, disrupt supply chain and freight logistics, damage electric power infrastructure, and result in huge economic costs. Agencies responsible for public infrastructure must address the risks associated with climate change.

Airports have many types of infrastructure assets from pavements to buildings, parking structures, pipelines, bridges, communications, and concessions. In a recent Airport Cooperative Research Program (ACRP) report [ACRP 12a] the information from a survey of U.S. and Canadian airport operators and available literature were synthesized for the risks associated with climate change and approaches to mitigate these risks. Some examples follow:

- At the Oakland International Airport in California, sea-level rise and post-hurricane dike standards are being considered as important in design changes.

- Climate change risk was considered for stormwater system reviews and water quality regulatory compliance activities at Canada's Toronto Pearson International Airport.

- Disruptions from extreme weather events and risk management systems were believed to be important for climate change adaptations, but this approach was not a formal practice on U.S. airports.

- A sustainability management plan for Hartsfield-Jackson Atlanta International Airport was in place in 2010 using Federal Aviation Administration (FAA) funding. A chapter on sustainability management and asset management is being developed for the airport's master plan.

20.2.2   Sustainability Rating Scales

A sustainability rating system quantifies sustainable practices on a common rating scale. An example of a point-based sustainability rating for roadway design and construction is Greenroads, developed by the Greenroads Foundation that also provides four certification levels (www.greenroads.org/, accessed December 12, 2012).

20.3 Environmental Stewardship

20.3.1 The Role of Leadership in Energy and Environmental Design

Leadership in Energy and Environmental Design (LEED), a comprehensive sustainability rating system, has become prominent in recent years, particularly in the design, construction, and operation of buildings. It is a process directly related to sustainability and environmental stewardship. Developed by the Green Building Council (USGBC) [LEED 08], LEED consists of a suite of rating measures, which has grown to encompass more than 7,000 projects in the United States and 30 other countries, at the time of writing this book. The highest rating of nine categories is platinum, which would represent a building with essentially zero "footprint" regarding energy usage, emissions, environment, etc. For example, the Student Services Building at the University of Texas, Dallas, is the first academic building in that state to receive LEED platinum status, in 2011. A directory of LEED-certified projects is provided online by the U.S. Green Building Council, and in Canada by the Canada Green Building Council [LEED 08].

It should be noted that LEED is a design tool, not a performance-measurement tool, and it is not modified to be climate specific. For example, a building in Maine would be rated the same as one in Arizona regarding water conservation, which is obviously inconsistent. While the initial capital cost of a LEED-certified building can be high, the evidence is mounting that resulting life-cycle operating-cost savings alone may, in fact, lead to a very positive return on investment.

To access the vast body of information on the LEED system and related projects to date, including numerous references, an Internet query will direct you to Wikipedia and other online sources.

20.3.2 Examples of Major Initiatives toward Conservation of Resources

The pavement field consumes large amounts of energy and other resources in producing and processing materials, hauling, and paving, etc. As a result, recycled asphalt pavement (RAP) has become an extensively reused material today. It has become cost-effective to use RAP in most areas; making it, in essence, a valuable resource itself. In addition, the economics and energy savings from the technology and engineering have advanced. More recently, another technology called "warm mixes" has been growing in use, which consumes less energy in mixing and can be compacted at lower temperatures [Hutschenreuther 10]. A summary of these two examples follows.

In the early years of recycling, agencies tried hot in-place recycling on major highways. But because of past weathering, processing issues, and old maintenance that had been performed, it was hard to predict the properties of the recycled mix in place. As a result, many RAP pavements failed rapidly at great expense. After many years of trial and error and testing, most highway agencies now limit the amount of recycled material used on major road reconstruction to 10- to 20 percent of the total mix, although larger percentages are often used on secondary roads. The experience has been that proper engineering of mixes and in related aspects is absolutely essential, even though, there is little proof that they will last as long as 100 percent virgin mixes. It takes time to validate such results. On the plus side there is a body of literature and experience now in existence, which can be accessed in relevant journals such as the *Journal of Asphalt Paving Technology* and in such organizations as the National Asphalt Pavement Association (NAPA) and the Asphalt Institute in the United States, the European Asphalt Pavement Association (EAPA), and the Australian Asphalt Pavement Association (AAPA).

In the case of warm asphalt mixes, the industry has determined over time that proper density of hot mixes is essential and can be done with the right equipment, procedures, mix design, temperature, etc. Warm mixes are harder to densify than hot mixes. They require additives now on the market to achieve reasonable density. This increases costs, and the additives may of themselves create an environmental problem in their own production. While the motivation for warm mixes is largely energy savings, and again while there is a growing body of information available, proof as to how warm mixes will actually perform in the long-term remains to be determined. In any case, comparing the trade-offs of cost, life-cycle performance, savings on energy and various emissions is needed. Only when these factors are adequately quantified can warm-mix performance be considered in an asset management system in the true sense.

20.4   Incorporating Sustainability and Environmental Concerns into Asset Management

New targets for reductions in GHG emissions are being proposed at state, city, and local levels for the infrastructure services offered by public and private agencies. Sustainability measures are being incorporated by states, cities and airports in new facilities, as well as retrofitting and replacement strategies. Future studies will be needed to determine the true value of such efforts. The following discussion is directed toward full-fledged incorporation of sustainability and environmental concerns into a quantitative infrastructure asset management system (IAMS) in the near future. Progress is being made, as we will see in some of the cited case studies.

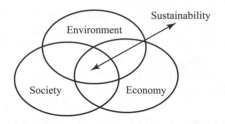

FIGURE 20-1 Fundamental sustainability model [after CH2M Hill 09].

20.4.1 Overview of Sustainable Practices

In 2011, an NCHRP study [NCHRP 11] evaluated sustainable practices in pavements. They worked from Figure 20-1 [CH2M Hill 09] as a framework. During their studies, the team surveyed all U.S. states and Canadian provinces. They identified eight case studies for elaboration as shown in Table 20-1. We do not have space to address all eight cases but will summarize two of them: (1) GreenPave from MTO Toronto, Ontario, and (2) the New York State Department of Transportation, GreenLITES. The following details were extracted from the NCHRP report and related references [MTO 10, NCHRP 11].

Other efforts can be cited. For example, in 2008 the state of Oregon installed 600 solar panels along the highway south of Portland, claiming to be a solar-powered highway. The renewable energy source is used to power road lighting, signs, and tunnel fans [Klinger 12]. The author further gives the following examples from Europe. Italy's 80,000 roadside solar panel system produces 12 million kWh per year, and Germany's 16,000 solar panel system along their roadways produces 3.2 million kWh per year. These examples show how sustainable practices are changing the landscape for the future.

20.4.2 Case Study: GreenPave, Ministry of Transportation, Ontario

GreenPave is a new tool that builds on more than 20 years of research and implementation of green initiatives at the Ministry of Transportation, Ontario, Canada. The system has recently been updated with a green-rating system for pavements. The program focuses on assessing the greenness of pavement designs and construction. Subsequent maintenance is not considered in the analysis. GreenPave takes some of its initiative from LEED certification, but it was also influenced by green roads, GreenLITES, and the Transportation Association of Canada's guide for green roads. Table 20-2 summarizes the key features and how they relate to environmental sustainability impact areas. GreenPave makes a direct effort to quantify environmental sustainability with MTO's long-standing green pavement initiatives (GPI) program. GPI has been in

Case Study	Agency/Location	Reason for Inclusion
Environmental Stewardship Program	Arizona DOT Phoenix, Arizona	Comprehensive system-wide framework for a DOT
Transportation Sustainability Plan	Oregon DOT, Salem, Oregon	Evaluation of air quality and water quality at a system level
Sustainable Highway Materials	Washington State DOT, Olympia, Washington	Evaluation using a practical and straightforward approach to assessing sustainability impact factor areas
GreenPave	Ministry of Transportation Ontario, Toronto, Ontario, Canada	Evaluation using a practical and straightforward approach to assessing sustainability impact factor areas
Dirt & Gravel Maintenance Program	Pennsylvania State Conservation Commission, Pennsylvania	Comprehensive assessment of dirt and gravel road maintenance
Greenworks Program	City of Philadelphia, Pennsylvania	Comprehensive city overview of sustainability impact factor areas
GreenLITES Program	New York State DOT Albany, New York	Utilizes a spreadsheet-based self-assessment of design that could be adapted to the design/selection of pavement preservation and maintenance treatments
Sustainable Land Transport	New Zealand Transport Agency Wellington, New Zealand	Provides specific guidance for low-volume roads

TABLE 20-1 Case Study Program Summary

place since 1988 and is credited with recycling more than 1.2 million tons of aggregate as well as the following reductions in GHGs:

- 88,400 tons of carbon dioxide
- 720 tons of nitrous oxide
- 15,400 tons of sulfur dioxide

These figures are reported in [Lane 10], but it is not clear how the results are broadly transferable. Certainly there has been improvement in GHGs using this program. GreenPave may be one of the most quantitative efforts yet to add sustainability to use in asset management.

Sustainability Impact Factor Area	Description/Relevance to Pavement Preservation and Maintenance Treatments
Virgin Material Usage/ Alternative Material Usage	• This is directed at optimizing usage/reusage of recycled materials and to minimize material transportation distances. • A total of 14 points is assigned to this category. • Directed at design, construction, and rehabilitation, but possibly could be modified for use with pavement preservation and maintenance treatments.
Program for Pavement In-Service Monitoring and Management	• Attempts to optimize sustainable designs to include long-life pavements, permeable pavements, noise-mitigating pavements, and pavements that minimize the heat island effect. • Total of 9 points is assigned to this category. • Directed at design, construction, and rehabilitation, but possibly could be modified for use with pavement preservation and maintenance treatments.
Noise	• Directs the use of means to minimize road noise.
Air Quality/Emissions	• Minimizes energy consumption and GHG emissions. • Total of 9 points is assigned to this category. • Directed at design, construction, and rehabilitation, but possibly could be modified for use with pavement preservation and maintenance treatments.
Energy Usage	• Directs consideration of fuel consumption in decision making.
Other: Innovation and Design Process	• Recognizes innovation and exemplary efforts to foster sustainable pavement designs. • Total of 4 points is assigned to this category. • Directed at design, construction, and rehabilitation, but possibly could be modified for use with pavement preservation and maintenance treatments.

TABLE 20-2   MTO GreenPave Case Study Factors

20.4.3   Case Study: New York State Department of Transportation GreenLITES

GreenLITES is a project design certification program for typical New York State Department of Transportation (NYSDOT) construction projects [NYSDOT 09]. The program is comprehensive and references pavement preservation and maintenance treatments such as diamond grinding, crack sealing, and liquid asphalt treatments.

In addition, it accords credit to practices that significantly build upon GreenLITES categories and objectives or that incorporate significant innovations. As of now, this provision is very generic but it potentially could be used to add practices to approved lists of green activities. The approach makes a first effort at quantifying some of the benefits that derive from "green practices" as shown in Table 20-3.

Sustainability Impact Factor Area	Description/Relevance to Pavement Preservation and Maintenance
Virgin Material Usage and Alternative Material Usage	• Emphasis on limiting virgin material usage and alternative material usage is stressed in the program • Quantification on pavement preservation and maintenance done through credits earned, use of recycled tire rubber, and recycled asphalt pavement
Noise	• Specific reference to inclusion of measures for reducing pavement noise • Quantification on pavement preservation and maintenance done through credits earned for diamond grinding
Air Quality/Emissions	• Emphasis is placed on reducing impacts to the air through proper maintenance of equipment and appropriate operations • Quantification on pavement preservation and maintenance done through credits earned
Water Quality	• Development of protocols to reduce sediment into streams and ensure proper storm water management • Quantification on pavement preservation and maintenance done through credits earned
Energy Usage	• Credit for reducing electricity and petroleum consumption • Quantification on pavement preservation and maintenance done through credits earned for documented analysis of design that reduces carbon footprint
Other: Innovation	• System encourages engineers to extend the scope of the program to previously unlisted techniques that comply with the spirit of the program • Quantification on pavement preservation and maintenance done through credits earned in innovative and/or unlisted category

TABLE 20-3 NYODOT GreenLITES [NYSDOT 09]

As of 2012, it is not a quantitative approach that can be incorporated into an asset management system. It is currently based on a spreadsheet rating system, but this could become the beginning of a tool to enhance environmental sustainability in road construction.

20.4.4   Case Study: Airports and Aviation Industry Efforts

Aviation infrastructure serves both passengers and supply chain stakeholders. It is estimated at any given time about 4,000 aircraft operate in U.S. national airspace. In the 1990s the United States produced about 1.5 million tons or 24 percent of the world's total carbon dioxide emissions from aviation sources. However, the aviation emission in the United States dropped in the 2000s [Uddin 12] due to better air traffic control and navigation, more fuel-efficient commercial aircraft engines, and efficient airline routing and scheduling. Airports are striving to introduce a range of sustainability measures from installing renewable energy production, introducing green roofs instead of traditional bituminous shingle roofs, recycling programs, and operating ground-side and air-side equipment and vehicles with fuel-efficient and "clean" engines [ACRP 12b].

As an example of sustainability in airline operations, the following data is provided by industry sources [Delta 12]:

- First comprehensive in-flight recycling program by Delta Airlines, 2007
- Reducing environmental footprint by managing the plane fleet; with today's aircraft a gallon of jet fuel flies as much as 120 percent farther than it did in 1978
- Verifying GHG inventory by a third party and posting on The Climate Registry, a nonprofit organization with established standards for calculating, verifying, and publicly reporting carbon footprints
- A place on the 2011 Dow Jones Sustainability Index for North America as a result of progress made by Delta Airlines on environmental fronts

20.4.5   Case Study: New York City Long-Term Sustainability Plan

The Mayor's Office of Long-Term Planning and Sustainability, New York City, set public policy goals to reduce citywide carbon dioxide emissions by 30 percent below 2005 levels by 2030. Figure 20-2 compares New York City with other large cities on the sustainability scale of per capita carbon dioxide equivalent (CO_2e) emissions. Other NYC sustainability program achievements follow [NYC 07]:

- Transportation in 2005 represented 23 percent of total emissions. An annual average 30 percent savings in a typical household's energy cost due to the use of public transportation is reported.

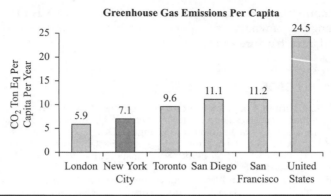

Figure 20-2   New York City compared with other cities on the sustainability scale [after NYC 07].

- New York's extensive mass transit system is estimated to transport 40 percent of all people traveling in NYC by motorized transportation, on a typical weekday, while producing only 12 percent of transportation CO_2 emissions, and 3 percent of overall CO_2 emissions.

- Other implemented measures that contribute to reducing NYC greenhouse gas emissions include energy reduction programs, street tree-planting programs, conversion of streetlights to more efficient technologies, landfill methane recovery, use of alternative fuel vehicles, and solid waste recycling.

20.5   Summary

Good asset management systems require use of equations, models, and rigorous mathematical and financial calculations. They require good engineering design, and life-cycle analysis of costs and benefits to optimize the expenditures of funds, not only immediately but in the long term. They also require trade-offs among assets in the portfolio, pavements, bridges, safety, congestion, etc. It is now time to add sustainability and environmental impacts. This chapter points out the need but shows that most efforts so far lack the quantitative detail for quantitative inclusion on an asset management system, except as additional criteria to use in final selection among relatively equal options. Case studies included herein point out that some agencies are close to quantifying the needed criteria and that broader AMS may be available in the next few years as additional information is gained and analyzed.

Aviation agencies and cities are incorporating sustainability in their facilities and operations by reducing wastes, adverse environmental impacts, emissions, and carbon footprint. Incorporating these environmental stewardship and sustainability goals beyond project level applications to a comprehensive AMS framework is the future challenge for owners and operators of both public and privately owned infrastructure assets.

20.6   References

[ACRP 12a] Airport Cooperative Research Program (ACRP), "Airport Climate Adaptation and Resilience," *ACRP Synthesis 33*, Transportation Research Board, The National Academies, Washington, DC, 2012.

[ACRP 12b] ACRP, "Guidebook for Incorporating Sustainability into Traditional Airport Projects," *ACRP Report 80*, Transportation Research Board, The National Academies, Washington, DC, 2012.

[Brundtland 87] G. H. Brundtland, "Our Common Future," Brundtland Commission, World Commission on Environment and Development, Oxford University Press, Oxford, U.K., 1987.

[CH2M Hill 09] CH2M Hill, "Transportation and Sustainability Best Practices Background," *Proceedings, AASHTO Sustainability Peer Exchange*, Washington, DC, 2009, p. 32.

[Delta 12] "Flying Greener Skies," deltaskymag.com September 2012, pp. 10–12.

[Haas 13] R. Haas, and L. C. Falls, "Realistic Long Term Warranty and Sustainability Requirements for Network Level PPP's," *Proceedings, First International Conference on Public-Private-Partnership*, Dalian, China, August, 2013.

[Hutschenreuther 10] J. Hutschenreuther, "New Pavement Technology Developed by the Use of Warm Asphalt Technologies," presented at the *2010 International Conference on Transport Infrastructure*, São Paulo, Brazil, August 4–6, 2010.

[Klinger 12] R. L. Klinger, "The Future Is Bright for Solar Energy and Asphalt Pavements," *Asphalt*, Asphalt Institute, Vol. 27, No. 3, Fall 2012, pp. 16–20.

[Lane 10] B. Lane, S. Chan, and T. Kazmierowski, "Pavement Preservation—A Solution for Sustainability," *Proceedings, Annual Conf. of the Transportation Association of Canada*, Halifax, Nova Scotia, September, 2010.

[LEED 08] "LEED Green Building Rating System for New and Major Renovations," Version 1.0, Canada Green Building Council, Ottawa, Ontario, 2008.

[MTO 10] Ministry of Transportation (MTO), "Ontario's Transportation Technology Transfer Digest," Vol. 16, No. 1, Ontario, Canada, 2010.

[NCHRP 11] National Cooperative Highway Research Program, "Sustainable Pavement Maintenance Practices," *Research Results Digest 365*, Transportation Research Board, Washington, DC, December 2011.

[NYC 07] New York City (NYC) Mayor's Office, "Inventory of New York City Greenhouse Gas Emissions," New York City Mayor's Office of Long-Term Planning and Sustainability, April 2007, http://www.nyc.gov/planyc2030, accessed September 15, 2010.

[NYSDOT 09] New York State Department of Transportation (NYSDOT), "GreenLITES Project Design Certificate Program," Albany, New York, 2009.

[Uddin 12] W. Uddin, "Mobile and Area Sources of Greenhouse Gases and Abatement Strategies," Chapter 23, *Handbook of Climate Change Mitigation*, (Editors: Wei-Yin Chen, John M. Seiner, Toshio Suzuki and Maximilian Lackner), Springer, 2012, pp. 775–840.

CHAPTER 21

Future Directions for Infrastructure Asset Management

21.1 Introduction

No technical area, including infrastructure asset management, can remain static. IAMS or AMS presents a kaleidoscope of opportunities for improvement and advancement through changing technology, changing needs, and changing user demands. Numerous ideas could be presented here by the authors and then immediately expanded upon by the readers. Rather than belabor intricate detail, we have chosen to outline vision and future directions in four categories, to provide a semblance of order:

1. Advances in technology to assist the AMS process and innovative funding mechanisms

2. Improvements in AMS resulting from its continued use (experience)

3. Advances made possible by the broader use of AMS methodology including environmental stewardship and sustainability issues

4. Better education, implementation, and adoption of AMS

There are other ways of organizing these ideas, and there may be overlap in these four categories, but this will suffice to outline the new directions needed here.

21.2 Advances in Technology to Assist the AMS Process

Technology has advanced greatly in the 45 years since modern asset management was initiated in the form of pavement management. Major developments include increased size and capability of computers and automated data-collection and storage functions. These changes will continue and accelerate, and during the next two decades the following specific technologies will have major impacts on AMS:

1. **Geographical Information Systems (GIS).** GIS is already used in asset management as illustrated herein. GIS and geo-spatial applications will greatly improve with the availability of remote sensing data from terrestrial, airborne, and space-borne technologies. It will have a dramatic impact in the coming decade as more user-friendly and generally applicable systems are developed and implemented.

2. **Global Positioning Systems (GPS).** GPS technologies are already being used in a variety of civil-engineering infra-structure. The accuracy of these satellite-based systems is increasing annually while the cost is decreasing rapidly. The ability of asset management staff to collect data in the field or perform maintenance and directly code it by location and identification through GPS will greatly enhance AMS.

3. **Faster and More Convenient Databases.** Fifty years ago, punched cards provided the best access to data storage and processing, along with mechanical sorting and high-speed linear magnetic tape (a useful medium, but one that provided no random access). Current random-access memory has increased its ability to store and process data by orders of magnitude. This improvement will continue as bubble memories and other new technologies are installed. A note of caution, however, is necessary, because it is tempting to invoke data pollution with such economical data-storage techniques. The collection, storage, and retention of vague or nonspecific data should be studiously avoided. One such example is the unnecessary storage of distress photographs. Photographs take hundreds of times more storage space than processed distress data, and distress photographs must be processed before they can be used. Thus, processing should take precedence over storage.

4. **Smart Systems.** So-called smart systems (in this context) are those that have high-quality, long-lived sensors that can detect meaningful response in an asset and feedback that response data to adjust the characteristics of the system. Such systems have broad use in military and space, in guidance and aiming devices. They can potentially be applied in asset management in certain types of maintenance feedback activities. Many other uses will also be developed as they become available and more reliable at a lower cost.

5. **Better Communications Technology.** Communications is a critical part of asset management. Communications technology will continue to improve and find applications in AMS.

Again, however, it will be critical to use the technology wisely. We all can cite examples of misuse of e-mail, the Internet, and the web just because it is there.

6. **As Yet Unknown Factors.** It is said that when digging in a gold mine, removing one nugget often uncovers several others. Such is true in the asset management field. Many new ideas and applications will continue to evolve that we have not even imagined.

21.3 Improvements in AMS Resulting from Its Continued Use

When an AMS is used by a specific agency, improvements will accrue within that agency itself, just as the ability of an athlete or musician improves with practice and experience. Pooling experience among agencies, improvements will accelerate this process, as shown with improvements in COTS systems.

1. **Performance Models.** At its inception, any AMS starts with approximate performance models. As the agency gains 5, 10, or more years of experience and history in the database for the segments or units within the management system, perfor mance-history data becomes available for developing improved performance models. Thus, the predictions from the AMS process improve with time and decision quality increases.

2. **Better Segmentation.** Within many AMSs, it is necessary to subdivide a network into sections and subsections. This segmentation is often approximate at the beginning of the process but improves with time and use of the system, including dynamic segmentation.

3. **Identification—Location.** Since most networks or portfolios of infrastructure are pre-existing at the outset of an AMS, it is not always possible to identify these separate parts correctly and completely. After the AMS has been used for several years, the identification and location information improves within the system providing more precise results.

4. **Add-On Subsystems.** As an AMS is used continuously, it is possible to add additional subsystems and therefore greatly improve the overall quality of the AMS. Begin, for example, with pavements then add bridges and congestion to produce a corridor analysis.

5. **Flexibility—User-Friendly.** Output reports and utility of AMS can be greatly improved over years of use, within an existing agency. It is not always possible to know precisely the user needs at the birth of the system, but clarity comes with continued use.

6. **Improved Quantification of Qualitative Factors.** A limitation on the use of AMS is the ability to account for qualitative factors, such as judgment and public opinion. These factors are critical to the AMS process, and the ability to quantify knowledge of the proper relationships is greatly improved by continued use of any system.

21.4   Advances Made Possible by the Broader Use of AMS

As AMS methodology has become more widely used in diverse areas, such as water and wastewater, it becomes possible to integrate new technologies and ideas into the process. This is an important way in which new directions will continue to develop.

1. **Evaluation and Use of New Materials.** New materials can be integrated into AMS in a rational way; however, if no AMS exists, there is often no way to access the relative value of the new material versus existing, less costly materials.

2. **New Technologies.** The same is true for new technologies. An AMS framework provides an ideal way to apply new technologies that have little chance of proving their worth under conventional design and maintenance operations, except by expensive trial and error.

3. **Incentive to Develop New Materials.** Experience has shown that once materials and technology experts know that AMS exists, they can speed up the development of new technologies and new materials because the developers and inventors have a better chance of seeing their new products and methods used effectively.

4. **Institutional Improvements.** The fact that AMS is used effectively in other agencies has encouraged institutional improvements among other potential users. That is, agencies that are not now using an integrated system will be able to see the benefit of such an approach among their peers in other similar agencies. This fact has led to wider use of AMS in US DOTs, for example.

5. **Integration of Additional Facilities.** An agency, such as an urban center or city, may start with a simple AMS involving its streets and bridges. Then such use can encourage the expansion and addition of water facilities, wastewater, sewer, and associated utilities in the right-of-way, such as phone land line and cable tv, electric, and gas.

6. **Application of AMS to Unitized Facilities.** Currently, not all airports, park agencies, and airport centers or port facilities

use the benefit of an integrated management system for their infrastructure. As practical use of such systems increases on similar facilities in the future, the adoption and use of such facilities by previous nonusers will be encouraged. This effectively ties in with the geographically central location of any unitized facility.

7. **Automated Data Collection**. As AMS is more commonly used, it will become easier to obtain automated data-collection services from service-providers and contractors. When the market becomes large enough, this service can be economically offered by sharing among several user groups, and the price will continue to drop and quality will continue to improve.

8. **Improved Resource Allocation between Existing and New Facilities**. As AMS is used more broadly, it will become a more effective tool for an agency to allocate its resources between maintaining existing assets and building new assets. This is often an ad hoc activity at the present time, with political forces often pushing for new facilities that offer strong political impact in lieu of much-needed, but less glamorous, maintenance funding.

9. **Improved Resource Allocation among Types of Infrastructure**. Improved AMS also will make it possible to better allocate resources among various types of infrastructure: roads versus sewers versus water, etc. Currently, such allocations are almost always politically affected.

10. **Improved Consideration of Benefits**. Broader use of AMS, combined with the desire to evaluate among and across asset types, will encourage better consideration of user benefits as part of the resource-allocation equation. This will allow agencies to move closer and closer to optimal use of funds, and offer an improved quality of life for the taxpayers and stakeholders who are the true owners of the infrastructure.

11. **Sustainability and Environmental Factors**. From 2000 to 2013 much has been discussed about these two factors. Asset management with multiyear, multifactor optimization provides a necessary framework for incorporating the external factors effectively.

12. **Public-Private Partnerships**. With reduced public funds available, innovative funding methods will continue to thrive. AMS provides the ideal framework for evaluating and allotting costs and benefits among private parties, public agencies, and users.

13. **Risk-Based Management**. As this book is being written, Hurricane Sandy is devastating the Mid-Atlantic and New England coast in the United States. This employed the need

for risk-based management. Some such AMS software is touted as of this writing (December 2012). However, there is a paucity of valid risk consideration since the analysis and management objectives are increased tenfold when considering risk. There will be, however in the next decade, significant improvement in the ability of AMS to quantify, predict, and deal with risk.

14. **Broader Use of AMS by Small Agencies**. During its first 30 years, asset management has largely been used in larger agencies such as state and provincial DOTs. The next two to three decades will see great advances in the use of AMS in cities, counties, and smaller private agencies as COTS software becomes more prevalent and thus cheaper, good data collection continues to improve, and knowledge of the benefits of good AMS spreads.

21.5   Better Education, Implementation, and Adaptation of AMS

AMS technology will lag in its use until potential users become fully educated about its benefits. Current college graduates are taught little or nothing about the need to integrate design, maintenance, and rehabilitation. In most colleges, they are taught design and perhaps construction quality, but little else about the total life cycle of facilities. Education in the form of short courses, college-level courses, and graduate work in AMS will encourage improvements in and wider use of AMS by engineers and managers. Educational workshops and short courses will also encourage the exchange of information among peer groups of practicing professionals and will foster the spread of AMS among agencies and from one type of asset to another.

At the administrative level, courses of one- or two-day duration including demonstrations to apprise elected officials and administrators of the benefits of AMS as a decision-making tool also will broaden its acceptability and adoption. Of particular value is the introduction of an undergraduate course in AMS in civil-engineering curricula. Such a course introduces young civil engineers to the need for improved maintenance and rehabilitation engineering. Currently, the emphasis is strongly on design, while at least 90 percent of civil-engineering graduates now deal with maintenance and rehabilitation in their job in some fashion without associated formal education.

21.6   Improvement in Technologies

The quality of implementation of AMS is increasing and will continue to increase. In the late 1900s, most AMS was homemade and used simple prioritization of one asset. By 2010 that was changing and will continue to change. Agencies that started with a simple pavement management system based on repairing cracks in worst-first sections, now have fully functioning PMSs and are moving into bridge management, etc. Commercial firms such as Deighton, AgileAssets, Exor, etc. that started by developing simple PMSs are now adding bridges and congestion management, etc. to their portfolios. Many state agencies that started with pavement management in the 1990s now have two, three, four, or more subsystems or silos integrated into their AMS. The knowledge of the benefits of such integration is spreading and will continue to spread. Likewise the knowledge of the benefits of AMS at state and provincial level is spreading to metropolitan areas and smaller agencies with whom they work and have contact.

21.7 Summary

Ultimately, the future of AMS lies with its users and potential users, like you, the readers of this book. Anyone who has read this book, or a major portion of it, and has now reached this section, is clearly a potential user of AMS technology. Hopefully, you can and will apply this technology and make it known to your peers and other potential users. There are many sources of additional AMS information in the literature, from vendors and at user conferences and meetings. This information may come in bits and pieces, but readers should take the opportunity to use such information as they continue their quest to improve the quality of infrastructure asset management of public and private facilities, both in their own agencies and throughout the world. Apply this knowledge, and share it with others.

Acronyms

3D	Three-dimensional
4GL	Fourth-generation language
4R	Reclaim, Reuse, Recycle, Reduce
AADT	Average annual daily traffic
AAPA	Australian Asphalt Pavement Association
AASHO	American Association of State Highway Officials
AASHTO	American Association of State Highway and Transportation Officials
ACRP	Airport Cooperative Research Program
ADA	Americans with Disabilities Act
AGCA	Associated General Contractors of America
AI	Asphalt Institute
AISC	American Institute of Steel Construction
AISI	American Iron and Steel Institute
AMS	Asset management system
ANN	Artificial neural network
ANOVA	Analysis of variance
APWA	American Public Works Association
ARIMA	Auto Regressive Integrated Moving Average
ARRA	American Recovery and Reinvestment Act of 2009
ASCE	American Society of Civil Engineers
ASQC	American Society for Quality Control
ASTM	American Society for Testing and Materials
ATC	Applied Technology Council
ATOA	Asset trade-off analyst
AWTTS	Advanced Water Technology Test Site
BART	Bay Area Rapid Transit
B/C	Benefit-cost ratio
BIM	Building information model
BLCC	Building life-cycle cost
BMDB	Building Maintenance, Repair, and Replacement Database
BMS	Bridge management system
BMWi	German Federal Ministry of Economics and Technology

BRB	Building Research Board
BS	British Standards
BTS	Bureau of Transportation Statistics
C-SHRP	Canadian Strategic Highway Research Program
CADD	Computer-aided drafting and design
CAF	Compound amount factor
CAFR	Comprehensive Annual Financial Reports
Caltrans	California Department of Transportation
CBI	Confederation of British Industry
CBR	California Bearing Ratio
C/E	Cost effectiveness
CE	Constructability enhancement
CEC	Commission for Environmental Cooperation
CEO	Chief executive officer
CERL	Construction Engineering Research Laboratory
CFC	Corrosion-fatigue cracking
CI	Condition index
CIA	Central Intelligence Agency
CII	Construction Industry Institute
CNN	Cable News Network
CO	Carbon monoxide
CO_2	Carbon dioxide
CO_2e	CO_2 equivalent
Cobb DOT	Cobb County Department of Transportation
COTS	Commercial off-the-shelf
CPAR	Construction Productivity Advancement Research
CPU	Central processing unit
CRCP	Continuously reinforced concrete pavements
CRF	Capital recovery factor
CSA	Canadian Standards Association
CSP	Corrugated steel pipe
DBMS	Database management system
DCP	Dynamic cone penetrometer
DHS	Department of Homeland Security
DOD	Department of Defense
DOE	Department of Energy
DOT	Department of Transportation
DSS	Decision support system
DWT	Dead weight tons
EAPA	European Asphalt Pavement Association
ECMS	Electrical/communication management system
EFMS	Engineering Facilities Management System
EIA	Energy Information Administration
EIS	Executive Information System
EM	Electromagnetic
EPA	Environmental Protection Agency
ERS	End-result specifications

ESAL	Equivalent single-axle loads
EU	European Union
FAA	Federal Aviation Administration
FCC	Federal Communications Commission
FEMA	Federal Emergency Management Agency
FHWA	Federal Highway Administration
FMEA	Failure mode and effects analysis
FOD	Foreign object damage
FRA	Federal Railroad Administration
FRP	Fiber-reinforced plastic
FWD	Falling-weight deflectometer
GAO	General Accounting Office
GASB	Government Accounting Standards Board
GASB 34	GASB Statement 34
GB	Gigabyte
GDP	Gross domestic product
GHG	Greenhouse gas
GIS	Geographical information system
GML	Geography markup language
GNP	Gross national product
GPI	Green pavement initiatives
GPR	Ground-penetrating radar
GPS	Global Positioning System
GPU	Graphics processing unit
GUI	Graphical user interface
HBRRP	Highway Bridge Replacement and Rehabilitation Program
HOV	High-occupancy vehicle
HRB	Highway Research Board
HTML	Hypertext markup language
HTR	Hard-time replacement
IAMS	Infrastructure Asset Management System
IATA	International Air Transport Association
IE	Impact Echo
IEA	International Energy Agency
IfSAR	Interferometric Synthetic Aperture Radar
IIAMS	Integrated infrastructure asset management system
IP	Internet Protocol
IQLs	Information quality levels
IR	(Bridge) Inventory rating
IR	Infrared
IRI	International roughness index
ISO	International Organization for Standardization
ISTEA	Intermodal Surface Transportation Efficiency Act
IT	Information technology
ITS	Intelligent transportation system
JRP	Jointed rigid pavement

KBES	Knowledge-based expert systems technology
kWh	Kilowatt-hour
LAN	Local-area network
LCA	Life-cycle assessment
LED	Light-emitting diode
LEED	Leadership in Energy and Environmental Design
LiDAR	Light detection and ranging
LNG	Liquid natural gas
LOS	Level of service
LP	Linear programming
LPR	Length of paved roads
LTPP	Long-term pavement performance
Maglev	Magnetic levitation
MAP-21	Moving Ahead for Progress in the 21st Century Act
MAU	Multi-attribute utility
MCE	Marginal cost-effectiveness
MDI	Mission Dependency Index
MDT	Mean downtime
MIAMS	Municipal infrastructure asset management systems
MLRS	Multiple Linear Referencing System
MMA	Methyl methacrylate acrylic
MMS	Maintenance management system
MPO	Metropolitan Planning Organization
M&R	Maintenance and rehabilitation
M,R&R	Maintenance, rehabilitation, and reconstruction
M,R&R	Maintenance, rehabilitation, and renovation, replacement, and reconstruction
MTBF	Mean time between failures
MTC	Metropolitan Transportation Commission
MTO	Ministry of Transportation Ontario
MTTR	Mean time to repair
NAFTA	North American Free Trade Agreement
NAPA	National Asphalt Pavement Association
NASA	National Aeronautics and Space Administration
NAT	National Aviation and Transportation
NAVFAC	Naval Facilities Engineering Command
NAVSEA	Naval Sea Systems Command
NBI	National bridge inspection
NBIS	National bridge inspection standard
NCAT	National Center for Asphalt Technology
NCHRP	National Cooperative Highway Research Program
NDE	Nondestructive evaluation
NDT	Nondestructive testing
NEPA	National Environmental Policy Act
NFESC	Naval Facilities Engineering Service Center
NGA	National Governors Association
NHPP	National Highway Performance Program

NHS	National Highway System
NIST	National Institute of Standards and Technology
NLA	New large aircraft
NLCD	National Land Cover Datasets
NO_2	Nitrogen dioxide
NOAA	National Oceanic and Atmospheric Administration
NPIAS	National Plan of Integrated Airport Systems
NPV	Net present value
NRA	National Recovery Act
NSF	National Science Foundation
NTAD-95	National Transportation Atlas Data Bases: 1995
NYC	New York City
O_3	Ozone
OCM	On-condition maintenance
OECD	Organisation for Economic Cooperation and Development
OMMS	Operations and maintenance management subsystem
OODBMS	Object-oriented database management systems
OOP	Object-oriented programming
OPI	Overall Priority Index
OR	(Bridge) Operating rating
ORE	Netherlands Office of Research and Experiments
OSHA	Occupational Safety and Health Administration
P-wave	Compression wave
P3s	Public-Private Partnerships
pc	Per capita
PC	Personal computer
PCA	Portland Cement Association
PCI	Pavement Condition Index
PCR	Pavement condition rating
PGNP	Per capita GNP
PI	Performance index
PIARC	Permanent International Association of Road Congresses
PM	Particulate matter
PMS	Pavement management system
PPB	Planning-programming-budgeting
PPPs	Public-Private Partnerships
PRT	Personal rapid transit
PS	Parallel seismic
PSAB 3150	Public Sector Accounting Board section 3150 of Canada
PSI	Present serviceability index
PSP	Problem-solving process
PSR	Present serviceability rating
PTAP	Pavement Technical Assistance Program

QA	Quality assurance
QC	Quality control
QI	Quality index
RAM	Random-access memory
RAP	Reclaimed asphalt pavement or recycled asphalt pavement
RCM	Reliability-centered maintenance
RCP	Reinforced-concrete pipe
RDBMS	Relational database management software
RFMS	Right-of-way features management system
ROW	Right-of-way
RP	Rigid pavement
RQI	Ride quality index
RTA	Road Transport Authority
S-wave	Shear wave
SAFETEA-LU	Safe, Accountable, Flexible, Efficient Transportation Equity Act—A Legacy for Users
SAM	Stress-absorbing membranes
SANS	Sanitary-sewer network management system
SASW	Spectral-analysis-of-surface waves
SCC	Stress-corrosion cracking
SCI	Structure condition index
SDHPT	State Department of Highways and Public Transportation
SDM	Spatial data management
SF	San Francisco
SHOALS	Scanning Hydrographic Operational Airborne LiDAR Survey
SHRP	Strategic Highway Research Program
SI&A	Structure inventory and condition appraisal
SMS	Safety management system
SN	Structural number
SQL	Structured query language
SR	Sufficiency rating
STMS	Storm-sewer network management system
STR	Sales tax revenue
TB	Terabyte
TDR	Time-domain reflectometry
TEA-21	Transportation Equity Act for the 21st Century
TIGER	Topological Integrated Geographic Encoding and Referencing
TIN	Triangulated irregular network
TMS	Transportation management system
TQI	Track quality index
TQM	Total quality management
TRB	Transportation Research Board
TRDF	Texas Research and Development Foundation

TSCI	Track Structure Condition Index
TVA	Tennessee Valley Authority
TxDOT	Texas Department of Transportation
UAMS	Unitized asset management systems
UPR	Unsaturated polyesters resins
UPV	Ultrasonic pulse velocity
URMS	Urban roadway management system
U.S.	United States
US	United States
USACE	U.S. Army Corps of Engineers
USDOT	U.S. Department of Transportation
USGBC	U.S. Green Building Council
USGS	U.S. Geological Survey
VA	Veterans Affairs
VE	Value engineering
VECP	Value engineering in construction project
VGA	Video graphic adapter
VNIR	Very near infrared
VOC	Vehicle operating cost
WIM	Weigh-in-motion
WMS	Work management system
WNS	Water Main Network Management System
WRMs	Waste and reclaimed materials
WSSC	Washington Suburban Sanitary Commission
WTC	World Trade Center
www	World Wide Web
XML	Extensible markup language

Index

Note: Figures are indicated by an *f* and tables by a *t*.

X